Dan Kelly

INTRODUCTION TO MOLECULAR THERMODYNAMICS

INTRODUCTION TO MOLECULAR THERMODYNAMICS

Robert M. Hanson
Susan Green
ST. OLAF COLLEGE

4   5
6.1-6.7

University Science Books
Sausalito, California

University Science Books
www.uscibooks.com

Production Manager: *Paul C. Anagnostopoulos, Windfall Software*
Copyeditor: *Lee A. Young*
Proofreader: *MaryEllen N. Oliver*
Text Design: *Paul C. Anagnostopoulos*
Cover Design: *George Kelvin*
Illustrator: *Lineworks*
Compositor: *Windfall Software*
Printer & Binder: *Maple-Vail Book Manufacturing Group*

This book was set in Times Roman and Helvetica and composed with ZzTEX, a macro package for Donald Knuth's TEX typesetting system.

This book is printed on acid-free paper.

ISBN: 978-1-891389-49-8

Library of Congress Cataloging-in-Publication Data

Hanson, Robert M., 1957–
  Introduction to molecular thermodynamics / Robert M. Hanson, Susan M.E. Green.
    p. cm.
  Includes index.
  ISBN 978-1-891389-49-8 (alk. paper)
  1. Thermodynamics. 2. Molecular dynamics. I. Green, Susan M. E., 1971– II. Title.
  QD504.H33 2008
  541'.369—dc22
                                                                    2007047842

Printed in the United States of America
10 9 8 7 6 5 4 3 2 1

Contents

The premise of this book is a simple one: You don't have to be a mathematical genius to appreciate the essentials of thermodynamics. All you need is a basic understanding of probability—how it can be used to predict the outcome of events and how, especially when dealing with a huge number of events, simple ideas of probability predict outcomes that are for all practical purposes totally certain.

In the first chapter of this book, we show that with just a basic understanding of probability you can understand why it is that chemical reactions approach equilibrium and why the natural unpredictability of random events makes equilibrium a dynamic state, constantly fluctuating, constantly testing the possibilities. You'll see how Le Châtelier's principle—that a disturbed equilibrium will spontaneously adjust to form a new one—is just a natural outcome of this fluctuation. And you will find that these fluctuations seemingly disappear as we go to more and more particles. But that is just an illusion. The fluctuations are always there and, basically, those fluctuations, tiny as they may be, are what allow for chemical reactions to take place at all.

But to understand what's really going on in the world, we have to add the idea of energy: countable energy. In Chapter 2, when we start to consider the simplest chemical system—with just a few particles and just a few "quanta" of energy—we find the most amazing result: The most probable distribution of that energy in a system (the Boltzmann distribution, what we call equilibrium) is easily predictable. Temperature plays a role here, not so much as the cause, but rather as the effect. Temperature, we will see, is simply a way of monitoring the distribution of energy in a system. It's really the energy and how it is distributed that makes all the difference.

The problem with energy is that it's conserved. It gets traded back and forth between a system and its surroundings, but it's a fair trade, and this in itself cannot explain why some chemical reactions occur and some do not. Enter entropy—a measure of the number of ways energy can be distributed. Entropy turns out to be the key to understanding what's going on in real chemical processes. Entropy is not conserved, and that fact alone drives natural processes in the direction we call "forward in time."

It's really quite amazing that just a few simple ideas can go so far. But they do, and that is why we are so excited to be sharing them with you at this introductory level. You are about to take part in an adventure into the inner workings of the molecular world. Be prepared to be amazed; be prepared to be surprised at the simplicity of it all. That's the beauty of molecular thermodynamics.

Bob Hanson, *hansonr@stolaf.edu*
Susan Green, *smgreen@comcast.net*
January, 2008

To the Instructor

It is a great privilege to share this book with you. We think you will enjoy teaching from this text as much as we do, and we look forward to hearing from you as you do so. We use this book in our first-year program here at St. Olaf College. It isn't an honors program at all—in fact, knowing that the majority of these students will not end up being chemistry majors is actually one of the most exciting aspects of using this book. We are convinced that having these discussions with first-year students, many of whom will never take another chemistry course, is one of the greatest services we can provide. The trick, we believe, is to have a book that hits the right level. And this book, in our opinion, does exactly that. It may seem at first blush that only students with very strong math backgrounds would thrive using this book, but our experience is that any student willing to put time into the course can do well. The mathematics in this book goes no further than algebra and the manipulation of logarithms—the background anyone taking a first-year chemistry course would have.

For those students who are headed toward being chemistry majors, this experience will prove to be an exciting challenge. They may have had an outstanding high school chemistry experience, but they still may not ever have thought of chemistry in terms of probability and energy the way they will in your class using this book. Time and again we have seen these students get more and more excited as they make their way through our course. The overwhelming reaction we have seen is something like, "Wow, I can't believe this all makes so much sense!" That's our goal, of course, and don't be surprised if you start saying that yourself as you teach it. It does make sense. This book is one single, sensible story. This approach pays big dividends later on in a chemistry major's coursework. In later courses students can use more math and focus on the details—they will already know the story and will be far less likely to lose sight of the forest for the trees.

The word *molecular* in the title of this book emphasizes that we are taking a distinctly chemical approach to thermodynamics. This is not the standard third-year physical chemistry approach, filled with calculus and differential equations. This is not an abstract introduction to the "pure" thermodynamics dear to physicists, which is wholly macroscopic and requires no understanding of the underlying structure of the system or the surroundings. Instead, by emphasizing always the molecular context and using a simple model that involves only as much mathematics as necessary, we offer students insight into the interaction of matter and energy at a level they can fully appreciate. The focus here is on chemical reactions—what makes them go or not go; how we can predict their course. We start with the simplest of reactions, just the isotope exchange between H_2 and D_2, and talk about probability. There are fluctuations in how the molecules are distributed, but the more particles we add, the less we are able to detect those fluctuations. We then add the idea of quantization of energy and find that one distribution that is most probable follows an amazingly simple mathematical law discovered by Boltzmann.

Chapter 3 introduces the four major ways energy is stored in real chemical systems—in the form of electronic, vibrational, rotational, and translational energy. Certainly this is an approximation, but we can use these differences, then, to talk about heat, work, bonding, and the effect of temperature on chemical equilibrium. Entropy is introduced as a way of counting possibilities, and the effect of concentration and pressure on molar entropy falls right out of this understanding. The approach to enthalpy involves focusing on the effect of energy transfer on the distribution of energy in the *surroundings*. Considerations of entropy in both the system and the surroundings lead us to Gibbs energy.

To relate this all to chemical reactions, we use the graphical method developed by Gibbs in the late 1870s, involving graphing Gibbs energy as a function of temperature. Although Gibbs used this method only for describing changes of phase (many such applications are given in Chapter 12), we find that it is also extremely valuable in the context of any sort of situation involving "reactants" and "products." In effect, we are using reaction Gibbs energies as a way of getting at reaction potentials.

In our experience, it takes about 20 days—roughly seven weeks, three days per week—to take this special "behind-the-scenes" look at what makes molecules do what they do. You can take a look at how we do this ourselves by checking the daily notes at *http://www.stolaf.edu/depts/chemistry/imt/days* or the syllabus at *http://www.stolaf.edu/people/hansonr/chem126*. In addition, be aware that there is a whole suite of helpful accessories awaiting you at *http://www.stolaf.edu/depts/chemistry/imt*. Take a look at the "Online Interactive Guide." We use this extensively in class on a daily basis, and students use it on their own, as well.

We encourage you to contact us to ask us more about how we make this work. We have a variety of lab experiences, for example, that cover all aspects of this book. We would be more than happy to share them with you if requested.

Above all, have fun!

To the Student: How to Study Thermodynamics

Learning anything new is a challenge, and this is going to be something new. Sometimes we professors forget how hard it is for students to remember all those new ideas. It's stressful. What's obvious for us is almost certainly not obvious for you. Here are a few tips that might relieve some of the stress:

- Check the web site *http://www.stolaf.edu/depts/chemistry/imt*, as there are several pages there you might find useful. The Concept Index page should be of great interest, for sure. We are very excited to offer this page, which provides an on-line overview of many of the chapters. For each chapter, there is a guided tour using slides, extensive commentary, searchable index, additional study questions, occasional animations and "tools," and many solved problems and quizzes.
- There's a *big* table in the back of this book (Appendix A) with all the many constants and symbols in alphabetical order. Refer to it when you forget what a symbol means.
- Equations are concise mathematical sentences. When you find an equation in the text that is especially important, it is labeled with a chapter equation reference such as 11.1. Highlight these and work hard not so much to memorize them as to understand what they are trying to tell you.
- As you read, try to say the mathematics to yourself in words (or better, if your roommate can stand it, say it out loud). For example, "$\Delta U = q + w$" should be pronounced, "The change in internal energy is the sum of the heat put in plus the work put in." If you find you can't do that, it's a flag that you probably don't understand it yet. Slow down, go to Appendix A, and check what those symbols mean.
- Read slowly, and take notes as you read. You can't do this stuff by looking at a homework problem and then figuring out where in the book is a good example just like it. Not here. No way. Forget that. Try to follow the progression of ideas in the chapter. You should be able to see the logic.
- Memorize now Hanson's Law of Thermodynamics: Learning requires work.
- Remember, no course lasts forever. Years from now, when you've gotten your dream job, you'll forget you ever read this. Still, we hope you will take something of this experience with you.

Even with these tips, this may be stressful for some of you. If you experience trouble, please talk it over with your professor or teaching assistant. They know lots of tricks and are probably more than happy to sit down with you and go over any problems you have with this material. The payoff to studying molecular thermodynamics will be a good sound understanding of what's happening around you and how you can (or cannot) control it. Good luck to you!

Acknowledgments

We wish to thank our many colleagues at St. Olaf College over the past seven years during the development of this text. Special thanks go to Paul Fischer (now at Macalester College), George Hardgrove, Gary Miessler, Pat Riley, Jeff Schwinefus, and Beth Abdella, who have taught from the draft and made many excellent suggestions for improvement over the years. It is a privilege to have such wonderful colleagues. Very special thanks go to Beth Abdella for test driving the first draft with only a week's notice. Way to go, Beth!

Well over 1500 students have gone through our program, either here at St. Olaf or at Macalester College. We appreciate all the feedback that has been offered.

Many of the ideas in this book are based on the pioneering work of others, namely Leonard Nash, *Elements of Statistical Thermodynamics* (Addison Wesley, 1968), Richard Dickerson, *Molecular Thermodynamics* (W. A. Benjamin, 1969), and especially William Davies, *Introduction to Chemical Thermodynamics* (Saunders, 1972).

RMH thanks Debbie, Ira, and "Daddy's stuck on the computer" Seth. Your understanding has meant the world to me. Let's go camping!

SG thanks Siri, Max, and especially Hans.

We both thank Bruce Armbruster, Jane Ellis, Paul Anagnostopoulos, Lee Young, MaryEllen Oliver, George Kelvin, and Tom Webster for making this textbook a reality.

INTRODUCTION TO MOLECULAR THERMODYNAMICS

CHAPTER 1

Probability, Distributions, and Equilibrium

If nature does not answer first what we want, it is better to take what answer we get.

—Josiah Willard Gibbs, Lecture XXX, Monday, February 5, 1900

1.1 Chemical Change

What makes chemicals react? Do substances react to give the more stable product? Is it the release of heat that drives chemical processes? Or is it something else? It isn't hard to demonstrate that although most reactions give off heat, some actually absorb heat as they occur. (Did you know that if you add salt to ice water, it gets *really* cold? Why is that?) In addition, many chemical reactions seem to stop before they're complete. They just seem to never use all of their limiting reactant. Why is that? More practically, how can we predict what is going to happen, and how can we make chemistry work in our favor?

Your body is an incredibly complex mixture of chemicals. Is there a special life force that governs the chemistry in your body, or is it, like all nonliving things, governed by the force of nature?

Why does a drop of food coloring in a glass of water spread until the entire solution is colored? Why does ice melt in a glass of lemonade? If your answer to either of these questions is, "It just does!" you aren't too far from the mark. There's something about diffusion and melting that common experience tells us is irreversible. Imagine watching a movie where a glass of green liquid suddenly separates into half blue on the left and half yellow on the right. Would you believe it? What if a puddle of water suddenly came together and formed an ice sculpture?

These are questions of *thermodynamics*, the study of the driving force behind chemical and physical change. We all know from our experience that certain things happen only in one direction. Anything else would be ridiculous and totally unbelievable. The premise of this entire discussion is that the force of nature driving all chemical reactions in all systems, living and nonliving, is really nothing more than the *statistics* of blind chance. *All chemical reactions proceed because they are going from a less probable state to a more probable one.* For some of you this is going to be a hard pill to swallow. What about free will? How can it be that all natural processes are simply a result of chance?

You may not be comfortable with the idea that chance is behind all natural processes. That's OK. Albert Einstein, a pretty smart guy, whose discovery of quantum statistical mechanics underlies all of what we are about to discuss, to his death refused to believe that simple chance could be the real driving force behind all of life and all of the beauty of nature. His famous assertion, "Surely, God does not play dice," says it all. Einstein simply could not believe that his own discovery was truly fundamental and not just a trick.

But, trick or not, over the 80 years since that discovery, no one—Einstein included—has been able to find any simpler, more powerful driving force than simple blind chance to explain and predict all of nature's wonders. Basically, the argument here is going to be that chemical reactions proceed not because they release energy, but, instead, because the chance of their doing anything *else* is incredibly small.

The interesting thing is that rather than just throwing up our hands and saying, "So what happens, happens—too bad," understanding chemistry in this way gives us the opportunity to direct it in ways we want it to go. Just because a chemical reaction *by itself* is headed in one direction doesn't mean we can't direct it otherwise. Sure we can. We can be *agents of change* if we want to, but in order to do that, we need to understand what it is that is causing the change (or lack thereof) in the first place. This, then, is what thermodynamics is all about.

The really exciting thing about getting a feeling for thermodynamics is that by understanding the underlying means by which chemical changes occur, we might be able to *direct* those reactions the way we want them to go (or to stop them altogether). This is true whether we are talking about the chemistry of cancer or the chemistry of ozone depletion.

1.2 Chemical Equilibrium

Much of this book is an attempt to help you understand why chemical reactions happen and why, once started, they always proceed toward a state we call *equilibrium*. We will say that a chemical system is at equilibrium if there is no tendency for it to change in any way *of its own accord*. Equilibrium is the end of the road, the state of "no observable reaction." We say that chemical reactions continue until they "reach equilibrium," at which point we can observe no further changes in pressure, volume, temperature, concentration, or color. Nonetheless, we shall see in this chapter that chemical equilibrium is very *dynamic*, much like the constant shuffling of a deck of cards.

We want to understand both what equilibrium is and how chemical reactions proceed toward it. The goal, then, of this chapter is to convince you of two ideas:

1. The equilibrium state (especially, *equilibrium constants*) can be predicted based on probability alone, and
2. The predictive nature of probability is especially well suited to chemical systems due to the sheer magnitude of the number of particles involved.

First, though, we need to discuss just the basics of probability.

1.3 Probability Is "(Ways of getting *x*)/(Ways total)"

Imagine throwing a standard six-sided die. What is the probability of rolling a 4? If you said, "one sixth," or "one in six," you were right. There is one way a die can land as a 4 out of a total of six possible ways it can land. Provided the die is not "loaded," that is, that it is not constructed or damaged in some way that makes it land one way more often than another, we can say

$$P_4 = \frac{\text{(Ways of getting 4)}}{\text{(Ways total)}} = 1/6$$

where "P_4" is the probability of rolling a 4. We say that the probability of rolling a 4 is "1/6" or "one chance in 6" or "16.66%" or 0.16666 . . . . In general, when each possibility is equally probable, we can just write that the probability of some outcome "*x*" is simply

$$P_x = \frac{\text{(Ways of getting } x)}{\text{(Ways total)}} \tag{1.1}$$

This simple definition of probability as "ways of getting something divided by ways total" is fundamental to everything in this book.

1.4 AND Probability Multiplies

Now imagine throwing that same die 10 times. What is the probability of rolling a 4 ten times in a row? That is, what is the probability of it landing a 4 the first time *and* a 4 the second time *and* the third time *and* the fourth time, etc.? How probable do you think that is? Your intuition probably says, "not very!" But *exactly* how (im)probable is it? The answer is fairly intuitive:

$$P = \frac{1}{6} \times \frac{1}{6} \times \frac{1}{6} \times \frac{1}{6} \times \frac{1}{6} \times \frac{1}{6} \times \frac{1}{6} \times \frac{1}{6} \times \frac{1}{6} \times \frac{1}{6} = \left(\frac{1}{6}\right)^{10}$$

This number is quite small, 1.65×10^{-8}. But probability is more than just a number, it is (ways of getting x)/(ways total). Let's work it out as a fraction:

$$P = \left(\frac{1}{6}\right)^{10} = \frac{1}{60466176}$$

In fact, there is *one* way to roll a 4 ten times in a row, and there are 60,466,176 ways of rolling a die ten times. There is about a 1 in 60,000,000 chance of rolling that 4 ten times in a row. No wonder it's not very likely to happen!

The sorts of events we are talking about here are called *independent* events, because the outcome of one does not influence the probability of the outcome of the next. No matter how many times we roll a 4, the probability that the *next* roll will produce a 4 is still 1/6.

But here's another example, where the outcome of the second event is not independent of the first: What is the probability of drawing two hearts off the top of a standard well-shuffled deck of cards (13 spades, 13 hearts, 13 diamonds, and 13 clubs)? Can you see how to do it? Here is the calculation:

$$P_{\heartsuit\heartsuit} = \left(\frac{13}{52} \times \frac{12}{51}\right) = \frac{156}{2652} = 0.0588$$

The first fraction, 13/52, is the number of ways of drawing a heart for the first card divided by the total number of cards; the second fraction, 12/51, is the number of ways of drawing a heart *once one is removed from the deck* divided by the total number of cards *now in the deck*. Our calculation shows that there are 156 ways of drawing two hearts off the top of the deck out of a total of 2652 ways of drawing two cards. This amounts to about a 6% chance.

1.5 OR Probability Adds

What is the probability of throwing one normal six-sided die and rolling a 2 *or* a 3? The answer is 2/6:

$$P_{2 \text{ or } 3} = P_2 + P_3 = \frac{1}{6} + \frac{1}{6} = \frac{2}{6}$$

The probability of throwing a 2 is 1/6, and the probability of throwing a 3 is also 1/6. There are two ways to throw "a 2 or a 3" out of 6 total ways, for a probability of 2/6. Another way of saying this is that there is a "33% chance" of throwing a 2 or a 3.

1.6 AND and OR Probability Can Be Combined

Probability starts getting more interesting when we consider somewhat more complicated examples. For example, throw two normal six-sided dice. What is the probability of rolling an 8? This is trickier and can be solved more than one way. Based on the idea that $P_8 =$ (number of ways to roll 8) / (total ways), we could work out all the ways to get 8 and divide by the total number of ways of rolling two dice:

(number of ways to roll 8) = five: (2, 6), (6, 2), (3, 5), (5, 3), (4, 4)

(total number of ways of rolling two dice) = $6 \times 6 = 36$

Thus, $P_8 = 5/36$. Another way of doing this, though, is by figuring out the different probabilities of all the independent possibilities and combining them. Thus, throwing an 8 is the same as throwing (a 2 *and* a 6) *or* (a 6 *and* a 2) *or* (a 3 *and* a 5) *or* (a 5 *and* a 3) *or* (two 4s):

$$P_8 = P_{2,6} + P_{6,2} + P_{3,5} + P_{5,3} + P_{4,4}$$

$$= \left(\frac{1}{6} \times \frac{1}{6} \right) + \left(\frac{1}{6} \times \frac{1}{6} \right) + \left(\frac{1}{6} \times \frac{1}{6} \right) + \left(\frac{1}{6} \times \frac{1}{6} \right) + \left(\frac{1}{6} \times \frac{1}{6} \right)$$

$$= \frac{5}{36}$$

Each of these five possibilities is independent and gives the desired outcome. Thus, we add them to give the overall probability of rolling 8.

This second way may seem more cumbersome at first, but consider this next example: What is the probability that *exactly one* of the top two cards of a standard well-shuffled deck of cards is a heart? In this case figuring out all the possibilities by hand (there are 1014) is much more difficult than using AND and OR.

We must calculate the probability of first drawing a heart (13/52) *and* then a card which is not a heart (39/51) *or* first drawing a card which is not a heart (39/52) *and* then the heart (13/51):

$$P = \left(\frac{13}{52} \times \frac{39}{51} \right) + \left(\frac{39}{52} \times \frac{13}{51} \right) = \frac{1014}{2652} = 0.382$$

Thus, there is a 38.2% chance of exactly one of the first two cards being a heart. Note that once again we have to consider the fact that after the first card is accounted for, there are only 51 cards remaining in the deck.

1.7 The Probability of "Not X" Is One Minus the Probability of "X"

There are two ways of thinking about determining the probability that when two cards are drawn off the top of the deck, neither card is a heart. The first approach is the same one we used to determine the probability that both cards would be hearts. We need to calculate the probability of first drawing a card that is not a heart (39/52) *and* then a card which is also not a heart (38/51):

$$P = \frac{39}{52} \times \frac{38}{51} = \frac{1482}{2652} = 0.559$$

Thus, there is a 55.9% chance that when two cards are drawn off the top of the deck, neither one of them will be a heart.

The second approach takes advantage of the fact that we already know the probability of all the outcomes except one. In this example of drawing two cards off the top of a deck, there are three possible outcomes: both of the cards will be hearts, one of the cards will be a heart, or neither of the cards will be a heart. This is illustrated with a pie chart in Figure 1.1. Each slice of the pie represents a possible outcome. The probability of drawing two hearts is a slice accounting for 5.9% of the total pie. The probability of drawing one heart and one non-heart accounts for another 38.2% of the pie. Since the entire pie represents 100%, the remaining 55.9% must be the probability that neither card drawn will be a heart.

A more mathematical way to say this is

$$P_{\text{not } x} = 1 - P_x$$

In this example, the probability that neither card drawn will be a heart is the probability that the cards drawn are *not* "one (0.382) or both (0.059) are hearts." So in this case we could have calculated

$$P = 1 - (0.382 + 0.059) = 0.559$$

We arrive at the same result, that the probability of not drawing a heart is 56%! In this case the second approach was the easiest since we already knew the probabilities of all the outcomes except the one we wanted.

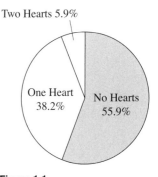

Figure 1.1
The three possible outcomes of drawing two cards off the top of a deck add up to 100%.

1.8 Probability Can Be Interpreted Two Ways

It's important to realize that in all of the above example we didn't have to do any multiplication or addition at all if we didn't want to. All we really had to do, for example, was count all the ways of getting a "neither card a heart" (A♠ 2♠, A♠ 3♠, etc., 1482 ways, believe it or not) and divide by all the possible ways of getting any two cards (of which there are 2652) to get our 55.9%. Similarly, for rolling a 4 ten times in a row, there is just *one* way to do it out of a total of 60,466,176 possibilities. Ultimately, probability always boils down to a fraction: the number of ways of x divided by the total number of ways possible. Everything else is just a trick to make the calculation of that fraction a little easier.

It's also important to realize that all of this works only when each "way" is equally probable. If the dice are loaded, or the cards are sticky, or the dealer is cheating, all bets are off! We'll see that this "fair game" sort of probability is all we need to understand the "force" behind chemical reactions.

The perspective we have been taking up to this point, counting all the ways of this or that outcome, is that of the gambler. From this perspective, probability gives some sort of indication of what to expect next: "If I roll this die or pick up that card, my chances of winning are . . . ." But there is another perspective, which turns out to be more appropriate to chemistry. This is the casino's perspective. From this perspective, probability gives an indication (a *strong* indication) of what to expect over the long haul: "If this game is played 10,000 times, we should expect a profit of . . . ."

Clearly in the case of casino gambling, the "odds are in favor of the house." For example, a slot machine is constructed to "deliver" a certain percentage of the time. You may get lucky and win the jackpot, but there are an awful lot of times that handle is pulled and no money comes out. As long as a casino operator can get someone to place a bet (and they do not mismanage their expenses), then over the long haul the casino is going to make a profit. It's not that they can't lose, just that it is *sooooo* unlikely that they will lose overall when the handle has been pulled 10,000 times. If they do, then something's wrong with that slot machine!

The casino operator's advantage is precisely why probability works to explain chemical reactions. If you think the odds are pretty likely to work out after 10,000 games are played, try 602,213,670,000,000,000,000,000 (a mole)!

1.9 Distributions

After shuffling cards, we often deal them out, creating *hands*, or sets of cards. If you take a deck of cards and deal them *all* out into hands, we say you have *distributed* the cards. For example, take a deck of just four cards, one of each suit: a heart ($\heartsuit$), a diamond ($\diamondsuit$), a club ($\clubsuit$), and a spade ($\spadesuit$). This deck of cards can be distributed into two hands of two cards each. We will see that there is an excellent analogy between such distributions of a deck of cards into hands and the distribution of a collection of atoms into molecules in a chemical reaction. To understand the likelihood of different molecules being produced in a chemical reaction, we first look at how probability applies to this simple, four-card deck, containing just one card of each of the four suits.

What is the probability, then, that, distributed into two hands of two cards each, one hand has both red cards ($\heartsuit\diamondsuit$) and the other has both black ($\clubsuit\spadesuit$)? First we will just work out all the possibilities. There are 24 in all, as shown in Table 1.1. The answer to our question is that only in eight cases (those indicated with asterisks) are both cards in the same hand the same color. There is an 8/24 chance of one hand having both red while the other has both black. All the other possibilities (16 of 24 total) have one red and one black card in each hand.

If we ignore the distinction between spades and clubs, just calling both black cards "B" and similarly consider both hearts and diamonds to be just red cards, calling them "R", we can say that there are only two distinct *distributions* of cards. One possible distribution might be characterized as "$B_2 + R_2$", where one hand is both black and the other is both red. The other distinct distribution is where there are two hands, both the same, having one black and one red card each: "2 RB". Overall we have

Distribution A:	$R_2 + B_2$	8 ways	$P_A = 8/24 = 0.33$
Distribution B:	2 RB	16 ways	$P_B = 16/24 = 0.67$

Clearly Distribution B is twice as likely as Distribution A. We expect, if we were to shuffle the deck and deal out all the cards in pairs over and over many times, that roughly 33% of the time we would end up with Distribution A and roughly 67% of the time we would end up with Distribution B.

Table 1.1
All 24 distributions of four different cards into two hands. Asterisks indicate "2 R + 2 B."

Hand 1	Hand 2	Hand 1	Hand 2
$\heartsuit\diamondsuit$ *	$\spadesuit\clubsuit$	$\diamondsuit\heartsuit$ *	$\spadesuit\clubsuit$
$\heartsuit\diamondsuit$ *	$\clubsuit\spadesuit$	$\diamondsuit\heartsuit$ *	$\clubsuit\spadesuit$
$\heartsuit\spadesuit$	$\diamondsuit\clubsuit$	$\spadesuit\heartsuit$	$\diamondsuit\clubsuit$
$\heartsuit\spadesuit$	$\clubsuit\diamondsuit$	$\spadesuit\heartsuit$	$\clubsuit\diamondsuit$
$\heartsuit\clubsuit$	$\diamondsuit\spadesuit$	$\clubsuit\heartsuit$	$\diamondsuit\spadesuit$
$\heartsuit\clubsuit$	$\spadesuit\diamondsuit$	$\clubsuit\heartsuit$	$\spadesuit\diamondsuit$
$\diamondsuit\spadesuit$	$\heartsuit\clubsuit$	$\spadesuit\diamondsuit$	$\heartsuit\clubsuit$
$\diamondsuit\spadesuit$	$\clubsuit\heartsuit$	$\spadesuit\diamondsuit$	$\clubsuit\heartsuit$
$\diamondsuit\clubsuit$	$\heartsuit\spadesuit$	$\spadesuit\clubsuit$	$\heartsuit\spadesuit$
$\diamondsuit\clubsuit$	$\spadesuit\heartsuit$	$\spadesuit\clubsuit$	$\spadesuit\heartsuit$
$\spadesuit\clubsuit$ *	$\heartsuit\diamondsuit$	$\clubsuit\spadesuit$ *	$\heartsuit\diamondsuit$
$\spadesuit\clubsuit$ *	$\diamondsuit\heartsuit$	$\clubsuit\spadesuit$ *	$\diamondsuit\heartsuit$

Surely there is a more sophisticated way of calculating these probabilities. With any more than four cards it could be pretty difficult to lay out all the possibilities. Doing the calculation using AND and OR goes like this:

- Consider the first pair of cards off the deck. Since there are two red cards and four total, the probability of the first card being red is 2/4:

$$P_{1R} = \frac{2}{4}$$

- The probability now of the second card being red is 1 in 3, or 1/3, because now there are only three cards, one red and two black. So the probability of having the two red cards in the first hand is

$$P_{RR} = \frac{2}{4} \times \frac{1}{3}$$

- Now, for the second pair, there are only two black cards left. The probability that the first is black, of course, is 2/2; after that card is taken there is only one card left and its chance of being black is 1/1. So the probability of getting first two red cards *and* then two black cards is

$$P_{RRBB} = \frac{2}{4} \times \frac{1}{3} \times \frac{2}{2} \times \frac{1}{1}$$

- Similarly, the probability of getting first two black cards *and* then two red cards is also

$$P_{BBRR} = \frac{2}{4} \times \frac{1}{3} \times \frac{2}{2} \times \frac{1}{1}$$

- Thus, the probability of drawing off two red cards *and* then two black cards *or* the other way around is

$$P_{R_2+B_2} = P_{RRBB} + P_{BBRR}$$

$$= \left(\frac{2}{4} \times \frac{1}{3} \times \frac{2}{2} \times \frac{1}{1} \right) + \left(\frac{2}{4} \times \frac{1}{3} \times \frac{2}{2} \times \frac{1}{1} \right) = \frac{8}{24} = 0.33$$

Either way you do it, you get the same result: There are 8 ways out of 24 to get "R_2+B_2", Distribution A, and (by default, then) 16 ways out of 24 to get anything else ("2 RB", Distribution B). Since Distribution B is more likely than Distribution A, we refer to it as the *most probable distribution*. (Technically, perhaps, we should say *more* probable, since there are only two possible distributions in this case.) We will find that it is important to identify which distribution is the most probable, because we will argue that the approach to equilibrium is simply the natural development of a system to this state.

1.10 For Large Populations, We Approximate

What if these were hydrogen atoms, and we had 1000 of them? Instead of red and black cards, we might talk about ^1H isotopes ("H") and ^2H isotopes ("D", for *deuterium*). Say we had 800 H atoms and 200 D atoms in a box. They would naturally combine to form three distinct *molecular* species: H_2, D_2, and HD. In total, we would expect these 1000 atoms to combine to form 500 diatomic molecules.

And if we could somehow put a catalyst in the box (some platinum metal would do the trick), maybe we could get the following two-atom–scrambling, or *isotope exchange*, reactions to take place:

$$H_2 + D_2 \longrightarrow 2\,HD$$

$$2\,HD \longrightarrow H_2 + D_2$$

What would you expect to find after these reactions went on for a while? (Perhaps we could monitor the situation and wait until it looks like the reaction is over or at least settled into some final state.) Surely there is going to be some of each type of molecule present in the end. Two important questions could now be considered based on the idea that these isotopes are just getting scrambled randomly by the catalyst.

First, what is the probability that, if we select one *atom* out of the box at random, it would be H? D? Easy. We just take the number of ways of getting an H or D atom divided by the total number of atoms:

$$P_H = \frac{800}{1000}$$

$$P_D = \frac{200}{1000}$$

Second, what is the probability that any one *molecule* we select will be H_2? D_2? HD? This is trickier. But basically, it is just like taking cards off the top of a deck:

$$P_{H_2} = \frac{800}{1000} \times \frac{799}{999} \approx \frac{800}{1000} \times \frac{800}{1000} = 0.64$$

$$P_{D_2} = \frac{200}{1000} \times \frac{199}{999} \approx \frac{200}{1000} \times \frac{200}{1000} = 0.04$$

$$P_{HD} = \frac{800}{1000} \times \frac{200}{999} + \frac{200}{1000} \times \frac{800}{999}$$

$$\approx \frac{800}{1000} \times \frac{200}{1000} + \frac{200}{1000} \times \frac{800}{1000} = 0.32$$

Our calculation predicts that it is twice as likely for a molecule to be H_2 as HD in the end, and there is only a 4% chance of pulling out a D_2 from the mix. Notice that to be absolutely correct here, we have to consider the first and second atoms to be dependent, as for the first two cards in a deck. Thus, technically, we have to take into consideration that the first atom is no longer in the collection in calculating the probability.

Here, though, we see a new trick. Due to the fact that we are dealing with the sampling of large populations, *we can just ignore the dependency of the sampling*. The values of "799/999" and "800/999" are so close to "800/1000" that we can just use "800/1000" anywhere we are considering the probability of an atom being H. (The difference is less than one part in one thousand.) Note that here we don't *have* to make any approximations, but it turns out that if we do, the solution takes a nice form:

$$P_{H_2} \approx P_H \times P_H$$

$$P_{D_2} \approx P_D \times P_D$$

$$P_{HD} \approx P_H \times P_D + P_D \times P_H$$

How does this analysis relate to what we would expect to observe? Those observations will be distributions—so many H_2, D_2, and HD present at a particular instant—and there will be exactly 101 distinct *distributions*, as shown in Table 1.2. We can't ever have fewer than 300 molecules of H_2 because at that point we run out of D_2 to react with it. We might refer to D_2 in this case as the *limiting reactant*.

Table 1.2
There are 101 distinct *molecular* distributions of 800 H atoms and 200 D atoms.

	H_2	D_2	HD
#1	400	100	0
#2	399	99	2
#3	398	98	4
⋮			
#101	300	0	200

But which of these 101 distributions is the most probable one? Now, *this* is a very interesting question. We determined the probabilities of fishing out each of the specific types of molecules, H_2 (0.64), D_2 (0.04), and HD (0.32). What does it mean that the approximate probability of fishing out a molecule of H_2 is 0.64? Doesn't it mean that about 64% of the molecules are H_2? Surely it does! Here the "ways" of getting H_2 corresponds to the actual number of H_2 molecules in the system, and the "ways total" corresponds to the total number of molecules, 500. (Note that we had 1000 *atoms* but only 500 *molecules*.) We can write

$$P_{H_2} = \frac{N_{H_2}}{N_{\text{molecules}}} \quad \text{or} \quad 0.64 = \frac{N_{H_2}}{500}$$

Solving this for N_{H_2} we get $0.64 \times 500 = 320$. Likewise, we find $N_{D_2} = 0.04 \times 500 = 20$, and $N_{HD} = 0.32 \times 500 = 160$. We conclude that the *most probable* distribution in this case is 320 H_2 + 20 D_2 + 160 HD.

Finally, how probable is this most probable distribution? This is a difficult question. It can be calculated, but it's not a pretty sight. One way to do it is to list all the possible distributions (101 in this case) and determine the number of ways of getting each one. Adding all these ways together, we have the total number of ways possible. Then all we have to do is divide the number of ways of getting this particular distribution by the total number of ways possible. For example, it turns out that the number of possible ways of getting 800 H atoms and 200 D atoms to arrange themselves as 320 H_2 + 20 D_2 + 160 HD is roughly 7.35×10^{214}. And the total number of ways there are to distribute 800 H atoms and 200 D atoms in any way turns out to be 6.6×10^{215}. Thus, the probability of actually getting this particular distribution is only about 11.1%. Still, this is the most probable distribution. The next most probable, characterized by 319 H_2 + 19 D_2 + 162 HD, has a probability of just 10.9%.

1.11 Relative Probability and Fluctuations

Although we have little hope in general of determining exactly how probable the most probable distribution is, still valuable will be how probable this distribution is in relation to some other distribution. If we divide the probabilities of the various distributions by the probability of the most probable one, we obtain what are called *relative probabilities*, P_{rel}, as shown in Table 1.3. Note that just these eleven distributions account for a full 88% of the possible ways of distributing the atoms. For example, the most probable distribution in this case is #81, consisting of 320 H_2, 20 D_2, and 160 HD molecules, with 7.35×10^{214} possible ways of finding this arrangement of atoms. This is more ways than any other possible distribution can be found, and we assign this distribution a relative probability of 1.

Table 1.3
Percent HD, number of ways, absolute probabilities, and relative probabilities for the eleven most probable molecular distributions of 800 H atoms and 200 D atoms. Distribution #81 is the *most probable distribution*, but not by much.

	H_2	D_2	HD	% HD	Ways	P	P_{rel}
#76	325	25	150	30.0%	2.64×10^{214}	0.040	0.36
#77	324	24	152	30.4%	3.74×10^{214}	0.057	0.51
#78	323	23	154	30.8%	4.94×10^{214}	0.075	0.67
#79	322	22	156	31.2%	6.07×10^{214}	0.092	0.83
#80	321	21	158	31.6%	6.93×10^{214}	0.105	0.94
#81	**320**	**20**	**160**	**32.0%**	**7.35×10^{214}**	**0.111**	**1.00**
#82	319	19	162	32.4%	7.21×10^{214}	0.109	0.98
#83	318	18	164	32.8%	6.54×10^{214}	0.099	0.89
#84	317	17	166	33.2%	5.47×10^{214}	0.083	0.74
#85	316	16	168	33.6%	4.20×10^{214}	0.064	0.57
#86	315	15	170	34.0%	2.96×10^{214}	0.045	0.40

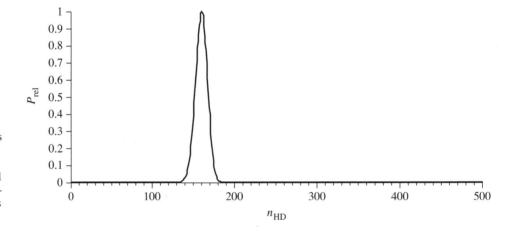

Figure 1.2
Relative probabilities of specific number of HD molecules in the equilibration of 800 H atoms and 200 D atoms into 500 molecules of H_2, D_2, and HD. The most probable distribution has 160 HD molecules (32% HD).

The slightly different distribution #82, characterized by 319 H_2 + 19 D_2 + 162 HD, is only slightly less probable than the most probable distribution, with "only" 7.21×10^{214} possible ways of being found. We could say that this distribution is 7.21×10^{214} / 7.35×10^{214} = 0.98 (that is, 98%) as probable as #81. A graph of the relative probabilities vs. number of HD molecules for all 101 possible distributions is shown in Figure 1.2. Notice that even our most probable distribution really has a probability of turning up only about 11% of the time. If we check our box of 500 molecules now and then, we will not always observe the most probable distribution any more than we would expect a flip of 100 pennies to always end up 50 heads and 50 tails. That is, we expect to see some *fluctuation* in the observations. Sometimes we will see more than 160 HD molecules, and other times we will see fewer.

However, a very interesting characteristic of all probability-based systems is that as the number of particles grows, our capability of *discerning* those fluctuations decreases exponentially. The reason this happens has to do with how we would measure the identity of the distributions. We wouldn't really count any molecules. Instead, we might weigh them, or titrate them, or somehow measure a property such as light absorbance that is unique to only one of the molecules in the mix.

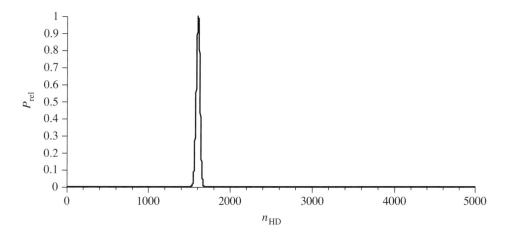

Figure 1.3
Relative probabilities of specific number of HD molecules in the equilibration of 8000 H atoms and 2000 D atoms into 5000 molecules of H_2, D_2, and HD. The most probable distribution is still 32% HD, but now the relative probability of finding 30% HD (1500 HD molecules) in the mix is only about 1 in 12,000.

In all such measurements there is a limit to our precision. For example, in the case we have been studying, Figure 1.2 shows that the most probable distribution could be characterized as being 32% HD ($160/500 \times 100\% = 32\%$). However, the relative probability of instead finding 30% HD (150 HD molecules) is 0.36. So that distribution is 36% *as probable* as the most probable distribution and will be seen a significant fraction of the time.

On the other hand, if we were to go to a system having ten times as many atoms—8000 H and 2000 D (now 5000 molecules)—then the most probable distribution would have the same percent H_2, D_2, and HD, and would be characterized as $3200 \, H_2 + 200 \, D_2 + 1600$ HD, still with 32% HD. But the relative probability of finding $3250 \, H_2 + 250 \, D_2 + 1500 \, HD$ (30% HD) instead would be only about one chance in 12,000. This is seen in Figure 1.3.

And if we had 50,000 molecules, our chances of finding only 30% HD instead of 32% HD when we measured would be only about one chance in 10^{43}. With a *mole* of molecules, our relative chances of finding even 31.999999999% HD instead of 32% HD turns out to be only one part in 10^{105}.

What this means is that on a *mole* scale, the way real chemistry is carried out, we could not possibly expect to see any fluctuation at all. The fluctuations would be there of course, it's just that we couldn't *detect* them even with the most sensitive devices we could ever build. Our graph would be essentially a sharp spike at 32% HD, the most probable distribution.

Importantly, we could still use the technique described above to calculate the most probable distribution of molecules in our system. For example, if we started with 4.0 mol of H_2 and 1.0 mol of D_2, we would have 8.0 mol of H atoms and 2.0 mol of D atoms in the box, for a total of 10.0 mol of atoms, and we could calculate the *atomic* probabilities as

$$P_H = \frac{8.0 \text{ mol}}{10.0 \text{ mol}} = 0.80$$

$$P_D = \frac{2.0 \text{ mol}}{10.0 \text{ mol}} = 0.20$$

From that, we could calculate the approximate *molecular* probabilities:

$$P_{H_2} \approx P_H \times P_H = 0.80 \times 0.80 = 0.64$$

$$P_{D_2} \approx P_D \times P_D = 0.20 \times 0.20 = 0.04$$

$$P_{HD} \approx P_H \times P_D + P_D \times P_H = (0.80 \times 0.20) + (0.20 \times 0.80) = 0.32$$

Since we know that we have a total of 5.0 mol of *molecules* (H_2, D_2, and HD), we conclude that in this case the most probable distribution is characterized by:

$$N_{H_2} = P_{H_2} \times N_{total} = 0.64(5.0 \text{ mol}) = 3.2 \text{ mol } H_2$$

$$N_{D_2} = P_{D_2} \times N_{total} = 0.04(5.0 \text{ mol}) = 0.2 \text{ mol } D_2$$

$$N_{HD} = P_{HD} \times N_{total} = 0.32(5.0 \text{ mol}) = 1.6 \text{ mol } HD$$

These are the amounts of H_2, D_2, and HD we expect to find in the system any time we looked, even with fluctuations. Sure these are approximations, but they are very *good* approximations. Even with fluctuations, our calculation for the most probable distribution gives precisely what, with great certainty, *must* be observed.

1.12 Equilibrium and the Most Probable Distribution

Now imagine starting with this same "deck" of 8 mol H atoms and 2 mol D atoms already "stacked" in the form of 4.0 mol H_2 and 1.0 mol D_2 molecules. Start shuffling. What would you expect to happen? The relative probability of this starting distribution, with absolutely no HD, is essentially 0. Think it will last long? The most probable distribution we found above to be 3.2 mol H_2, 0.2 mol D_2, and 1.6 mol HD. How will that arise? Surely we will see some HD formed from what we might call the *forward* reaction:

$$H_2 + D_2 \longrightarrow 2 \text{ HD}$$

The amount of HD will build up, and then we will probably start to see the *reverse* reaction taking place:

$$2 \text{ HD} \longrightarrow H_2 + D_2$$

At some point, the rate of the forward reaction and the rate of the reverse reaction will become equal: Roughly as many HD will split up in any shuffle as are made. Isn't this what we call *equilibrium*? But we have also learned that the outcome of this shuffling is most certainly going to be the most probable distribution, or at the very least, with these kind of numbers, something very close to it. Thus, what we call the *equilibrium state* is really just the most probable distribution, along with the generally undetectable set of fluctuations very similar to it. That is, chemical equilibrium is a dynamic, fluctuating state centered around the most probable distribution.

1.13 Equilibrium Constants Describe the Most Probable Distribution

Consider again our small system containing just 800 H atoms and 200 D atoms, with a most probable distribution of 320 H_2, 20 D_2, and 160 HD molecules. For this equilibrium we might write

$$H_2 + D_2 \rightleftharpoons 2 \text{ HD} \qquad K = \frac{[HD]^2}{[H_2][D_2]}$$

If equilibrium is really the most probable distribution, then we should be able to substitute our values for the most probable distribution number of H_2, D_2, and HD molecules for equilibrium concentrations. If we do that, we get

$$K = \frac{160^2}{(320)(20)} = 4$$

OK, if that is really an equilibrium *constant*, then we should be able to start with *any* initial conditions (number of molecules, % H, and % D) and still get 4. Let's check it out by calculating this value, which we call the *reaction quotient*, Q, for the most probable distribution in each case:

N_{atoms}	% H	% D	Most Probable Distribution	Q
2000	80	20	$640\ H_2 + 40\ D_2 + 320\ HD$	4
2000	50	50	$250\ H_2 + 250\ D_2 + 500\ HD$	4
2000	10	90	$10\ H_2 + 810\ D_2 + 180\ HD$	4

That is,

$$\frac{320^2}{(640)(40)} = \frac{500^2}{(250)(250)} = \frac{180^2}{(10)(810)} = 4$$

Note that each of these distributions is determined exactly as above, by first determining the atomic probabilities, P_H and P_D, then the molecular probabilities, P_{H_2}, P_{D_2}, and P_{HD}, and finally multiplying those probabilities by the total number of molecules (1000). Sure enough! No matter how much H or D we start with, we always get the same value for Q!

1.14 Le Châtelier's Principle Is Based on Probability

If a system at equilibrium is disturbed, a reaction will occur which will reduce (but not eliminate) that disturbance and create a new, shifted, equilibrium. This is Le Châtelier's principle, and it is supposed to apply to all chemical equilibria. Let's see if it works here. In Figure 1.4 we see three distributions. Distribution A is once again the most probable distribution for 500 molecules arising from mixing 800 H atoms with 200 D atoms. There are 320 H_2, 20 D_2, and 160 HD in this distribution. Note that it has $P_{rel} = 1$, because it is the most probable distribution.

Upon adding 300 D_2 molecules (600 D atoms), the equilibrium is disturbed, giving Distribution B. Notice that P_{rel} drops tremendously even though the number of ways, W, increases. This is because now we are comparing this distribution to others having many more atoms than before. Once the system starts scrambling, it is very unlikely that it will ever be found in this distribution again. Rather, the most probable distribution after this disturbance, Distribution C, is characterized by 200 H_2, 200 D_2, and 400 HD molecules.

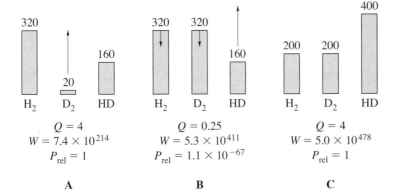

Figure 1.4
Three distributions illustrating Le Châtelier's principle. In each case, the reaction quotient, Q, the number of ways of that distribution being found, W, and the relative probability of this distribution relative to the most probable for that number of H and D atoms, P_{rel}, are indicated. Distribution A is the most probable distribution before the disturbance. Adding 300 D_2 molecules creates Distribution B, which, though having more ways of being found relative to A is still incredibly less probable to be found with continued shuffling *relative to Distribution C*. It is inconceivable that with scrambling the system will ever be found again in Distribution B.

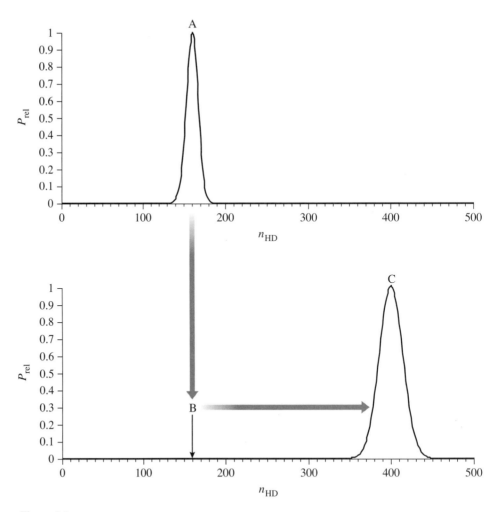

Figure 1.5
Relative probabilities vs. number of HD molecules in the mixture of 800 H and 200 D atoms before (above) and after (below) adding an additional 300 D_2 molecules. The equilibrium "shifts to the right," producing more HD, illustrating Le Châtelier's principle. Distribution A from Figure 1.4 is at the center of the curve on the top; Distribution C is at the center of the curve on the bottom. Distribution B is at the same place as Distribution A (160 HD), but its probability *relative to Distribution C* is essentially 0.

The reaction that occurs in response to the disturbance is the forward reaction

$$120 \ H_2 + 120 \ D_2 \longrightarrow 240 \ HD$$

which gets rid of some of that added D_2 and ends up with 200 H_2 + 200 D_2 + 400 HD. The resultant most probable distribution still has $Q = 4$, but we say that the equilibrium has *shifted* to the right. The shift is easily seen on a graph of relative probability vs. number of HD molecules (Figure 1.5). As another example, let us now add 400 HD molecules to our new equilibrium mix, giving 200 H_2 + 200 D_2 + 800 HD initially (Figure 1.6). The system responds to the added HD product with the reverse reaction

$$100 \ H_2 + 100 \ D_2 \longleftarrow 200 \ HD$$

shifting the equilibrium to the left and re-establishing $Q = 4$ and a much more probable distribution, 300 H_2 + 300 D_2 + 600 HD.

Note that there is no magic here. The shift in the equilibrium is simply the establishment of a new most probable distribution based entirely on the idea that *that* distribution has *soooo*

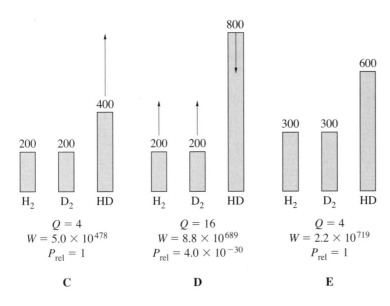

Figure 1.6
Adding 400 HD molecules to Distribution C (the third distribution in Figure 1.4) results in Distribution D. With continued scrambling, this distribution is not expected to last long, because it has a probability of only 4.0×10^{-30} relative to Distribution E. Random exchange will lead ultimately to a new most probable distribution, Distribution E, or something very close to it.

many more ways of being found. Each individual way is just as likely as another, but this final most probable distribution constitutes a whopping 2.2×10^{719} ways. With a few thousand atoms it might be possible to observe slightly different distributions now and then, but with mole quantities, we don't have to worry about them. (The curves in Figure 1.5 collapse to sharp spikes.) It's just too unlikely that the system will be found in any significantly different distribution. Again, there will be fluctuations around this distribution. We just won't be able to detect them.

1.15 Determining Equilibrium Amounts and Constants Based on Probability

Given the following information:

$$H_2 + D_2 \rightleftharpoons 2\,HD \qquad K = 4$$

and specific initial concentrations, you may already have learned how to determine the equilibrium concentrations of H_2, D_2, and HD. For example, say you had 4.0 mol of H_2 and 1.0 mol of D_2 in a 1-L flask to start with. The technique involves setting up a table something like the following, where $2x$ is the number of moles of HD formed at equilibrium:

	H_2	D_2	HD
[ ]$_0$	4.0	1.0	0
Δ	$-x$	$-x$	$+2x$
[ ]$_{eq}$	$4.0-x$	$1.0-x$	$2x$

and substituting these values into the equilibrium expression:

$$4 = K_c = \frac{[HD]^2}{[H_2][D_2]} = \frac{(2x)^2}{(4.0 - x)(1.0 - x)}$$

Solving this equation using either algebra or an equation-solving calculator gives $x = 0.8$. Substituting in on the bottom line, we get $[H_2]_{eq} = 3.2\ M$, $[D_2]_{eq} = 0.2\ M$, and $[HD]_{eq} = 1.6\ M$.

This method works just fine when there is only one equilibrium equation and we know the value of its equilibrium constant. However, as we have seen, there is another way, based on probability, which does NOT depend upon knowing the value of K initially and works for all isotope exchanges. In summary, the method goes like this:

1. Determine the number of moles of each of the exchanging atom types ($N_H = 8.0$ mol, $N_D = 2.0$ mol, for example).
2. Add these to get the total number of exchanging atoms ($N_{atoms} = 10.0$ mol).
3. Determine the probability of a randomly selected atom being of each type:

$$P_H = \frac{N_H}{N_{atoms}} = \frac{(8.0 \text{ mol})}{(10.0 \text{ mol})} = 0.80$$

$$P_D = \frac{N_D}{N_{atoms}} = \frac{(2.0 \text{ mol})}{(10.0 \text{ mol})} = 0.20$$

Note that probabilities properly have no units, and the sum of these atomic probabilities is 1.

4. From this information, determine the approximate probability of each *molecule* type:

$$P_{H_2} = P_H \times P_H = (0.80)(0.80) = 0.64$$

$$P_{D_2} = P_D \times P_D = (0.20)(0.20) = 0.04$$

$$P_{HD} = P_H \times P_D + P_D \times P_H = (0.80)(0.20) + (0.20)(0.80) = 0.32$$

Note that, here too, the values are unitless, and the sum of the probabilities is 1.

5. Determine the total number of *molecules*. In this case we are starting with 4.0 mol of H_2 and 1.0 mol of D_2, for a total of 5.0 mol of molecules. Note that in isotope exchange reactions the total number of molecules does not change, so in the end we still have 5.0 mol of molecules.
6. Finally, using the idea that $N_x = P_x \times N_{molecules}$, determine the number of each type of molecule at equilibrium:

$$N_{H_2} = P_{H_2} \times N_{molecules} = (0.64)(5.0 \text{ mol}) = 3.2 \text{ mol } H_2$$

$$N_{D_2} = P_{D_2} \times N_{molecules} = (0.04)(5.0 \text{ mol}) = 0.2 \text{ mol } D_2$$

$$N_{HD} = P_{HD} \times N_{molecules} = (0.32)(5.0 \text{ mol}) = 1.6 \text{ mol HD}$$

So we have that the most probable distribution in this case is characterized by 3.2 mol H_2, 0.2 mol D_2, and 1.6 mol of HD. Now, from *this* result we can actually calculate the value of the equilibrium constant, K:

$$K = \frac{[HD]^2}{[H_2][D_2]} = \frac{(N_{HD}/V)^2}{(N_{H_2}/V)(N_{D_2}/V)} = \frac{(1.6/1)^2}{(3.2/1)(0.2/1)} = 4$$

Note that we could have ignored the volume, V, altogether here, as it drops out from the equation in this and all isotope exchange reactions. In fact, we can easily show that Q is always 4 for this system regardless of how much H_2, D_2, or HD we start with. Based on the idea that $N_x = P_x N_{molecules}$, we have:

$$K = \frac{N_{HD}^2}{N_{H_2} N_{D_2}} = \frac{(P_{HD} N_{molecules})^2}{P_{H_2} N_{molecules} P_{D_2} N_{molecules}} = \frac{(P_{HD})^2}{P_{H_2} P_{D_2}}$$

(Here P stands for probability, not pressure!) So we can use P_x just as we could use n_x or [X] in these isotope exchange reactions. Substituting in the equivalent *atomic* probabilities we calculated in Step 4, all the probabilities cancel, and we get the value 4:

$$K = \frac{(P_H P_D + P_D P_H)^2}{(P_H P_H)(P_D P_D)} = \frac{(2 P_H P_D)^2}{P_H^2 P_D^2} = 4$$

Thus, the reaction quotient, Q, is a constant at equilibrium, regardless of the initial number of H or D atoms in the system. We say that at equilibrium, $Q = K$, which is 4 in this case.

One of the main objectives of this book is to convince you that not only in simple isotope exchange reactions but in *all* reactions, equilibrium amounts, equilibrium expressions, and the whole concept of an equilibrium constant are all based solely on probability. You may not be convinced yet that this method is easier than the tabular method of determining equilibrium concentrations involving setting up and solving an equation involving x, but remember that here we do not have to worry about there being more than one equilibrium at work, and we didn't even have the value of K to start with. Using probability not only gives us the equilibrium amounts of all reactants and products, it also gives us the *value* of the equilibrium constant itself.

1.16 Summary

Applying just the very basics of probability theory, namely

$$P_x = \frac{\text{ways of getting } x}{\text{ways total}}$$

we have derived four basic ideas useful for calculating the probability of various outcomes:

- AND probability multiplies
- OR probability adds
- AND and OR probability can be combined
- $P_{\text{not } x} = 1 - P_x$

With small systems we have to be careful about dependencies and count our cards carefully. However, as the number of cards or atoms in a system increases, we can make some very good approximations that remove the effect of these dependencies. In particular, when we want to know the *most probable distribution* of atoms, we can base our entire calculation on simple atom probabilities calculated as $P_x = N_x/N_{\text{atoms}}$.

For simple isotope exchange reactions, even with relatively few atoms, the idea of an equilibrium constant is justified. The equilibrium state in such an atom-scrambling reaction is characterized by the most probable distribution of atoms. Just as for flips of a coin, we cannot guarantee a result, but due to the incredible number of atoms involved in real chemical reactions, the *fluctuations* we might imagine observing are just too small to detect.

In reality, of course, things aren't quite so simple. The big difference between what we have been doing and real chemical reactions is that in real chemical systems there is something else besides atoms that is being distributed, namely energy. In Chapter 2 we will start to look at the rules relating to how energy is distributed in real chemical systems.

Problems

The symbol ▣ *indicates that the answer to the problem can be found in the "Answers to Selected Exercises" section at the back of the book.*

Basic Probability

1.1 ▣ A jar contains one blue ball, one green ball, one red ball, and three orange balls. What is the probability that a ball drawn at random (a) will be green? (b) will be orange? (c) will not be blue?

1.2 A CD player holds three CDs. The first disk by U2 has 10 tracks, the second disk by Pink Floyd has 9 tracks, and the final disk by Yo-Yo Ma has 16 tracks. If the CD player is set to pick a track at random from any CD, but will not repeat any one track until all tracks have played on all CDs, what is the probability that (a) the first track played will be by U2? (b) the first three tracks will be by U2? (c) the first song will be my favorite Pink Floyd song? (d) none of the first 5 tracks played will be by Yo-Yo Ma?

1.3 Your dresser has three shirts, one black, one blue, and one green; three pairs of pants, also one black, one blue, and one green; and three pairs of socks, also one black, one blue, and one green. You randomly pick out one shirt, one pair of pants, and one pair of socks. What is the probability (a) that your shirt will be green? (b) that you won't have to wear the green pants? (c) that your outfit will be monochromatic—that is, that the shirt, pants, and socks will be the same color? (d) that you will look like Johnny Cash—that is, that the shirt, pants, and socks will all be black?

1.4 ▣ A four-sided die (shown below) is a tetrahedron rather than a cube. All of the sides of the die are regular equilateral triangles rather than squares as in a six-sided die. The faces of the four-sided die are labeled 1, 2, 3, and 4. (a) Assuming there is an equal probability of landing on any side, what is the probability of rolling a 2 with this die? If a pair of four-sided dice is rolled, what is the probability that the sum of the two dice (b) will be 4? (c) will be 4 or 6?

1.5 An eight-sided die is a regular octagon rather than a cube. All sides of the die are labeled 1 to 8. A ten-sided die is a regular decahedron rather than a cube. All sides of this die are labeled 1 to 10. For both of these dice there is an equal probability of landing on any side. For each die (a) what is the probability of rolling an 8? Why is there a greater probability of rolling an 8 with one die? (b) what is the probability of not rolling a 2? (c) what is the probability of rolling an 8 or a 9?

1.6 (a) For the single roll of a normal six-sided die, show that the sum of the probabilities of all possible outcomes adds up to 1. (b) Show that this is true also for the roll of two four-sided dice. (c) Provide a simple nonmathematical explanation of why the sum of all the probabilities of all of the outcomes of an event *has* to add up to 1.

1.7 ▣ What is the probability that the first card drawn from a well-shuffled standard deck of cards (a) is a diamond? (b) is the jack of diamonds? (c) will not be a face card—a jack, queen, or king? (d) What is the probability that the first two cards drawn will be diamonds?

1.8 Powerball™ is a lottery game played in many states. To win the jackpot you need to match 5 numbers, in any order, to the numbers drawn from a field of 49 and match one number, the powerball, to a number drawn from a field of 42. You can think of it as drawing five numbered balls from a jar with 49 balls numbered 1–49 and then drawing one ball from a second jar that has 42 balls numbered 1–42. (a) What is the probability of winning the jackpot? (b) What is the probability of not winning the jackpot?

Chemical Distributions

1.9 Simple probability governs the equilibrium positions of chemical systems and the number of heads and tails you can toss with a coin. Why does a "normal" chemical system never seem to fluctuate from its equilibrium position, but if you flip a coin 100 times you expect that you might not get exactly 50 heads and 50 tails?

1.10 ▣ A system initially containing 2.0 mol of H_2 and 1.0 mol of D_2 is allowed to equilibrate, forming HD. Determine the amounts of H_2, D_2, and HD expected at equilibrium.

1.11 A system initially containing 2.0 mol of H_2O, 1.0 mol of D_2O, and 2.0 mol of HDO is allowed to exchange hydrogen for deuterium. Determine the amounts of H_2O, D_2O, and HDO expected at equilibrium. (Note that the oxygen atom should be ignored because it is not exchanging.)

1.12 ▣ Chlorine naturally occurs with two major isotopes ^{35}Cl and ^{37}Cl. On earth, approximately 76% of all chlorine atoms are ^{35}Cl and 24% are ^{37}Cl. Given 1.5 mol of chlorine atoms in their natural abundances, what is the most probable distribution of $^{35}Cl_2$, $^{37}Cl_2$, and $^{35}Cl^{37}Cl$?

1.13 ▣ Starting with the system in the previous problem, what is the new most probable distribution if (a) 0.5 mol of $^{35}Cl_2$ is added? (b) if 0.25 mol of $^{35}Cl^{37}Cl$ is added? (c) if 0.50 mol of $^{37}Cl_2$ and 0.25 mol of $^{35}Cl^{37}Cl$ are added?

1.14 The three natural isotopes of oxygen and their natural abundances are ^{16}O (99.757%), ^{17}O (0.038%), and ^{18}O (0.205%). What is the most probable distribution of O_2 molecules among the six different possibilities in a sample containing 2 mol of natural O_2?

1.15 Copper naturally occurs in two major isotopes, ^{65}Cu and ^{63}Cu. What is the number of moles of $^{63}Cu^{65}Cu$ you expect to have at equilibrium if you start with 1.75 mol of $^{65}Cu_2$ and 0.50 mol of $^{63}Cu_2$?

1.16   Starting with the system in the previous problem, what is the new most probable distribution if (a) 0.75 mol of $^{63}Cu_2$ is added? (b) 1.75 mol of $^{63}Cu^{65}Cu$ is added? (c) 0.25 mol of $^{63}Cu_2$ and 0.25 mol of $^{63}Cu^{65}Cu$ are added?

1.17 Ⓐ   Iridium occurs naturally in two isotopes ^{191}Ir and ^{193}Ir with natural abundances of 37.3% and 62.7% respectively. A 1.345-mol sample of natural Ir_2O_3 is analyzed. (a) What would be the expected amounts of $^{191}Ir^{193}IrO_3$, $^{193}Ir_2O_3$ and $^{191}Ir_2O_3$? (Note that the oxygen atom should be ignored since it is not exchanging.) (b) Using your answers from (a), calculate the equilibrium constant for the isotope exchange reaction between $^{191}Ir^{193}IrO_3$, $^{193}Ir_2O_3$, and $^{191}Ir_2O_3$ as written:

$$2\ ^{191}Ir^{193}IrO_3 \rightleftharpoons\ ^{191}Ir_2O_3 +\ ^{193}Ir_2O_3$$

1.18   A sample of 1.75 mol of N_2 is created by mixing 1 mol of $^{14}N_2$ and 0.75 mol of $^{15}N_2$. The molecules are allowed to react, and equilibrium is established. What is the most probable distribution of $^{14}N_2$, $^{15}N_2$, and $^{14}N^{15}N$ at equilibrium?

1.19 Ⓐ   Silver naturally occurs in two major isotopes, ^{107}Ag and ^{109}Ag. A 1.75-mol sample of natural silver atoms is allowed to react to form Ag_2. After equilibrium is achieved, the most probable distribution of Ag_2 molecules is found to be 0.235 mol $^{107}Ag_2$, 0.203 mol $^{109}Ag_2$, and 0.437 mol $^{107}Ag^{109}Ag$. Using this information and a bit of algebra, determine the natural abundance of ^{107}Ag, expressed in percent (%).

Brain Teasers

1.20   A normal six-sided die has been *loaded* so that the probability of rolling a 6 is twice as much as that of rolling any other number. What is the probability (a) of rolling a 6 with this die? (b) of rolling a 4? (c) How much more likely are you to roll two of these loaded dice and have the sum of the two dice be 8 than you are with the roll of two normal dice?

1.21   A mixture of ammonia isotopes containing 2.0 mol of NH_3 and 1.0 mol of ND_3 is allowed to equilibrate, producing a mixture of NH_3, ND_3, NH_2D, and NHD_2. (a) Ignoring the N atoms (because they are not exchanging), determine the most probable distribution once equilibrium is achieved. (b) Use this information to predict the value of the equilibrium constants for each of the following two isotope exchange reactions:

$$NH_3 + ND_3 \rightleftharpoons NH_2D + NHD_2$$
$$NH_3 + NHD_2 \rightleftharpoons 2\ NH_2D$$

1.22   Methane, CH_4, can undergo two types of isotope exchanges. The carbon atom can exchange ^{12}C for ^{13}C, and the hydrogens can exchange H for D. For example, one possibility would be:

$$^{13}CH_2D_2 +\ ^{12}CH_2D_2 \rightleftharpoons\ ^{13}CH_3D +\ ^{12}CHD_3$$

(a) List all possible isotope exchange reactions in this case (including simple switches of reactants for products). (b) Determine the value of K for each possibility based on probability.

CHAPTER 2

The Distribution of Energy

As often in natural science the picture changed quite suddenly. New fruitful concepts appeared, through the interplay and extension of which most of the darkness has been to a large extent dispelled in a single stroke.

—*Walther Nernst*

2.1 Real Chemical Reactions

The $H_2/D_2/HD$ equilibrium we looked at in Chapter 1:

$$H_2 + D_2 \rightleftharpoons 2HD \qquad K = \frac{[HD]^2}{[H_2][D_2]}$$

should have an equilibrium constant of 4 based on probability. However, this reaction has been studied carefully and found to have an equilibrium constant of only 3.25 at 25°C. It increases slowly with temperature until, around 2000 kelvins, it finally reaches the value 4. Why isn't K just 4 at all temperatures? Something is missing from our analysis! Shown in Figure 2.1 is a graph of K vs. T for four related reactions. Remember, the value of K is a measure of extent of a chemical reaction. The horizontal line where $K = 1$ is important. At a temperature where $K = 1$, neither reactants nor products will be favored. When $K > 1$, we say that products are favored in the reaction, while when $K < 1$, we say that reactants are favored.

In all of these cases, the value of K either increases or decreases as the temperature is raised. Why do two of these reactions have a very large value of K at low temperature, which decreases with increasing temperature? Why do the other two slope upward, increasing in value as the temperature is raised?

2.2 Temperature and Heat Energy

One way to interpret these temperature effects is to look at how the absorption and release of energy affects chemical reactions. Basically, reactions that *release* heat ("exothermic" reactions) are driven backward, toward reactants (K decreases), with increasing temperature. In contrast, reactions that *absorb* heat ("endothermic" reactions) are driven forward with increasing temperature (K increases). This is depicted in Figure 2.2. Note that at high temperature, whether reactants or products are favored is independent of whether the reaction is endothermic or exothermic. This is an important point we will return to in Chapter 6, where the focus is on the effect of temperature on chemical equilibrium.

Chemical reactions that absorb heat are sometimes depicted with "heat" as a reactant. For example, the reaction shown in Figure 2.1 involving the dissociation of N_2O_4 into NO_2

21

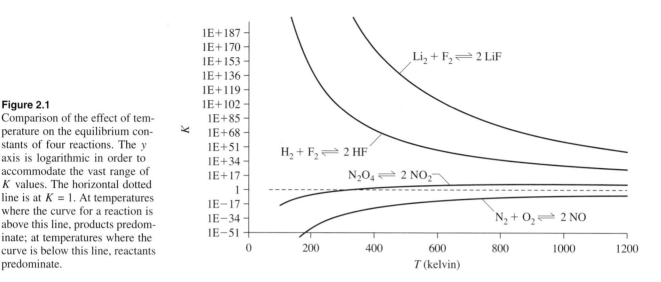

Figure 2.1
Comparison of the effect of temperature on the equilibrium constants of four reactions. The y axis is logarithmic in order to accommodate the vast range of K values. The horizontal dotted line is at $K = 1$. At temperatures where the curve for a reaction is above this line, products predominate; at temperatures where the curve is below this line, reactants predominate.

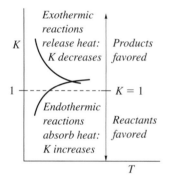

Figure 2.2
Exothermic reactions show a decrease in the value of K as temperature increases, while endothermic reactions show the opposite effect. Whether or not the value of K crosses the value of 1 as temperature increases is a totally different question.

might be written

$$N_2O_4 + heat \rightleftharpoons 2\,NO_2 \qquad K = \frac{[NO_2]^2}{[N_2O_4]}$$

In this reaction, energy is required to break the bond in N_2O_4; heat is absorbed. Although this reaction generally favors reactants, an increase in temperature results in an increase in K, shifting the equilibrium more to the right:

$$reactants + heat \longrightarrow products$$

On the other hand, reactions that are exothermic and thus release heat, such as the upper two reactions in Figure 2.1 and the upper curve in Figure 2.2, can be depicted as:

$$reactants \longrightarrow products + heat$$

In this case, increasing the temperature does not promote the formation of products, and instead drives the reaction backward in the direction of reactants. We will find that reactions of this type, exothermic reactions, always involve the *making* of bonds. For example, in the reaction:

$$H_2 + F_2 \longrightarrow 2\,HF + heat$$

two very strong H−F bonds are made at the expense of two much weaker bonds. Whenever stronger bonds are made than are broken, we will see that the reaction is exothermic, releases heat, and is driven *backward* by increasing the temperature.

Clearly temperature and/or the availability of heat plays a major role in determining the direction of real chemical reactions. But what is *heat energy* really, and why is it associated with equilibrium constants and temperature this way? Is heat a "substance" to be gained or lost? Before we can give definitive answers to these questions (in Chapter 4), we must consider how exactly energy and matter interact (in this chapter and Chapter 3). We will see that, once again, probability has *everything* to do with it.

2.3 The Quantized Nature of Energy

The key to understanding what is going on here is in understanding what energy is and how molecules interact with it. Just as we can talk about distributing atomic isotopes among a set of molecules, as was done in Chapter 1, so, too, can we talk about distributing energy among a set of molecules or atoms.

We'll refer to molecules or atoms as *particles* in this chapter. In a system of HCl(*g*), for example, the particles are molecules; in a system of Ne(*g*), the particles are atoms. In a complex reaction, the particles might be of many sorts. The idea is to figure out what the *most probable distribution of energy* will be within whatever system we are investigating. By doing that, we will be able to say what the system should look like at equilibrium. Will all the energy go to one particle or one sort of molecule? Will it be evenly spread around? We can only guess at this point, but the answer is not far off.

Fundamental to this discussion is the idea that energy (light and heat radiation) comes in "packets" called *quanta* or *photons*. Some photons are very energetic (X rays and ultraviolet rays, for example) while some are not very energetic at all (infrared and radio waves, for example). All chemical substances interact with light and heat by absorbing some of the energy that they encounter while allowing the rest to pass through.

The discoveries by Planck and Einstein that light is quantized and interacts with matter as discrete photons completely revolutionized the fields of chemistry and physics around the turn of the 20th century. Einstein realized that in order to understand this interaction, statistics and the laws of probability would have to be applied.

What Einstein's perspective teaches us is that all chemical reactions can be thought of as an exchange of energy within a system of particles (atoms and molecules) in the presence of that system's surroundings. For now, we'll focus on the system and leave the surroundings for later. What we will discover is that when a chemical system absorbs energy, it does so in very predictable ways, much as gambling in a casino results in predictable profits, due to the nature of randomness and probability when applied to large systems.

2.4 Distributions of Energy Quanta in Small Systems

We start with a very small system—just three particles and three "quanta" of energy. It's not important what a "quantum" of energy is, only that quanta are discrete packets of energy. A particle will be allowed to have and exchange only a countable number of quanta of energy. In this imaginary system, we will allow each particle to have only 0, 1, 2, or 3 units of energy. We'll depict the particles as follows:

3 quanta: (((●)))

2 quanta: ((●))

1 quantum: (●)

0 quanta: ●

You can think of the parentheses as representing shivering. The more quanta a particle has, the more it is shivering. (This really isn't very far from the truth!) Since there are only 3 quanta total in our system, clearly there are no other possibilities. A particle can't have 1½ quanta, nor can it have 4.

OK, the idea is to figure out what it will look like when we look into the box. We must determine what all the different ways of distributing our three quanta are and then group them into distinct sets and decide which set is most probable. Here goes. It turns out there are ten ways in all to distribute three quanta of energy among three particles, and these ten can be divided into three distinct sets, or *distributions:*

Distribution A: One particle has all the energy (three possibilities):

Distribution B: One particle has two units, one has one, and one has none (six possibilities):

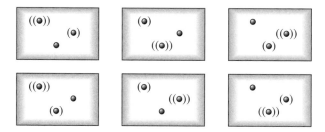

Distribution C: Each particle has one unit of energy (one possibility):

These are the only ways to distribute three quanta of energy among three particles. In all, we have $3 + 6 + 1 = 10$ possible ways. Each of these ways we call a *microstate*.

We make an important assumption here that no microstate is any more likely than any other. Whether this is an appropriate assumption is a good question. We're assuming, for example, that it's no better or worse to have two particles next to each other jiggling with the other resting than having two further apart jiggling with one in the middle doing nothing. This is like assuming, when dealing with cards, that having two red cards next to each other is no different fundamentally from having two black cards next to each other. Only a very strange deck of cards would be otherwise.

In addition, we must assume that there is some mechanism (like bumping into each other) that allows particles to continually trade energy. This property of molecular systems is called *continuous weak coupling,* and it is analogous to the shuffling of a deck. To make predictions based on probability in a card game you have to maintain a shuffled deck; to make predictions in chemistry, you have to have continuous weak coupling.

So we have three particles in a box, and we have given them a total of three units of energy. What should we observe when we look into the box? Provided no microstate is any more likely than another, three out of ten times we look we will find a single particle moving and the other two resting (Distribution A). Most of the time (60%) we will see one particle with 2 units, one with 1 unit, and one with no energy at all (Distribution B), and 10% of the time we look we will find that each particle has a single-unit share of the energy (Distribution C). Clearly the "most probable distribution" is Distribution B, with 6 possibilities.

To keep track of these microstates, we will use the letter W, for *ways*. In this example, $W_A = 3$, $W_B = 6$, and $W_C = 1$. In addition, we could say $W_{total} = 10$. Based on these microstates being equally probable, we conclude that the probability of finding the system with a particular distribution of energy is again (ways of x) / (ways total):

$$P_x = \frac{W_x}{W_{total}}$$

Thus, the probabilities of finding the individual distributions in this case are calculated to be:

Distribution A: $W_A = 3$ $P_A = 3/10 = 0.3$
Distribution B: $W_B = 6$ $P_B = 6/10 = 0.6$ (most probable)
Distribution C: $W_C = 1$ $P_C = 1/10 = 0.1$

Another name for W is *thermodynamic probability*. To be a real probability, we have to divide W_x by W_{total}, but for comparing two distributions we never have to determine W_{total}. Let's see why.

Notice that the *relative probability* of distribution A with respect to the most probable distribution (B) is 0.5, since $P_A / P_B = 0.3/0.6 = 0.5$. But we could have gotten this relative probability using thermodynamic probabilities just the same, because $W_A / W_B = 3/6 = 0.5$ as well. So thermodynamic probabilities W are just as good as "real" probabilities if all we are interested in is comparing two possible distributions. It turns out that is exactly what we will be doing!

After you practice a little figuring out all of the microstates of a few simple systems, you will realize that it doesn't take too many particles or too much energy to start getting very complicated. For example, just giving 5 units of energy to 5 particles involves seven distinctly different distributions and a full 2002 microstates! We need to learn how to calculate W without going to all the trouble of writing all these pictures depicting boxes containing shivering particles. The idea is to organize our information not in terms of distinct particles, but instead in terms of distinct amounts of energy.

Let's look again at our first way of distributing 3 units of energy among 3 particles:

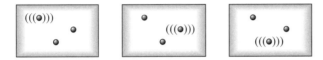

All three of these microstates can be depicted in one drawing if we focus on the fact that each has one particle with 3 quanta of energy and two with 0 quanta:

Quanta	Particles		
3	●	1	one particle with 3 quanta
2			
1	or		
0	● ●	2	two particles with 0 quanta

Distribution A

Now instead of looking at individual *microstates* we are looking at individual *distributions*. You can draw two particles as "oo" or just write the number "2." It's your choice. It takes a little practice, but once you get the hang of it, it is very handy.

All three possible distributions together can be written in one neat little package:

Quanta	Distinct Distributions		
3	●		
2		●	
1		●	● ● ●
0	● ●	●	
	A	B	C

This says there are three possible *distinct* distributions. In each case there are three particles and three units of energy total. In the first distribution, only one particle has any energy, all 3 units. In the second, one particle has 2 units, one has 1 unit, and one has no energy. Finally in the third way of distributing 3 units of energy among 3 particles, each has 1 unit.

It's important to realize that these simplified depictions say nothing about *which* particle has *what* energy. These depictions are simply tricks to quickly map out all the possible *distinct* distributions of energy. Indeed, they represent *all possible ways* by which the system can be found in a state involving a certain number of particles having a certain amount of total energy. In the next section we'll see a quick method of figuring out the number of possible ways or *thermodynamic probability*, W, for each distinct distribution.

2.5 Calculating *W* Using Combinations

Say we had the state shown below, which is one of several ways to distribute 7 units of energy among 9 particles:

2	● ●	2 particles have two units of energy; $n_2 = 2$
1	● ● ●	3 particles have one unit of energy; $n_1 = 3$
0	● ● ● ●	4 particles have zero units of energy; $n_0 = 4$

We'll just call this "Distribution M" (for *Mountainous*). The number of ways of doing this can be thought of as the product of two *independent* terms:

$$W_M = W_2 \times W_1$$

where

$W_2 = $ the ways of giving 2 units of energy to 2 of 9 particles

$W_1 = $ the ways of giving 1 unit of energy to 3 of the *remaining* 7 particles

The process of distributing this energy could thus be laid out as follows:

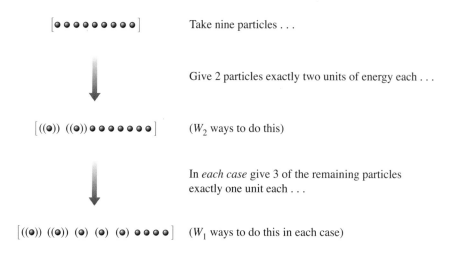

$\left[● ● ● ● ● ● ● ● ● \right]$ Take nine particles . . .

Give 2 particles exactly two units of energy each . . .

$\left[((●))\ ((●)) ● ● ● ● ● ● ● \right]$ (W_2 ways to do this)

In *each case* give 3 of the remaining particles exactly one unit each . . .

$\left[((●))\ ((●))\ (●)\ (●)\ (●) ● ● ● ● \right]$ (W_1 ways to do this in each case)

So if we could calculate W_2 and W_1, we would just write $W_M = W_2 \times W_1$, and we would be done. Mathematically what we are doing is calculating the "number of combinations of n things taken p at a time," C_p^n.[1] Thermodynamic probability W_2 is the number of combinations of 9 things taken 2 at a time, C_2^9. Thermodynamic probability W_1 is the number of combinations of 7 things taken 3 at a time, C_3^7. It can be shown that

$$\mathrm{C}_p^n = \frac{n(n-1)(n-2)\cdots(n-p+1)}{1\cdot 2\cdot 3\cdot\ \cdots\ \cdot p} \tag{2.1}$$

Note that the number of combinations of n things taken 1 at a time, C_1^n, is n:

$$\mathrm{C}_1^n = \frac{n}{1} = n$$

That is, there are n ways to take n things one at a time. You can take the first one or the second one or the third, etc.—n ways.

And the number of combinations of n things taken n at a time, C_n^n, is 1:

$$\mathrm{C}_n^n = \frac{n(n-1)(n-2)\cdots 1}{1\cdot 2\cdot 3\cdot\ \cdots\ \cdot n} = 1$$

This means that there is only one way you can take all n things all at once. You just take them all! Here we are interested in

$$\mathrm{C}_2^9 = \frac{9\cdot 8}{1\cdot 2} = 36 \quad \text{and} \quad \mathrm{C}_3^7 = \frac{7\cdot 6\cdot 5}{1\cdot 2\cdot 3} = 35$$

There are 36 ways to give two units of energy to 2 of 9 particles. You could give it to the first and the second, or the first and the third, or the first and the fourth, and so on. And there are 35 ways to give one unit of energy to 3 of 7 particles. You could give it to the first and the second and the third, or the first and the second and the fourth, etc.

So we have that $W_{\mathrm{M}} = 36 \times 35 = 1260$. There are 1260 ways to distribute 7 units of energy among 9 particles in this manner, with two units going to each of 2 particles and three units going one to each of the three of the remaining 7 particles. Note that we could also say that we are finally giving no energy to each of the remaining 4 particles and define for the lowest level, $W_0 = \mathrm{C}_4^4 = 1$. Thus, we can write:

$$W_{\mathrm{M}} = W_2 \times W_1 \times W_0 = \mathrm{C}_2^9 \times \mathrm{C}_3^7 \times \mathrm{C}_4^4$$

$$= \frac{9\cdot 8}{1\cdot 2} \times \frac{7\cdot 6\cdot 5}{1\cdot 2\cdot 3} \times \frac{4\cdot 3\cdot 2\cdot 1}{1\cdot 2\cdot 3\cdot 4} = \frac{9!}{2!\,3!\,4!} = 1260$$

where we introduce here the *factorial* notation "$n!$" which means $1\cdot 2\cdot 3\cdot\ \cdots\ \cdot n$. (Look for it on your calculator.) If you look closely, you will see that what we have here is the factorial of the total number of particles (9) in the numerator and the product of the factorials of the number of particles of each energy type (2 with two units, 3 with one unit, and 4 with no units) in the denominator. The general form of the equation for calculating the number of possible ways of a system being in a particular distribution looks like this:

$$W = \frac{n!}{n_0!\,n_1!\,n_2!\cdots} \tag{2.2}$$

where now n is the total number of particles, n_0 is the number of particles with no units of energy, n_1 is the number of particles with one unit of energy, and so forth.

---

[1] In some statistics texts the combination of n things taken p at a time is written

$$\binom{n}{p} \quad \text{or} \quad {}_nC_p$$

For another example, consider a different distribution of seven units of energy among 9 particles:

3	___•___	1 particle has three units of energy; $n_3 = 1$
2	_____	0 particles have two units of energy; $n_2 = 0$
1	_•••• _	4 particles have one unit of energy; $n_1 = 4$
0	_•••• _	4 particles have zero units of energy; $n_0 = 4$

In this case, which we will call "Distribution K" (for *Kingly*, because it looks like a crown), we have:

$$W = \frac{9!}{4!\,4!\,0!\,1!} = 630$$

Note that for Level 2, we use 0!, which is defined to be the number 1. The Kingly distribution or "state" is only half as probable as the Mountainous one, since $630/1260 = 0.5$.

2.6 Why Equations 2.1 and 2.2 Work

To see *why* these equations work, consider the following example: Say you had 10¢ in the form of two nickels to distribute among three people (Fred, Mary, and John) such that no one person got both nickels. It turns out there is just one distinct distribution:

Nickels

3	___
2	___
1	••
0	•

Two people will each get a nickel and one person will be left empty handed. How many ways are there of doing this? The calculation can be done based either on Equation 2.1 or on Equation 2.2:

$$W = C_2^3 \times C_1^1 = \frac{3(2)}{2} \times \frac{1}{1} = 3$$

$$W = \frac{3!}{1!\,2!} = \frac{1 \cdot 2 \cdot 3}{(1)(1 \cdot 2)} = 3$$

The factor 3(2) in the numerator in either method just gives the number of ways of lining up the first two of the three people:

first:	Fred		Mary		John	
second:	Mary	John	Fred	John	Mary	Fred
	#1	#2	#3	#4	#5	#6

But this isn't quite what we want. We plan to give a nickel to each of two people, but we do not care who gets the first nickel and who gets the second. In each case there is a duplication:

Fred/Mary (#1)   Mary/Fred (#3)
Fred/John (#2)   John/Fred (#6)
Mary/John (#4)   John/Mary (#5)

Thus, in just listing all the possible arrangements of two people (3×2), we have overcounted the possibilities, and we have to divide by 2. There are really only three combinations of three people, two at a time. This is now the number of combinations of 3 things taken 2 at a time, $C_2^3 = (3 \times 2)/2 = (3)(2/2) = 3$. Generalizing, to calculate the number of ways of taking n things 2 at a time, C_2^n, we multiply n by $(n-1)/2$.

Had we distributed *three* nickels, one to each person, we would have ended up with just one possibility: $C_3^3 = (3 \times 2 \times 1)/(2 \times 3) = (3)(2/2)(1/3) = 1$, because all six possibilities are the same:

first: Fred   Mary   John
second: Mary John   Fred John   Mary Fred
third: John Mary   John Fred   Fred Mary
            #1   #2   #3   #4   #5   #6

and to calculate the number of ways of taking n things 3 at a time, C_3^n, we have to multiply n by both $(n-1)/2$ and $(n-2)/3$.

Generalizing, to calculate the number of ways of taking n particles p at a time, we need to multiply n by $(n-1)/2$ and $(n-2)/3$ and $(n-3)/4$, and so on, until we get to $(n-p+1)/p$. In each case, we are dividing so as to account for the fact that we do not care about the order in which items are combined. It does not matter who got the nickel first, nor does it matter which of two particles got their energy first.

2.7 Determining the Probability of a Particular Distribution of Energy

If we want to know the actual *probability* of the system being found in Distribution M, then we have to know W for each and every possible distribution. We can then add all these ways together to get W_{total}. The actual probability of Distribution M is W_M / W_{total}. Unfortunately, there is no simple way of figuring this out. We have to generate all possible distributions and calculate W for each.

However, there is a systematic way of getting all possible distributions that works quite well. The idea is to keep the energy the same by "colliding" two particles at a time in a process by which one gains a unit of energy and one loses. One goes up a level while one goes down a level, until all possible distributions have been found (Figure 2.3).

U	1	0	0	0
$U-1$	0	1	0	0
$U-2$	0	0	1	1
...	...	...	...	...
2	0	0	1	...
1	0	1	0	2
0	$n-1$	$n-2$	$n-2$	$n-3$
	A	B	C	D

Figure 2.3
A systematic way of generating all possible distributions of U units of energy among n particles. Start with all the energy in one particle, and list all possible ways that particles can trade energy to give new distributions. Distribution B gives both C and D. Only the first four are shown.

For the general case of U units of energy among n particles, the scheme works like this:

1. Put all the energy U into one particle and call this Distribution A:

$$W_A = \frac{n!}{(n-1)!\,1!} = \frac{(n-1)!\,n}{(n-1)!} = n$$

2. Next, move this particle down one level and move a particle from Level 0 up one level to get Distribution B, which still has the same total energy, U, but is now considerably more probable:

$$W_B = \frac{n!}{(n-2)!\,1!\,1!} = \frac{(n-2)!(n-1)n}{(n-2)!} = (n-1)n$$

3. Repeat Step 2, but consider *both* possibilities—either the particle from Level 1 could move up or another particle from Level 0 could become energized. This thinking generates Distributions C and D.

4. Continue in this way, always moving the highest particle down one level while moving all possible particles below it up one level until no new distributions are found. There isn't any need to actually calculate W at this stage, but be careful not to label the same distribution twice!

For example, for 7 units of energy distributed among 9 particles you should get the 15 distributions listed in Figure 2.4. The two distributions we have been considering already, K and M, are shown in this figure. The most probable distribution is J, with 1512 microstates. The total number of ways, W_{total}, is $9 + 72 + 72 + 252 + 72 + 504 + 504 + 252 + 252 + 1512 + 630 + 504 + 1260 + 504 + 36 = 6435$. Thus, the probability of the system being found in State K is $630/6435 \times 100\% = 9.8\%$ and in State M is $1260/6435 \times 100\% = 18.6\%$. Notice that the *most probable* state isn't K or M. It's J, since $W_J = 1512$. The probability of the system being found in that state is 23.5%.

This is a lot of work! You will be happy to know that we will never find it necessary to determine all of the possible distributions of any real amount of energy among any real number of molecules. Instead, we will always be interested only in the probability of one state being found relative to one other. For example, in this case we can say that it is twice as probable for the system to be found in State M as in State K. For that we need not total up or even know all of the different thermodynamic probabilities. All we will need to know is W for each of the two states of interest.

Level	A	B	C	D	E	F	G	H	I	J	K	L	M	N	O
7	1														
6		1													
5			1	1											
4					1	1	1								
3					1			2	1	1	1				
2			1			1			2	1		3	2	1	
1		1		2		1	3	1		2	4	1	3	5	7
0	8	7	7	6	7	6	5	6	6	5	4	5	4	3	2
	A	B	C	D	E	F	G	H	I	**J**	K	L	M	N	O
W	9	72	72	252	72	504	504	252	252	**1512**	630	504	1260	504	36

Figure 2.4
All 15 possible distributions of 7 units of energy among 9 particles. The most probable distribution is distribution J, with 1512 microstates.

2.8 The Most Probable Distribution Is the Boltzmann Distribution

When a chemically significant number of particles are in a system—on the order of a mole—the number of possible distributions of energy is astronomical. We have no hope of even starting to list all of them. What we really need instead is an understanding of just the *most probable distribution* which, although not particularly probable in and of itself, represents an average of a very narrow band of observable distributions. These distributions in real systems are so tightly packed that they are for all practical purposes indistinguishable from each other.

So what can we possibly know about the most probable distribution of energy in a system? This question was answered in 1869 by Ludwig Boltzmann. He was able to derive an expression relating specifically to the most probable distribution of energy in a system where particles can only have specific amounts of energy.

These systems in general can be described as we have been doing already, as having a set of levels which are numbered 0, 1, 2, 3, . . . (or sometimes just 1, 2, 3, . . .) from lowest to highest in energy. We will refer to these levels as "Level i" or "Level j" where i and j are integers. Each Level i has an energy associated with it we will call ε_i ("epsilon sub eye") as well as a number of particles, n_i ("en sub eye"). This general case is depicted in Figure 2.5. Now, Levels i and j can be any two levels in the system. Generally we will consider i to be the lower level, in which case $\varepsilon_j > \varepsilon_i$, and so the difference $\varepsilon_j - \varepsilon_i > 0$.

What Boltzmann discovered *for the most probable distribution* was the relationship between the ratio of the number of particles in one level to those in another (n_j/n_i), the difference in energy between the two energy levels ($\varepsilon_j - \varepsilon_i$, also written as $\Delta\varepsilon_{ij}$), and the temperature. **The *Boltzmann law* is perhaps the most important equation in this entire book:**

$$\frac{n_j}{n_i} = e^{-(\varepsilon_j - \varepsilon_i)/kT} = e^{-\Delta\varepsilon_{ij}/kT} \qquad (2.3)$$

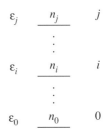

Figure 2.5
The variables associated with a general system. Levels are identified with a subscript such as 0, i, or j. Each level has an associated number of particles n and energy ε. We define $\Delta\varepsilon_{ij}$ as $\varepsilon_j - \varepsilon_i$.

The parameter T is the absolute temperature, in kelvins. The e is the base of the natural logarithm system, 2.71828182845905 . . . . The parameter k is called *Boltzmann's constant* and has units of energy/temperature. The value of k is 1.381×10^{-23} joules per kelvin (J/K).

The importance of this little equation cannot be overemphasized. It is the key to linking the microscopic world of atoms and energy to the macroscopic world of pressures, volumes, and temperatures with which we are more familiar. Boltzmann actually derived this equation from other laws of physics. We will just show here that it is reasonable. First, though, let's look carefully at the Boltzmann law and see what we can make of it.

First of all, consider the following example: Say we had a system with evenly spaced energy levels which were 2.0×10^{-20} joules apart. If the temperature were 298 K, and we knew we had 1 mol of particles in the lowest level, Level 0, then we would calculate for the number of particles in the next-to-lowest level, n_1, to be

$$n_1 = n_0 e^{-\Delta\varepsilon_{0,1}/kT} = (6.022 \times 10^{23})e^{\frac{-2.0\times 10^{-20}\,\text{J}}{(1.381\times 10^{-23}\text{J/K})298\,\text{K}}} = 4.668 \times 10^{21}$$

In effect, the Boltzmann law says that less than 1% of the particles are going to have any energy at all at this temperature!

We could then get the next level's population from n_1, since our system is evenly spaced:

$$n_2 = n_1 e^{-\Delta\varepsilon_{1,2}/kT} = (4.668 \times 10^{21})e^{\frac{-2.0\times 10^{-20}\,\text{J}}{(1.381\times 10^{-23}\text{J/K})298\,\text{K}}} = 3.619 \times 10^{19}$$

Note that because the exponential term is the same in both cases, the ratio of particles n_2/n_1 is the same as the ratio n_1/n_0, which is less than 1%.

Alternatively, we could get the population of Level 2 directly from the population of Level 0 using the fact that $\varepsilon_2 - \varepsilon_0 = 4.0 \times 10^{-20}$ J:

$$n_2 = n_0 e^{-\Delta \varepsilon_{0,2}/kT} = (6.022 \times 10^{23}) e^{\frac{-4.0 \times 10^{-20} \text{ J}}{(1.381 \times 10^{-23} \text{J/K}) 298 \text{ K}}} = 3.619 \times 10^{19}$$

Or, say we wanted to know the temperature at which this system would have a ratio of $1/2$ going from one level to another up the energy ladder. Putting 0.5 in for n_1/n_0 (or putting in 1 for n_1 and 2 for n_0) we could solve for $T = 2090$ K.

Finally, if we know, for example, that at 298 K the ratio $n_j/n_i = 0.0100$, then we can solve this equation for the energy difference between Level i and Level j, $\Delta \varepsilon_{ij}$. In this case it would be 1.9×10^{-20} J.

Four important characteristics of the Boltzmann distribution should be noted:

1. The most probable distribution will have more particles at the lowest level than at any other level, and fewer and fewer particles will be in each level as we go up in energy. This is because $\Delta \varepsilon_{ij}$ is positive, and since T and k are also positive, the right-hand side of Equation 2.3 takes the form e^{-a}, where $a > 0$. So the right side is between 0 and 1, and so is n_j/n_i. Then

$$n_j/n_i < 1 \quad \text{and} \quad n_j < n_i \quad \text{when } \varepsilon_j > \varepsilon_i$$

That is, the number of particles in a higher level (n_j) is always less than the number of particles in a lower level (n_i).

2. If the levels are evenly spaced, like going up a ladder of energy, then the sequence $n_0, n_1, n_2, \ldots$ forms a *geometric progression*. That is,

$$\frac{n_1}{n_0} = \frac{n_2}{n_1} = \frac{n_3}{n_2} = \frac{n_4}{n_3} = \cdots = e^{-\Delta \varepsilon_{0,1}/kT} < 1$$

The fact that all the ratios are the same for evenly spaced energy levels can be very handy. Once we calculate this single ratio, we can use it along with the number of particles in Level 0 to get the number of particles in Level 1. From that we can get the number of particles in Level 2, and from that, the number in Level 3, etc.

3. Often our lower level i will be the very lowest or *ground state*, and we will assign the lowest energy to have "no" energy, $\varepsilon_0 = 0$. Then the Boltzmann law takes the following alternative form:

$$\frac{n_j}{n_0} = e^{-(\varepsilon_j - \varepsilon_0)/kT} = e^{-\varepsilon_j/kT} \tag{2.4}$$

This is a popular form of the law, but be careful! It works only when the lowest energy level has been assigned 0 for its energy.

4. Let's think about what these numbers and lines mean. The numbers count particles, and their vertical position (their "level") denotes how much energy each particle has. We'll see in the next chapter, when we consider the particles to be actual molecules, that it is quite possible for two molecules to have exactly the same amount of energy but still be doing two different things. That is, the two are really in different molecular states. When this occurs, the energy level is said to be *degenerate*, and we indicate this degeneracy by drawing two (or more) separate lines at the same vertical position. In the example shown in Figure 2.6, system B has an energy level that is doubly degenerate. The number of particles in two such states i and j will be identical in the most probable distribution. That is, $n_i = n_j$ whenever $\varepsilon_i = \varepsilon_j$. We can see how this arises from the Boltzmann law, because if $\Delta \varepsilon_{ij} = 0$, then $n_i/n_j = e^{-0} = 1$, and $n_i = n_j$. In effect, each state gets a full share of particles as though there were no degeneracy.

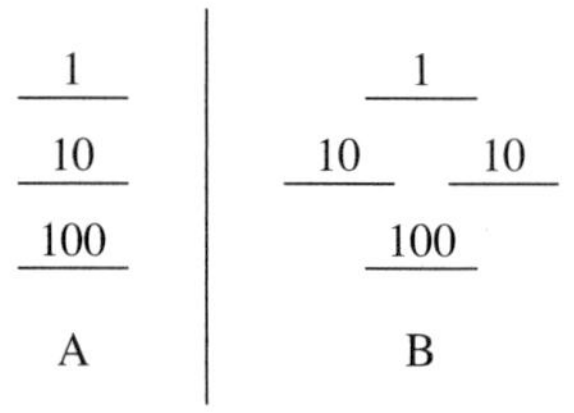

Figure 2.6
Boltzmann distributions for two different systems, both at the same temperature. Because of the degeneracy in B, there are twice as many particles of intermediate energy in B as in A.

We will see that the phenomenon of degeneracy is very important. It shows up specifically when the energy can be put into the system aligned somehow in more than one direction. For example, electronic energy can be put into a system in such a way that an electron ends up with "spin up" or "spin down" (a double degeneracy). Or we may say that an electron in an atom is excited into a p_x, p_y, or p_z orbital, all of which have the same energy (leading to a triple degeneracy). Or if a molecule is flying through space, it may be directed in the x, y, or z direction (also a triple degeneracy).

The value of Boltzmann's constant, $k = 1.381 \times 10^{-23}$ J/K, is not just any number. Amazingly, it is the ideal-gas constant, R, divided by Avogadro's number:

$$k = \frac{R}{N_{av}} = \frac{8.31451 \, \text{J/mol} \cdot \text{K}}{6.02214 \times 10^{23}/\text{mol}} = 1.38066 \times 10^{-23} \, \text{J/K} \tag{2.5}$$

You may be more familiar with R in units of L·atm/mol·K, where its numerical value is 0.082054, but R can be written in many different units. This relationship will have to be left for now as a bit of a mystery. It arises from two ideas:

1. All gases are always observed with their most probable distribution of energy.
2. The ideal-gas equation *itself* ($PV = nRT$) can be derived from the Boltzmann law!

Notice that k and R are really the same thing, just expressed in different units. Both have units of "energy/temperature" and, you will find, both often appear multiplied by T. In effect, k and R are really just "conversion factors" that allow us to relate units of absolute temperature (kelvins) to units of energy (calories, joules, L·atm, etc.). The constant k is useful when we are referring to individual particles, atoms, or molecules; the constant R is preferred when we are dealing with *moles* of things.

Thus, still another form of the Boltzmann law is sometimes used where the energy of levels is written in units of "energy-units/mol" (which is admittedly rather strange!) and R is used in place of k:

$$\frac{n_j}{n_i} = e^{-(E_j - E_i)/RT} = e^{-\Delta E_{ij}/RT} \tag{2.6}$$

Note that capital E is used here instead of ε to help us remember that our energy units are per mole. When solving problems using the Boltzmann law, just make sure you check your units carefully. A good idea might be to pick your own personal standard unit for k and R, such as joules, and stick with it. Some calculators have k as a built-in constant with units J/K and "Rc" built in with units of J/mol·K.[2]

Anytime you have a situation where energy is expressed in units of calories (handy when heat is involved) or ergs (an older energy unit), consider doing a conversion to joules first. In any case, *be careful*! For example, the two lowest vibrational energy levels in gaseous HCl molecules are reported to be "8.25 kcal/mol" apart. If $n_0 = 6.022 \times 10^{23}$, then we can calculate n_1 at 25°C using

$$n_i = 6.022 \times 10^{23}$$

$$\Delta \varepsilon_{ij} = 8.25 \, \text{kcal/mol}(1000 \, \text{cal/kcal})(4.184 \, \text{J/cal})(1 \, \text{mol}/6.022 \times 10^{23})$$

$$k = 1.381 \times 10^{-23} \, \text{J/K}$$

$$T = 298 \, \text{K}$$

---

[2] If you are using an equation-solving calculator, you will probably want to take a quick look at *http://www.stolaf.edu/depts/chemistry/imt/js/tinycalc/ti-calc.htm*. Always list the information for a problem on paper first, indicating all units. *But there is no need to do any algebra on paper*.

In this case $j = 1$ and $i = 0$. Solving for n_j, we get 5.38×10^{17}. Thus, $n_1/n_0 = 8.9 \times 10^{-7}$, and only about one molecule in a million in gaseous HCl is excited vibrationally at 25°C. All the rest are in the ground state. Alternatively, we could have used the following information:

$$n_i = 6.022 \times 10^{23}$$
$$\Delta E_{ij} = 8.25 \,\text{kcal/mol}(1000 \,\text{cal/kcal})$$
$$R = 1.987 \,\text{cal/mol·K}$$
$$T = 298 \,\text{K}$$

We would have gotten exactly the same result. The choice is yours and will probably depend upon the units inherent to the problem you are solving.

2.9 The Effect of Temperature

Now we are ready to consider the effect of temperature, at least on individual substances, if not whole reactions. Consider again the Boltzmann law:

$$\frac{n_j}{n_i} = e^{-(\varepsilon_j - \varepsilon_i)/kT} = e^{-\Delta\varepsilon_{ij}/kT}$$

Let's look at the effect of changing the temperature on a system where all the levels are evenly spaced. Specifically, consider just the lowest two levels, which we will call 0 and 1 here. Note that as T approaches 0, we have:

$$\frac{n_1}{n_0} = e^{-\Delta\varepsilon_{0,1}/kT} \Rightarrow e^{-\Delta\varepsilon_{0,1}/0} = e^{-\infty} = 0 \quad \text{as } T \Rightarrow 0$$

Thus, the Boltzmann law predicts that at the limit of absolute zero $n_1 = 0$, there are no energized particles, and only the very lowest energy level is populated.

On the other hand, as T approaches infinity, we have:

$$\frac{n_1}{n_0} = e^{-\Delta\varepsilon_{0,1}/kT} \Rightarrow e^{-\Delta\varepsilon_{0,1}/\infty} = e^{-0} = 1 \quad \text{as } T \Rightarrow \infty$$

As the temperature is raised, more and more particles get energized until, at the limit of infinitely high temperature, all levels will have an equal number of particles.

Shown in Figure 2.7 are the most probable distributions of vibrational energy at five different temperatures for gaseous HCl, which has energy level separations of 5.77×10^{-20} J. Three general trends should be clear:

1. At all temperatures fewer particles are at higher energy than at lower energy.
2. At very low temperatures, essentially all the particles are at the lowest possible energy.
3. As the temperature increases, more and more levels become populated. This appears to be at the expense of only the lowest level. (Only at about 40,000 K in this case does Level 1 start to *decrease* in population along with Level 0.)

Of course, neither the extreme of 0 K nor the extreme of infinite K can be reached in practice. Nonetheless, these three features are characteristic of all Boltzmann distributions.

i	10 K	100 K	200 K	400 K	1000 K
12					98
11					6430
10					4.20×10^5
9					2.74×10^7
8					1.79×10^9
7					1.17×10^{11}
6					7.64×10^{12}
5				12	4.99×10^{14}
4				426000	3.26×10^{16}
3				1.46×10^{10}	2.13×10^{18}
2			426000	5.07×10^{14}	1.40×10^{20}
1	1	426000	5.07×10^{14}	1.75×10^{19}	9.07×10^{21}
0	6.022×10^{23}	6.022×10^{23}	6.022×10^{23}	6.022×10^{23}	5.93×10^{23}

Figure 2.7
The effect of temperature on the distribution of energy among a mole of particles in a system having energy level separations all equal to 5.77×10^{-20} J.

2.10 The Effect of Energy Level Separation

Now consider the effect on n_i of different energy level separations, $\Delta\varepsilon_{ij}$. This separation is an intrinsic property of each different substance and also depends hugely upon the energy type and in some cases even the volume. Several examples are given in Table 2.1. Note that vibrational energy levels are about 100 times closer together than electronic levels. Likewise, rotational levels are about 100 times closer together than vibrational levels, and translational levels are about another 10^{20} as close as that! We'll learn much more about these four types of energy levels in Chapter 3.

Shown in Figure 2.8 is a comparison of four hypothetical systems all at 300 K and all containing one mole of particles, but with successively wider energy level separations. The wider the separation between levels, the smaller the ratio n_1/n_0. Notice that the population of Level 1 consistently decreases as we go to wider and wider energy level separations. As the energy level separation increases, more and more particles are trapped in the ground state. It isn't that they *can't* get to higher levels, only that it is *less likely* for this to happen.

Table 2.1
Examples of energy level spacings for a variety of particles and conditions ordered from largest to smallest.

Substance	Energy Type	$\Delta\varepsilon_{0,1}$ (joules)
H atom	electronic	1.6×10^{-18}
H_2	vibrational	8.2×10^{-20}
HCl	vibrational	5.7×10^{-20}
HCl	rotational	4.2×10^{-22}
Cl_2	rotational	9.6×10^{-24}
H_2	translational ($V = 1\,\text{L}$)	5.0×10^{-39}
H_2	translational ($V = 1000\,\text{L}$)	5.0×10^{-41}

$\Delta\varepsilon$:	$1 \times 10^{-20}\,\text{J}$	$2 \times 10^{-20}\,\text{J}$	$3 \times 10^{-20}\,\text{J}$	$4 \times 10^{-20}\,\text{J}$
	$n_5 = 3.15 \times 10^{18}$			
	$n_4 = 3.52 \times 10^{19}$	$n_2 = 3.83 \times 10^{19}$		$n_1 = \mathbf{3.86 \times 10^{19}}$
	$n_3 = 3.93 \times 10^{20}$		$n_1 = \mathbf{4.31 \times 10^{20}}$	
	$n_2 = 4.93 \times 10^{21}$	$n_1 = \mathbf{4.78 \times 10^{21}}$		
	$n_1 = \mathbf{4.91 \times 10^{22}}$			
	$n_0 = 5.48 \times 10^{23}$	$n_0 = 5.97 \times 10^{23}$	$n_0 = 6.02 \times 10^{23}$	$n_0 = 6.02 \times 10^{23}$

Figure 2.8

The effect of energy level separation on the distribution of energy among a mole of particles in a system of evenly spaced energy levels at 300 K. Note how the population of the first excited state (Level 1, in bold) decreases with increasing energy level separation. As the energy level separation increases, more and more particles are stuck in the ground state, until at sufficiently large energy level separations, we can just say that the ground state population is approximately one mole.

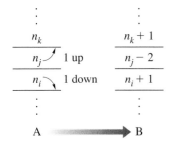

Figure 2.9
When particles exchange energy, one goes up, and one goes down.

2.11 Why Is the Boltzmann Distribution the Most Probable?

Boltzmann's genius in deriving his law can only be alluded to here. All we will do is see that even the slightest change in a system that is in the distribution described by his law leads to a distribution with lower overall probability. In addition, we'll see that once again, when large numbers of particles are involved, any significant fluctuation from the most probable distribution is inconceivably improbable.

We start by imagining just three levels of a much larger energy level system. To keep this simple, let's assume that the three levels are evenly spaced and that, furthermore, they are adjacent (Figure 2.9). In the real proof of the Boltzmann law neither of these is required, but that proof is way beyond what we need here. What we are going to do is compare the thermodynamic probability of this state (A) to another (B) where two particles from the same level j have collided. One has moved up in energy one level; the other has moved down so that there is no change in total energy. Note that W_A and W_B are very similar, differing only in three terms:

$$W_A = \frac{n!}{n_0!\,n_1!\,n_2!\,\ldots\,\boldsymbol{n_i!\,n_j!\,n_k!}\,\ldots}$$

$$W_B = \frac{n!}{n_0!\,n_1!\,n_2!\,\ldots\,\boldsymbol{(n_i + 1)!\,(n_j - 2)!\,(n_k + 1)!}\,\ldots}$$

Our goal will be to show that $W_B < W_A$. In addition, we get some important practice manipulating factorials. In particular, we will use the following two relationships of factorials:

$$(n - 2)! = \frac{n!}{(n - 1)n} \quad \text{and} \quad (n + 1)! = n!\,(n + 1)$$

From the Boltzmann law and the fact that these levels are evenly spaced, we can write:

$$\frac{n_j}{n_i} = \frac{n_k}{n_j} \quad \text{or} \quad \frac{n_j^2}{n_i n_k} = 1 \tag{2.7}$$

This relationship among n_i, n_j, and n_k will turn out to be the key to our proof.

We start by writing W_B in terms of W_A:

$$W_B = \frac{n!}{n_0! \, n_1! \, n_2! \ldots (n_i + 1)! \, (n_j - 2)! \, (n_k + 1)! \ldots}$$

$$= \frac{n!}{n_0! \, n_1! \, n_2! \ldots n_i! \, (n_i + 1) \frac{n_j!}{(n_j - 1) n_j} n_k! \, (n_k + 1) \ldots}$$

$$= \frac{n!}{n_0! \, n_1! \, n_2! \ldots n_i! \, n_j! \, n_k! \ldots} \left(\frac{(n_j - 1) n_j}{(n_i + 1)(n_k + 1)} \right) = W_A \left(\frac{(n_j - 1) n_j}{(n_i + 1)(n_k + 1)} \right)$$

Now, we know something important about the numerator and denominator. Simply from the fact that $n - 1 < n$ and $n + 1 > n$:

$$(n_j - 1) n_j < n_j^2$$

and

$$(n_i + 1)(n_k + 1) > n_i n_k \quad \text{or} \quad \frac{1}{(n_i + 1)(n_k + 1)} < \frac{1}{n_i n_k}$$

Applying these relationships to both the denominator and numerator, above, we can write:

$$W_B < W_A \left(\frac{n_j^2}{n_i n_k} \right)$$

The ratio in parentheses is just 1 (based on Equation 2.7), so we have that $W_B < W_A$. That is, distribution B is less probable than distribution A (our Boltzmann distribution).

Thus, the Boltzmann distribution is more probable than any other distribution made from it by even the slightest change. (More extensive changes just lead to a messier analysis but with the same overall result.) We can be confident that **the Boltzmann distribution is the most probable distribution.**

Now the only question is, how probable is this distribution, especially in relation to all the others that might be possible? Will we really observe only the Boltzmann distribution? Of course there will be fluctuations. Will they be observable? It can be shown that for a system of n particles in a Boltzmann distribution with thermodynamic probability W_A, making a change that, on average, changes the populations of levels by just p percent of their original values, results in a new distribution with thermodynamic probability W_B such that:

$$\frac{W_B}{W_A} \approx e^{-n(p/100)^2} \tag{2.8}$$

For example, in a system containing, say, 1×10^{20} particles, changing the populations on average by just one millionth of one percent of their original values (that is, $p = 1 \times 10^{-6}$) results in a new distribution that is not only less probable than the original Boltzmann distribution, but less probable by a factor of about $e^{-10000} = 1.5 \times 10^{-4342}$. This is unimaginably improbable. **With any realistic number of molecules involved, once again we do not expect to observe even the slightest fluctuation!**

2.12 Determining the Population of the Lowest Level

Although the Boltzmann law gives the ratio of populations throughout an energy system, it's not always easy to know where to start. In the case of evenly spaced energy systems such as those illustrating this chapter, there is a trick that can be used to get the lowest-

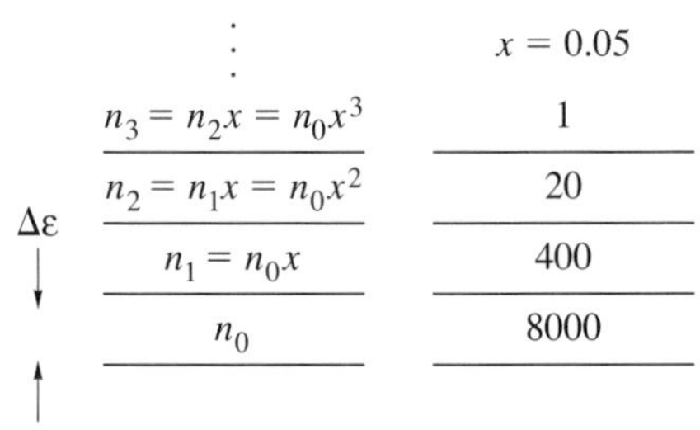

Figure 2.10
An evenly spaced energy system behaves as a geometric progression. In the example on the right, $n_0 = 8000$, $n_1 = 400$, $n_2 = 20$, and $n_3 = 1$, so $x = 0.05$. In this case the total number of particles in the example, n, is 8421. The number of particles in Level 1 or above is 421, which is very close to being nx, which equals 421.05. In addition, the number of particles in the lowest level, n_0, is very close to being $n(1-x)$, which equals 7999.95.

	$x = 0.05$
$n_3 = n_2 x = n_0 x^3$	1
$n_2 = n_1 x = n_0 x^2$	20
$n_1 = n_0 x$	400
n_0	8000

level population. This is a result of evenly spaced energy systems behaving as geometric progressions (Figure 2.10). If we define:

$$x = \frac{n_1}{n_0} = e^{-\Delta\varepsilon/kT}$$

for a system of n particles, then we can write for this special case:

$$n_1/n_0 = x, \quad n_2/n_1 = x, \quad n_3/n_2 = x, \quad \text{etc.}$$

Or, putting all in terms of the population of the lowest level, n_0:

$$n_1/n_0 = x, \quad n_2/n_0 = x^2, \quad n_3/n_0 = x^3, \quad \text{etc.}$$

or

$$n_1 = n_0 x, \quad n_2 = n_1 x^2, \quad n_3 = n_2 x^3, \quad \text{etc.}$$

Then we can write for the total number of particles:

$$n = n_0 + n_1 + n_2 + n_3 + \cdots = n_0(1 + x + x^2 + x^3 + \cdots)$$

In fact, this is just a specific case of a more general relationship. The number of particles *at or above* level j must be:

$$n_{j \text{ or above}} = n_j + n_{j+1} + n_{j+2} + n_{j+3} + \cdots = n_j(1 + x + x^2 + x^3 + \cdots)$$

Dividing $n_{j \text{ or above}}$ by n, we get the fraction of particles at or above a particular level j in an evenly spaced Boltzmann distribution:

$$\frac{n_{j \text{ or above}}}{n} = \frac{n_j(1 + x + x^2 + x^3 + \cdots)}{n_0(1 + x + x^2 + x^3 + \cdots)} = \frac{n_j}{n_0}$$

which can be rearranged to give us the number of particles at or above level j:

$$n_{j \text{ or above}} = \frac{n_j}{n_0}n \tag{2.9}$$

This is actually a very significant and useful finding. Specifically for our purposes here, we can say that the number of particles not in the ground state, $n - n_0$, is in fact the number of particles at or above Level 1:

$$n - n_0 = n_{1 \text{ or above}} = \frac{n_1}{n_0}n$$

Replacing n_1/n_0 with $e^{-\Delta\varepsilon/kT}$ and rearranging gives the number of particles in the ground state:

$$n_0 = n\left(1 - e^{-\Delta\varepsilon/kT}\right) \tag{2.10}$$

For example, in the first case in Figure 2.8, where $\Delta\varepsilon = 1 \times 10^{-20}$ J and $T = 300$ K, we calculate $\Delta\varepsilon/kT = 2.414$, and $n_0 = (6.022 \times 10^{23})(1 - e^{-2.414}) = 5.48 \times 10^{23}$. We can then calculate $n_1 = n_0 e^{-2.414}$ and $n_2 = n_1 e^{-2.414}$, etc.

And in the last case in Figure 2.8, where $\Delta\varepsilon = 4 \times 10^{-20}$ J, we calculate $\Delta\varepsilon/kT = 9.657$ and $n_0 = (6.022 \times 10^{23})(1 - e^{-9.657}) = 6.022 \times 10^{23}$. **As a general rule, if $\Delta\varepsilon/kT$ is greater than about 4, we can say that virtually all of the particles are in the lowest level. We can just write $n_0 \approx n$, where n is the total number of particles.** From there we can use the Boltzmann law to get n_1, n_2, n_3, and so on, which will all be very small relative to n_0. Although only shown for evenly spaced systems, this result is quite general.

2.13 Estimating the Fraction of Particles That Will React

The Boltzmann law tells us that at any given point in time, there is a broad range of energies that particles in a system might have. Although the particles themselves are exchanging energy rapidly, the *fraction* of particles with any particular energy is constant as long as the system is at thermal equilibrium. Chemists often use the Boltzmann law to predict the rate of chemical reactions—or at least the relative rate expected for two different temperatures. The idea is quite simple: The fraction of particles above a given "activation energy" are assigned a certain probability of reacting (perhaps based on the probability that they will collide in a productive fashion and not just bounce off each other, or that they will jiggle just the right way and tear themselves apart). Any particle below that energy is considered stable (for this particular moment in time). Going back to how we derived Equation 2.9, we have

$$f_{\text{reactive}} = \frac{n_{j \text{ or above}}}{n} = \frac{n_j}{n_0} = e^{-E_a/RT} \tag{2.11}$$

where f_{reactive} is the fraction of particles that are reactive—that have enough energy—at any given time (Figure 2.11). The difference in energy between level j and level 0 would be our activation energy, usually written E_a and pronounced "EE-sub-AY". Notice that we are using R here, as in Equation 2.6, simply because chemists usually consider energy on a per mole basis.

Since the number of particles in a given level changes with temperature, we expect that a different fraction of the particles will be reactive, depending on the temperature. The higher the temperature, the more particles there will be at or above level j, and the fewer there will be in the ground state. As the temperature rises, we expect the fraction $n_{j \text{ or above}}/n$ to *increase*. The rate of the chemical reaction, which is proportional to f_{reactive}, should increase with temperature as well (Equation 2.12). We can predict that

$$\text{rate } \alpha \ e^{-E_a/RT} \tag{2.12}$$

This simply says that the rate is proportional to the fraction of particles that are reactive. We can't know what the proportionality constant is, but it turns out we often don't need it anyway. If we have two particular temperatures that we are interested in, say 20°C and 100°C, we can use the Boltzmann law to predict how much faster the reaction will be at the higher temperature:

$$\text{relative rate} = \frac{\text{rate}_{100}}{\text{rate}_{20}} = \frac{e^{-E_a/RT_{100}}}{e^{-E_a/RT_{20}}} = e^{-(E_a/R)(1/T_{100} - 1/T_{20})} \tag{2.13}$$

So, for example, if we have a reaction that has an activation energy of 20 kJ/mol, we would calculate:

$$\text{relative rate} = e^{-\frac{20000 \text{ J/mol}}{8.3145 \text{ J/mol} \cdot \text{K}}\left(\frac{1}{373 \text{ K}} - \frac{1}{293 \text{ K}}\right)} = 5.8$$

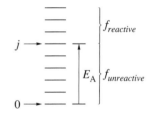

Figure 2.11
The fraction of particles that will react at any given temperature depends upon how many particles have at least the energy of Level j.

That is, we would estimate that this particular reaction might go about six times faster at 100°C than at 20°C. Conversely, if we know the relative rate of two reactions at two different temperatures, we can use that information to estimate the activation energy for the reaction.

2.14 Estimating How Many Levels Are Populated

With a few more tricks, it is also possible to estimate how many levels are populated for a given system of n particles at a given temperature, at least for systems with evenly spaced energy levels, $\Delta\varepsilon$. The trick is to determine at which Level j the population is roughly 1 particle. That level, then, can be considered the top level.

Here's how it works: First we use the above result to get n_0. Now, since the levels are evenly spaced, we can just call the difference in energy between Level 0 and Level 1, $\Delta\varepsilon_{0,1}$, "$\Delta\varepsilon$." Then the difference in energy between Level 0 and Level 2, $\Delta\varepsilon_{0,2}$, is $2\Delta\varepsilon$, and the difference in energy between Level 0 and Level 3, $\Delta\varepsilon_{0,3}$, is $3\Delta\varepsilon$, etc. In general, we have

$$\Delta\varepsilon_{0,j} = j\Delta\varepsilon \quad \text{and} \quad \frac{n_j}{n_0} = e^{-\Delta\varepsilon_{0,j}/kT} = e^{-j\Delta\varepsilon/kT}$$

Substituting in $n_j = 1$ and $n_0 = n(1 - e^{-\Delta\varepsilon/kT})$, we get

$$\frac{1}{n(1 - e^{-\Delta\varepsilon/kT})} = e^{-j\Delta\varepsilon/kT}$$

And, using the fact that $\ln 1/x = -\ln x$, this becomes

$$-\ln\left[n\left(1 - e^{-\Delta\varepsilon/kT}\right)\right] = -j\Delta\varepsilon/kT$$

which, upon rearrangement, gives the value of j for the highest level populated (the level with at most 1 particle):

$$j = \frac{\ln\left[n\left(1 - e^{-\Delta\varepsilon/kT}\right)\right]}{\Delta\varepsilon/kT} \tag{2.14}$$

For example, for the first case in Figure 2.8, where $n = 6.022 \times 10^{23}$ and $\Delta\varepsilon/kT = 2.414$, we get $j = 22.6$. Now, since j must be an integer, we just round down and say that the highest populated level is Level 22. (So there are 23 levels populated including Level 0.) We expect no particles to have more than the amount of energy associated with that level. (In Level 23 there would be less than one particle!) In the last case in Figure 2.8, where the energy level separations are much larger and $\Delta\varepsilon/kT = 9.657$, we get that the highest populated level is Level 5.

2.15 Summary

This chapter covered a lot of very important territory. The really interesting aspect of chemistry that makes it different from just a random scrambling of atoms comes with the introduction of energy to the picture. The fact that energy is quantized leads to the idea that there are countably many possibilities for distributing this energy among the particles in the system. These *microstates* can be grouped into sets called *distributions*, which are the observable states we expect to find. For small numbers of particles and small amounts of energy, we can enumerate all the possible distributions of energy in a system without too much trouble.

When just a few particles and just a few quanta of energy are involved, it is possible to lay out all of the possible distributions and figure out which is the most probable. In each case, we determine the *thermodynamic probability*, W, using Equation 2.2:

$$W = \frac{n!}{n_0! \, n_1! \, n_2! \ldots}$$

Although not giving the actual probability of a particular distribution of energy, the thermodynamic probability is useful in that it allows us to measure the *relative* probability of two distinct distributions of energy.

When lots of particles are present, once again due to the astronomical numbers of particles in real chemical systems, we get to focus on just the most probable distribution of energy in the system. This distribution is referred to as the *Boltzmann distribution* and is characterized by the Boltzmann law, Equation 2.3:

$$\frac{n_j}{n_i} = e^{-\Delta \varepsilon_{ij}/kT}$$

which relates the relative number of particles in two energy levels to the difference in energy between the two levels. We haven't proven this law by a long shot, but what we have been able to do is argue that it seems to describe the most probable distribution, because even the slightest change leads to a less probable distribution.

This then is the key: If we can describe the energy level systems of real molecules in detail, then we could use the Boltzmann law to learn what to expect for the most probable distribution. Since that distribution is the one we expect to find at equilibrium, even with a few fluctuations, that means we could predict the equilibrium energy distribution in a system of real chemical particles. It will turn out that we can apply this understanding to derive equilibrium-constant expressions for chemical *reactions* and predict the direction of reaction needed to reach equilibrium, the effect of temperature on equilibrium constants, and even how chemical energy can be stored and turned into work and electrical energy.

In Chapter 1 we learned how to make predictions based on probability and how equilibrium constants for isotope exchange reactions can be derived based on probability alone. No prediction based on probability is going to be perfect, though, due to the nature of random fluctuations. Nonetheless, with the astronomical numbers of particles we are dealing with in real chemical systems, we find that the fluctuations are undetectable, and we can focus entirely on the most probable distribution—the *Boltzmann distribution*. This distribution will be the one that characterizes what we call "equilibrium."

In Chapter 3, the four ways that molecules and atoms can absorb energy—electronic, vibrational, rotational, and translational—are introduced so that we can apply our understanding to real chemical systems. In Chapters 4 and 5, we take a close look at the factors that can affect the distribution of energy in a system. We'll see that there are actually just two fundamental ways of doing this: adding (or removing) heat and applying (or doing) work. All this is done with the ultimate goal of understanding the effect of temperature on chemical equilibria (Chapter 6), and discovering how the driving force behind all chemical reactions (blind chance, Chapters 7 and 8) underlies everything we know about equilibrium (Chapters 9–12) and electrochemistry (Chapter 13). We shall see that everything in this book is based on probability and the Boltzmann law, Equation 2.3.

Problems

The symbol Ⓐ indicates that the answer to the problem can be found in the "Answers to Selected Exercises" section at the back of the book.

Microstates and Thermodynamic Probability

2.1 Ⓐ (a) Write out all the ways (microstates) to distribute 2 quanta of energy to 5 particles, grouping these microstates into distinct distributions. (b) How much more probable is it to find two particles each with one quantum of energy than finding a single particle with both quanta? For example, one microstate might look like the box below.

2.2 Consider a system involving 2 quanta of energy distributed among 3 particles. (a) List the two possible distributions in a manner similar to Figure 2.4. (b) For each distribution, show each of its microstates using a box with dots and parentheses as illustrated above for Problem 2.1. (c) Calculate W and W/W_{total} for each. Which distribution is most probable?

2.3 Ⓐ Write out the distinct distributions of 4 quanta of energy among 5 particles in a manner similar to Figure 2.4. Calculate W and W/W_{total} for each. Which distribution is most probable?

2.4 Write out the distinct distributions of 4 quanta of energy to 6 particles in a manner similar to Figure 2.4. Calculate W and W/W_{total} for each. Which distribution is most probable?

2.5 Ⓐ Write out all of the distinct distributions of 4 quanta of energy among 7 particles. Calculate W and W/W_{total} for each. Which distribution is the most probable? What do you notice about this distribution that makes it different from the others?

2.6 (a) Write out all of the microstates for each of the two distributions shown below. (b) Which is the more probable distribution?

Quanta	Particles	
3	——	——
2	•	——
1	•	• • •
0	• • •	• •
	A	B

2.7 (a) Show that Equation 2.1 on page 27 can be written as

$$C^n_p = \frac{n!}{p!\,(n-p)!}$$

(b) Evaluate C^6_0, C^6_1, C^6_4, and C^6_2.

2.8 (a) What is the largest factorial, $n_{max}!$, that your calculator can handle? (b) Evaluate:

$$\frac{1000!}{998!\,2!}$$

[Hint: What is $n!/(n-2)!$?] (c) What system (number of units of energy and number of particles) might this calculation be used for?

2.9 Ⓐ (a) Determine W for each of the distributions shown below and verify that each has 8 units of energy. (b) Which distribution is most probable? How much more probable is the most probable distribution vs. (c) the next most probable? (d) the least probable?

Quanta	A	B	C	D
5	1			
4			2	
3	1			1
2				1
1		8		3
0	6		6	3

2.10 The following systems each have 12 particles. Distribution A has one particle with five units of energy, two particles with two units of energy, one particle with one unit of energy, and the remaining eight particles have zero energy. Distribution B has two particles with five units of energy and the remaining ten particles have zero energy. Distribution C has ten particles with one unit of energy each and two with no energy. Distribution D has one particle with four units of energy, one particle with three units of energy, one particle with two units of energy, one particle with one unit of energy, and eight particles with zero energy. (a) Sketch each of the distributions and verify that each has 10 units of energy. Determine W for each of the distributions. (b) Which distribution is the most probable? How much more probable (twice as probable? three times as probable?) is the most probable distribution vs. (c) the next most probable? (d) the least probable?

The Boltzmann Law

Note: If you have a "graphing" calculator, it almost certainly also has an equation solver. Be sure to take a look at *http://www.stolaf.edu/depts/chemistry/imt/js/tinycalc/ti-calc.htm* before approaching these problems. If you spend a little time learning more about your calculator now, you will save an immense amount of time doing homework and (if your instructor allows it) taking exams.

2.11 State the two major conditions required for the Boltzmann distribution to be the most probable distribution.

2.12 Ⓐ In a system with energy levels evenly spaced by 1.5×10^{-19} J at 500 K, n_0 is 1×10^{10}. What is n_1?

2.13 The ratio $n_j/n_i = 0.40$ for a system at 175 K. What is $\Delta\varepsilon_{i,j}$ for this system?

2.14 Ⓐ In a system with energy levels evenly spaced by 1.5×10^{-19} J at 250 K, n_0 is 1×10^{10}. What is n_1?

2.15 In a system with energy levels evenly spaced by 3.0×10^{-19} J at 250 K, n_0 is 1×10^{10}. What is n_1?

2.16 In a system with energy levels evenly spaced by 3.0×10^{-19} J at 500 K, n_0 is 1×10^{10}. What is n_1?

2.17 Ⓐ The ratio $n_j/n_i = 0.20$ for a system at 175 K. What is $\Delta\varepsilon_{i,j}$ for this system?

2.18 The ratio $n_j/n_i = 0.40$ for a system at 350 K. What is $\Delta\varepsilon_{i,j}$ for this system?

2.19 The ratio $n_j/n_i = 0.20$ for a system at 350 K. What is $\Delta\varepsilon_{i,j}$ for this system?

2.20 Each of the systems shown below is at 1000 K and has level spacings of 4.0×10^{-20} J. In each case the population of the ground state, n_0, is 1×10^9 particles. Complete the diagram by filling in the number of particles in each energy level. (Note the degeneracy in B, Level 2.)

(b)		(e)	
(a)		(c)	(d)
1.0×10^9		1.0×10^9	
A		B	

2.21 Ⓐ Relative populations of energy levels can sometimes be determined by examining intensities of bands in certain types of spectroscopy. It was determined by examining a photoelectron spectrum of a compound that the ratio n_1/n_0 was 0.20 at 175 K. (a) What is $\Delta\varepsilon_{0,1}$? (b) At what temperature would the system need to be for the ratio n_1/n_0 to reach 0.25?

2.22 Ⓐ It was determined from spectroscopy that $n_j/n_i = 0.15$ and $\Delta\varepsilon_{i,j} = 7.8 \times 10^{-21}$ J. (a) What is the temperature of the system? (b) If the temperature of the system was found to be 400 K, what would the observed n_j/n_i be?

2.23 Consider the evenly spaced equilibrium energy level system below. (a–c) Fill in the three missing populations. (d) Provided the smallest amount of energy that a particle can absorb is 5×10^{-22} J, what is the temperature?

	40	
(b)		(c)
	4000	
	(a)	

2.24 It was determined from spectroscopy that $n_j/n_i = 0.80$ and $\Delta\varepsilon_{i,j} = 8.0 \times 10^{-25}$ J. (a) What is the temperature of the

system? (b) If the temperature of the system was found to be 298 K, what would the observed n_j/n_i be?

2.25 It was determined from spectroscopy that $n_j/n_i = 0.30$ and $\Delta\varepsilon_{i,j} = 2.5 \times 10^{-21}$ J. (a) What is the temperature of the system? (b) If the temperature of the system was found to be 400 K, what would the observed n_j/n_i be?

The next three questions relate to the highly unusual compound Al_3O,[3] *for which the vibrational energy level spacings have been found to be* 3.08×10^{-21} J.

2.26 Ⓐ What is the ratio n_1/n_0 for vibrational excitation of Al_3O at (a) 175 K? (b) 500 K?

2.27 (a) What fraction of Al_3O molecules are vibrationally excited (that is, in Level 1 or above) at 300 K? (b) What fraction of Al_3O molecules is electronically excited at 300 K? (Assume the system has evenly separated energy levels with energy separation of 6.41×10^{-20} J.) (c) Calculate the fraction of H_2 molecules excited vibrationally (with energy level separation 8.2×10^{-20} J) at 300 K and compare this number to your answers for Parts (a) and (b).

2.28 In a sample of 1 μmol of Al_3O at 300 K, how many vibrational energy levels are populated? What is the most energy a particle in this system is expected to have?

2.29 Ⓐ If Distribution A is the Boltzmann distribution, and Distribution B is a distribution having energy level populations differing by 0.005% from those of A, what is W_B/W_A if the system contains (a) 10,000 particles? (b) 0.5 mol particles?

2.30 If Distribution A is the Boltzmann distribution, and Distribution B is a distribution having energy level populations differing by 0.015% from those of A, what is W_B/W_A if the system contains (a) 10,000 particles? (b) 0.5 mol particles?

2.31 If Distribution A is the Boltzmann distribution and Distribution B is a distribution having energy level populations differing by 0.01% from those of A, what is W_B/W_A if the system contains (a) 10,000 particles? (b) 0.5 mol particles?

2.32 Ⓐ For a certain reaction to occur, a particle in a system has to have an energy of at least 3.0×10^{-20} J above the ground state. What fraction of the particles will have enough energy to react (a) at 100 K? (b) at 500 K?

2.33 For a certain reaction to occur the system has to have on average an energy of 20 kJ per mole of particles. (a) What fraction of the particles will have enough energy to react at 300 K? (b) At what temperature does the system need to be for 2% of the particles to have enough energy to react?

2.34 Ⓐ For the system in Problem 2.32, if a mole of particles are trading energy in packets of 3.0×10^{-21} J, what is the highest

---

[3] The formula Al_3O is not a typographical error! Al_3O belongs to a very interesting class of compounds called *hyperstoichiometric* compounds. In these compounds, there are more metal atoms per nonmetal atom than you would normally predict. The most stable form of solid aluminum oxide, Al_2O_3, has two aluminum atoms for every three oxygens. But in Al_3O the ratio is 3:1. With only 15 valence electrons, Al_3O violates the octet rule and is very reactive.

number of quanta with which we can expect to find a particle (a) at 100 K? (b) at 500 K?

2.35 Ⓐ For the system in Problem 2.32, if a mole of particles are trading energy in packets of 3.0×10^{-21} J, what is the highest energy relative to the ground state with which we can expect to find a particle (a) at 100 K? (b) at 500 K?

2.36 For the system in Problem 2.33, if 0.50 mol of particles are present, and they are trading energy in packets of 5.0×10^{-21} J, how many particles should we expect to find in the ground state at any given time at 300 K?

Brain Teasers

2.37 Consider the following game: Thirty people stand in a circle, each holding a $1 bill. Half are poised to walk clockwise, half counterclockwise, so that each person has a partner. At the designated time, one of the following three possibilities is played out:

(a) If neither person has any money, both just wait.
(b) If only one person has money, that person gives a $1 bill to the person with no money.
(c) If both have money, they play "rock/paper/scissors" to determine the winner, who receives exactly $1 from the other person.

Once all exchanges are complete, each person takes a step forward and to the right around their partner to find a new partner. The process is repeated indefinitely. What is the most probable outcome of this game? That is, roughly how many people should end up with no money, how many with $1, how many with $2, etc.? Is this a fair game? Is this a model for the fair distribution of wealth?

2.38 Prove that the formula for determining the number of distinct ways of distributing red (R) and black (B) cards into pairs, as was discussed in Chapter 1, is

$$W = \frac{n_{\text{pairs}}!}{n_{\text{RR}}! \, n_{\text{BB}}! \, n_{\text{BR}}!} 2^{n_{\text{RB}}}$$

Show that for a deck of two red cards and two black cards the probability of dealing out two hands each having one red and one black card is twice as probable as dealing out two hands where one hand has both red cards and one has both black cards. Given a deck of 24 red cards and 24 black cards, (a) how many ways are there of dealing out $6\,R_2 + 6\,B_2 + 12\,RB$? (b) How many ways are there of dealing out $7\,R_2 + 7\,B_2 + 10\,RB$? (c) Fill in the blank: "Distribution (a) is _____ times more probable than Distribution (b)."

CHAPTER 3

Energy Levels in Real Chemical Systems

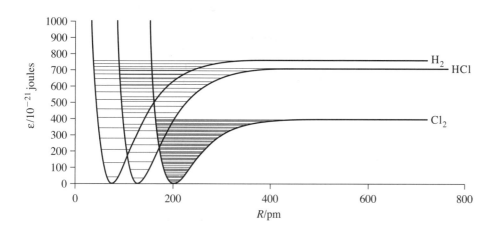

3.1 Historical Perspective

Boltzmann published his law governing the populations of energy levels in 1869. This set the stage for the redefining of chemistry in terms of probability. It is rather awesome to realize that in 1869:

- Experiments dispelling the idea of a "luminiferous ether" through which light traveled and composing the vacuum would not be carried out for another 18 years (by Michelson and Morley, in 1887).
- The electron wouldn't be discovered for another 28 years (by Thompson, in 1897).
- The idea of *quanta* of energy would not be proposed for another 31 years (by Max Planck, in 1900).
- The proton wouldn't be discovered for another 42 years (by Rutherford, in 1911).
- The neutron wouldn't even be *postulated* for another 51 years (also by Rutherford, in 1920) or identified for another 63 years (by Chadwick, in 1932).
- Albert Einstein wouldn't be born for another 10 years.
- Although the equation itself was published in 1869 by Boltzmann, the term "Boltzmann law" would not be coined for another 36 years (by Einstein, in 1905). Nor would the "k" in the equation be used for another 40 years (again, by Einstein, in 1909).

Clearly Boltzmann was far ahead of his time. He would hardly live long enough to see his theories applied, much less to see his theories widely accepted. (He succumbed to chronic depression and committed suicide in 1906.)

3.2 The Modern Viewpoint

To be sure, we have little real knowledge of what is going on in an atom, much less in a molecule. However, it is now completely accepted that molecules and atoms are real entities. The key to understanding what follows is to imagine atoms and molecules as dynamic, nervous things, always on the move, always shivering, always exchanging energy with their neighbors and with the electromagnetic radiation field they encounter from all directions at all times.

Something about these atoms and molecules, something extremely fundamental but not really understood, requires that when these particles absorb energy, they do so based on certain quantum rules. The rules we will use in this book are as follows:[1]

1. A *particle*[2] can absorb only one single quantum unit of energy at a time. (Using a laser it is possible for a particle to absorb two or more quanta almost simultaneously, but we won't consider that here.)
2. Simple collisions can be thought of as the means by which energy is traded among particles. One particle gains energy while the other loses.
3. Particles also gain and lose energy by the absorption and emission of light. When light energy is absorbed by a particle, what happens is completely dependent upon the frequency of the light, not its amplitude or "strength." The strength of the light simply increases the chance of absorption, much as having a lake stocked with fish makes hooking a fish more likely when fishing.
4. The motion of electrons can be largely decoupled from the motion of the nuclei of atoms. We say, as we did in Chapter 1 regarding dice, that electrons and nuclei are largely *independent*. This allows us to divide up the ways energy is absorbed, to talk about them separately, and most importantly to multiply their probabilities independently.

The model we will use here includes four general ways that atoms and molecules can absorb energy:

Electronic energy excitation, as its name implies, involves electrons. It takes place mainly as a result of interaction with light in the visible, ultraviolet, and X-ray frequency ranges. These are by far the largest energy jumps that can be accomplished by a particle, often enough to blow the particle into smaller pieces. All atoms and all molecules can become excited electronically. In most cases, electronic energy absorption requires lots of energy.

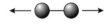

Vibrational energy excitation involves the motion of atomic nuclei *relative to one another* in the same molecule. Bond stretching, bending, and twisting are all examples of vibrations and require energy input. Vibrational excitation is not possible for isolated atoms, only for molecules, which are by definition groups of atoms. Generally, vibrational

---

[1] More sophisticated models have more complex rules. But we will see that our simple model with just these four rules will suffice for an introduction to molecular thermodynamics.

[2] In this chapter we will refer to atoms and molecules as *particles*. Although electrons, protons, and neutrons are also "particles," here we mean only whole atoms and discrete molecules when we say *particles*.

excitation requires less energy than electronic excitation and takes place in response to the absorption of light in the infrared region of the spectrum. Vibrational energy changes very commonly occur due to collision with a wall or with another particle in the system.

Rotational energy excitation involves motion that changes only how the entire molecule is tumbling through space. Like vibration, rotation implies motion of one nucleus relative to another in the same particle. Thus, rotation is not possible in atoms, only molecules. Rotational excitation generally requires substantially less energy than vibrational excitation and is typically associated with light in the far infrared, microwave, and radio regions of the spectrum. Generally a molecule has to be in the liquid or gas phase to rotate. Molecules in solids tend to be locked together too tightly by intermolecular forces to allow individual molecules to rotate.

Translational energy excitation always involves changes in the speed or direction a particle is traveling through space and usually occurs through collision with a wall or another particle. Translational excitation requires far less energy than any of the other categories of excitation and can involve molecules or isolated atoms such as those of the noble gases He, Ne, and Ar. Importantly, translation is possible only with gases, never liquids or solids.[3]

If you think about it, this pretty much accounts for anything an atom or molecule can do, although it is also possible for the nucleus itself to absorb energy. For example, the important medical technique of *Magnetic Resonance Imaging*, or *MRI*, is based on the absorption of radio waves by H atoms of H_2O when exposed to a strong magnetic field, and of course nuclear fission and fusion involve the absorption of energy by atomic nuclei. We will ignore nuclear excitation here simply because it is not important to the discussion of chemical reactions.

Everything that follows in this chapter is designed to help you understand more clearly how energy in molecules is quantized, thus leading to the sorts of energy level pictures presented in Chapter 2. You will see that there are several similarities and differences among electronic, vibrational, rotational, and translational energy excitation. To accomplish this, we are going to have to work with a few mathematical equations.

If you are used to just memorizing a mathematical equation such as the ideal-gas equation, $PV = nRT$, this is going to be a challenge for you. We are not after memorization here, rather *understanding*. The equations simply help us get a handle on this rather complex issue of energy, allowing us to come to some general points which will be summarized at the end of the chapter. Try not to get bogged down in the math, but also try not to ignore it, either. See if you can find what is important in each equation presented. What would be the effect of increasing this or decreasing that? How is this equation similar to that one? It's going to be a challenge, but if you give it a try, you really will learn something!

3.3 Planck, Einstein, and de Broglie

The equations that we are about to discuss all can be interpreted in terms of probability. Each can be thought of as the extension of an idea developed in the mid-1920s by Louis de Broglie (pronounced de-BROY-lee), a French physicist who received his PhD in 1924 and the Nobel Prize just five years later, at the age of 37, for his discovery of the wave nature of electrons. What in the world was he talking about?

---

[3] *Diffusion*, the random movement of molecules in a liquid, is basically the result of translation in a really tiny box. It won't be considered here because it is so totally different from translation in the gas phase. Technically, *throwing* a solid through space might be thought to involve translational energy changes, but we won't be considering that type of translation here.

Discoveries in the early 1900s by Planck and Einstein had established that light is quantized and is in some way equivalent to mass:

$$E = h\nu \text{ (Planck, 1900)} = mc^2 \text{ (Einstein, 1905)} \tag{3.1}$$

Here E is the energy content of light (Planck) and matter (Einstein), h is Planck's constant, $6.6260755 \times 10^{-34}$ joule-seconds, and ν (nu; sounds like "new") is the frequency of light, a variable that depends upon the sort of light involved. The constant m is the "equivalent mass" of the *photon*, and c is the speed of light, 2.998×10^8 meters per second.

Note that it isn't that the photon actually has mass, but when light energy is absorbed by a particle, the particle actually gains this amount of mass; when light is emitted, this is the mass the particle loses. So it's *almost* as if the photon has mass. Prior to Einstein it was a fundamental rule of nature that mass is conserved and energy is conserved, but *no!* It's more as if matter is some sort of "condensed energy" that can be added to or taken from by the absorption or emission of photons. Isn't that interesting? (And you thought mass was conserved!) Frequency ν (nu) and wavelength λ (lambda) of light are related as

$$\lambda\nu = c \quad \text{or} \quad \nu = c/\lambda \tag{3.2}$$

You can refer to the frequency or the wavelength of light. Your choice. Given one, we can always determine the other. For example, the wavelength of visible light is in the range 400 nm to 700 nm. This translates to a frequency range of 7.5×10^{14} s^{-1} to 4.3×10^{14} s^{-1}. (*Longer* wavelengths are associated with *smaller* frequencies and *lower* energy.)

Substituting c/λ for ν in the Planck/Einstein equation provides the wavelength of light in terms of its equivalent mass, and equivalent mass in terms of wavelength:

$$h\left(\frac{c}{\lambda}\right) = mc^2 \quad \text{therefore} \quad \lambda = \frac{h}{mc} \quad \text{and} \quad m = \left(\frac{h}{c}\right)\frac{1}{\lambda} \quad \text{(Einstein, 1905)} \tag{3.3}$$

That is, when a particle absorbs a single photon of light of wavelength λ, then its mass is increased by the amount m.

De Broglie's radical hypothesis was that electrons are wave-like in as strange a way as light is particle-like. As recently as 1963 de Broglie recalled, "After long reflection in solitude and meditation, I suddenly had the idea, during the year 1923, that the discovery made by Einstein in 1905 should be generalized by extending it to all material particles and notably to electrons."[4] Specifically, de Broglie's generalization amounted to replacing Einstein's speed of light c with the electron's velocity v_e (the letter "vee" here, not frequency, *nu*, which looks like this: ν) and Einstein's "loss of mass" with mass of the electron, m_e:

$$\lambda = \frac{h}{m_e v_e} \quad \text{or} \quad v_e = \frac{h}{m_e \lambda} \quad \text{(de Broglie, 1924)} \tag{3.4}$$

Now, the energy due to the motion of an electron (*kinetic* energy) can be written as

$$\varepsilon = \frac{1}{2}m_e v_e{}^2 \quad \text{("velocity" here)} \tag{3.5}$$

Putting all these together and rearranging, we get the energy of de Broglie's wave-like electron:

---

[4] L. de Broglie, preface to his reedited 1924 PhD thesis, *Rescherches sur la théorie des quanta*, p. 4, Masson, Paris 1963, as found in A. Pais, *'Subtle is the Lord . . . ': The Science and the Life of Albert Einstein*, Oxford University Press (1982) p. 436.

$$\varepsilon = \frac{1}{2}m_e v_e^2 = \frac{1}{2}m_e \left(\frac{h}{m_e \lambda}\right)^2 = \frac{h^2}{2m_e \lambda^2} \qquad (3.6)$$

By itself, this may not look like much. But an essential characteristic of all of the energy equations in this chapter arise from this one equation: **The energy of the electron is inversely related to mass.** In addition, this was the first relationship ever attributing wave-like character (in the form of a wavelength) to matter (in the form of an electron). Notice that this is different from the relationship for light: $E = h\nu = hc/\lambda$.

So far, this doesn't explain why atoms must absorb discrete quanta of energy. To add quantization, de Broglie had to add a *constraint* (or *boundary condition*) to his system. The constraint he used was to say simply that the electron particle had to be in some sort of "box" such that the wavelength, λ, matched the box, as shown in Figure 3.1. That is, the particle's wave had to have a *node*, or 0-point, at both ends of the box.

In order to make this work out using a simple sine wave, which has 0-points every half wavelength, this box constraint requires that

$$\frac{\lambda}{2} = \frac{d}{n}$$

where d is the length of the box, and n is the **quantum number**, which can be any positive integer, 1, 2, 3, etc. That is, $\lambda/2$ must equal $d/1$, or $d/2$, or $d/3$, etc.

By putting this restriction for λ into his energy equation, de Broglie had an equation for the *quantization of energy* for an electron in a one-dimensional box:

$$\varepsilon_n = \frac{h^2}{2m_e \lambda^2} = \frac{h^2}{2m_e (2d/n)^2}$$

$$= n^2 \frac{h^2}{8}\left(\frac{1}{m_e d^2}\right) \qquad \text{where } n = 1, 2, 3, \ldots \qquad (3.7)$$

This, then, is the famous contribution of de Broglie. The subscript "n" in "ε_n" refers to the fact that there are infinitely many possible energies that a particle can take, depending upon what is put in for the parameter n. An electron (or any particle, for that matter) constrained to a one-dimensional box can only have certain *specific* energies, and its motion can be described in terms of one or another "harmonic waves." In fact, there is no such thing as a "one-dimensional box." In this case, it's the *idea* that gets you the Nobel Prize. The four types of energy distribution—electronic, vibrational, rotational, and translational—can all be treated as variations on this theme. Although all of the boxes will be different in detail, they all share the important characteristic of requiring some sort of *wave equation* to describe the system and a set of *constraints* to set the boundary conditions for the problem. In all cases, the constraints will be two: a mass constraint and a "size-of-the-box" constraint. The fact that both m and d appear in the denominator of this equation tells us two very important relationships:

1. An increase in mass or size of the box results in a decrease in energy for the particle.
2. This decrease in energy of each possible energy the particle can take means that the difference in energy level *spacing*, $\Delta\varepsilon$, will be less as well.

The situation is exactly as for musical instruments (Figure 3.2). The larger instruments of the orchestra—tubas, bassoons, basses, and cellos, for example—have larger, more massive "boxes." They produce the deeper (low-energy) sounds. Smaller, less massive instruments such as piccolos and violins produce the high-frequency (high-energy) pitches.

This is more than just an analogy. Essentially the same mathematical equation for de Broglie's energy levels governs the pitch of musical instruments. Each string on a violin

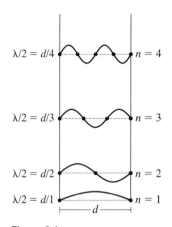

Figure 3.1
The one-dimensional box and the relationship between λ and d that is necessary to create a node (•) at the wall. Sine curves have nodes every $\lambda/2$. A node must come at integral divisions of the box width, d. That is, there must be a node at $x = d/1$, or at $x = d/2$, or at $x = d/3$, etc.

Figure 3.2
The range of pitches that can be produced by a musical instrument depends upon its mass and size. Larger, more massive instruments produce the deeper (lower-energy) sounds.

can play an infinite number of different notes, called *harmonics*. Go find a string player and get them to show you how to play a harmonic. If you look closely at the string while the harmonic is played, it will look precisely like the sine waves of Figure 3.1. If you listen really closely to a hydrogen atom as it goes from one energy state to another, it's singing to you!

For our purposes, we will consider each of the four energy types to be an independent system of energy levels. Thus, we will talk about the total electronic energy of a molecule or the vibrational energy level spacings of a system. We aren't trying to learn everything there is to know about these levels, just a few salient points related to how closely spaced the levels are, what molecular constraints affect these separations, and how this translates into the number of populated levels at the typical "room" temperature of 25°C (298 K).

3.4 The "Wave" Can Be Thought of in Terms of Probability

But what is the wave? De Broglie borrowed from Boltzmann and claimed that his matter-wave describes the probability of finding the particle in any particular place in the box. (His boundary condition of requiring a node at the wall simply means the particle has no probability of being found *at* the wall.) The problem Boltzmann had was that although his energy level ideas seemed to work, he had no real physical basis for them. De Broglie allowed a physical interpretation. Boltzmann's level numbers became de Broglie's quantum numbers 1, 2, 3, etc. Boltzmann's energies became de Broglie's quantum energies, ε_n. De Broglie described for the first time how particles could really have the quantized energy levels that Boltzmann was referring to, and, doing so, set the stage for others to clarify what exact form those levels should take.

3.5 Electronic Energy

The fact is, we know the actual relationship for the energy levels only in the simplest of atoms, hydrogen, consisting of a single proton with +1 charge and a single electron with

-1 charge. Anything else you have heard is an approximation based on various assumptions and subject to revision by future Nobel Prize winners. De Broglie's box for the hydrogen atom is *spherical*, and that changes the picture in three important ways:

- First, rather than having walls, the box is a loop. Imagine wrapping the box in Figure 3.1 into a cylinder, so that the two walls get glued together. The result is that we have a slightly different boundary condition. Now half of the sine waves don't work, because only the even ones make a continuous wave all the way around the loop. This changes the quantum dependence of the wavelength on d to $\lambda = d/n$ instead of $\lambda/2 = d/n$.
- Second, being spherical meant that the box itself "expands" as the quantum number n increases. In fact, it expands as $d = n^2(2\pi a_0)$, where a_0 is the radius of the first-level box (so $2\pi a_0$ is the circumference of a circle at that radius).
- Third, because of the peculiarities of a central force acting on the electron,[5] we have to add a minus sign to the energy, making it $\varepsilon = -\frac{1}{2}m_e v_e^2$ instead of just $\frac{1}{2}m_e v_e^2$.

Adding these changes to de Broglie's box, we get the following result for the H atom:

$$\varepsilon_n = -\frac{1}{2}m_e v_e^2 = -\frac{1}{2}m_e \left(\frac{h}{m_e\lambda}\right)^2 = -\frac{h^2}{2m_e\lambda^2}$$

$$= -\frac{h^2}{2m_e(d/n)^2} = -n^2\frac{h^2}{2}\left(\frac{1}{m_e d^2}\right) = -n^2\frac{h^2}{2}\left(\frac{1}{m_e(n^2 2\pi a_0)^2}\right)$$

$$= -\left(\frac{1}{n^2}\right)\frac{h^2}{8\pi^2}\left(\frac{1}{m_e a_0^2}\right) \quad \text{where } n = 1, 2, 3, \ldots$$

Well, that's a lot of substitutions, but the final result is the important thing. We have for the hydrogen atom:

$$\varepsilon_n = -\left(\frac{1}{n^2}\right)\frac{h^2}{8\pi^2}\left(\frac{1}{m_e a_0^2}\right) \tag{3.8}$$
$$\text{where } n = 1, 2, 3, \ldots \text{ (electronic; H atom only)}$$

Notice the mass constraint, m_e, and a constraint related to the size of the box, a_0, are both in the denominator again. The value of *the Bohr radius*, a_0, is 5.2918×10^{-11} m. This is the most probable distance between the nucleus and the electron in the ground state hydrogen atom.

Now, it turns out that the mass constraint m_e and the box constraint a_0 are both so small—the constraints on the system are so tight—that only the very lowest energy level has any particles in it! Let's see how that works out using the Boltzmann law. Using the

---

[5] When there is a central force, such as when a nucleus attracts an electron, the total energy is not just the kinetic energy, as before. The *total* energy in such a system, ε, is a combination of a negative potential energy, $\varepsilon_{\text{potential}}$, and a positive kinetic energy, $\varepsilon_{\text{kinetic}}$. (The potential energy is negative because we assign it to be 0 when the electron and proton are infinitely separated, and it has to drop as the two get closer and closer together.) In order to be a stable system, the *virial theorem* requires that $\varepsilon_{\text{potential}} = -2\varepsilon_{\text{kinetic}}$ and thus $\varepsilon = \varepsilon_{\text{potential}} + \varepsilon_{\text{kinetic}} = -\varepsilon_{\text{kinetic}} = -\frac{1}{2}m_e v_e^2$.

definition of a joule, 1 joule = 1 kg·m²/s², the constants in Equation 3.8 can be combined to give[6]

$$\varepsilon_n = (-2.1798736 \times 10^{-18} \text{ joules}) \left(\frac{1}{n^2} \right)$$

(3.9)

where $n = 1, 2, 3, \ldots$ (electronic; H atom only)

Notice that the energy is *negative*, and the lowest level is called Level 1, not Level 0. Solving for the first two levels, we get

$$\varepsilon_2 = -0.545 \times 10^{-18} \text{ joules}$$

$$\varepsilon_1 = -2.180 \times 10^{-18} \text{ joules}$$

Using the Boltzmann law (Section 2.8), we can now calculate the ratio n_2/n_1 for the hydrogen atom at 298 K as follows:

$$\frac{n_2}{n_1} = e^{-(\varepsilon_2 - \varepsilon_1)/kT} = e^{\frac{-1.635 \times 10^{-18} \text{ J}}{(1.381 \times 10^{-23} \text{ J/K})298 \text{ K}}} = 2.608 \times 10^{-173}$$

That's unimaginably small! (It has been estimated that there are only about 10^{70} atoms total in our whole *galaxy*.) We can be quite certain that no matter how many hydrogen atoms we have, at least around room temperature, there is *absolutely no chance* of finding hydrogen atoms in *any* electronic energy level above the first, at least not as part of the most probable distribution.

In 1926, Erwin Schrödinger redesigned de Broglie's box, writing an equation that gave not only the energies of the states, but also the probabilities of finding the electron in various relationships to the proton. *Schrödinger's Equation*, framed in the language of probability, had three constraints:

1. that the proton has a +1 charge attracting the electron's −1 charge,
2. that the electron must have no probability of being found right at the proton, and
3. that the electron must have no probability of being found infinitely far from the proton, either.

You're probably already aware that electrons in atoms and molecules can be thought of as residents of *orbitals*, like s, p, and d for atoms and σ, π, σ^*, and π^* for molecules (Figure 3.3). These orbitals are the solutions to Schrödinger's Equation. They represent

---

[6] This numerical value had been known to be correct as far back as 1885, when Johann Balmer found the pattern in the hydrogen spectrum. It also matched the result of Niels Bohr, who got the same number in 1911 (15 years earlier) using classical ideas of circular orbits.

Figure 3.3
Two different representations of the mathematical solution to Schrödinger's Equation for atoms and molecules. On the left is a depiction of a $2p$ orbital of the hydrogen atom. Density of dots represents probability. On the right is a depiction of one of the orbitals in CH_4. Here the "cage" represents the surface inside which the electron can be expected to be found 95% of the time.

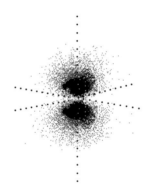
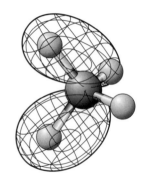

the probability of finding the electron in different relationships to the nuclei. Orbitals are simply the resultant mathematical functions, analogous to de Broglie's sine wave, which satisfy Schrödinger's set of constraints. When the hydrogen atom absorbs electronic energy, its electron shifts to a new pattern of movement. This changes its probability of being found at various points around the nucleus.

For example, when a hydrogen atom in its ground state (in Level 1) encounters a photon having energy that matches the energy difference between Level 1 and Level 2, it is quite possible that the photon will be absorbed by the hydrogen atom. That energy difference, $\Delta\varepsilon_{1,2}$, is precisely $\varepsilon_2 - \varepsilon_1 = (-0.545 \times 10^{-18}) - (-2.180 \times 10^{-18}) = 1.635 \times 10^{-18}$ joules. This corresponds to light of a frequency of $2.47 \times 10^{15}\ \text{s}^{-1}$ or a wavelength of 121 nm. This is in the ultraviolet part of the electromagnetic spectrum. Note that this new distribution of atoms, with one atom up in a higher level, is *not* the most probable distribution of energy. Not surprisingly, the atom will very quickly (in a matter of picoseconds) dispense of its newfound energy and return to the ground state.

In fact, if even more energy is given to the hydrogen atom, it may jump to a very high energy level. Upon returning to the ground state, it may take several smaller jumps to release all its energy. The result of this return to the ground state is the emission of light at many different energies, as depicted in Figure 3.4. In this figure, all of the possible emissions returning from Levels 6, 5, 4, 3, and 2 are depicted. The length of the arrow represents the amount of energy of each photon that is released corresponding to any given jump. Four of these emissions are visible and are called the *Balmer series*. These four emissions correspond to jumps from Levels 6, 5, 4, and 3 to Level 2. All other emissions are either too low in energy (Paschen series) or too high in energy (Lyman series) to make it into the visible region.

The extension of Schrödinger's box to multinuclear systems (*molecules*) involves adding more nuclei and more electrons to the box, effectively making the box bigger and the constraints less tight. Actually solving the equation turns out to lead to a rather intractable

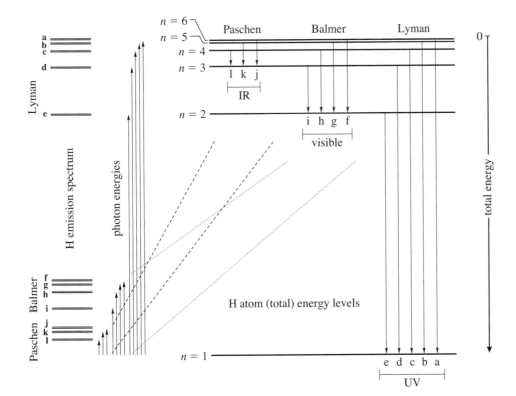

Figure 3.4
All of the possible return paths for a hydrogen atom returning to its ground state from Levels 6, 5, 4, 3, and 2. Some pathways are more probable than others, but all are observed. Each emission ($\Delta\varepsilon$) correlates with an observed line in the hydrogen spectrum. Only four of these lines are visible. The others are either too low or too high in energy. Note that some paths require more than one emission, for example $6 \rightarrow 4 \rightarrow 2 \rightarrow 1$.

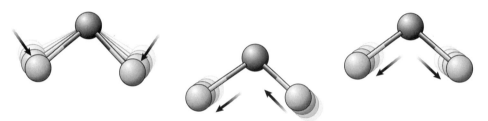

Figure 3.5
Typical sorts of vibrations involving H_2O. They are, from left to right in order of increasing energy, the bend, the asymmetric stretch, and the symmetric stretch. In each case, the relative positions of the atoms in the molecule change. It is almost as though the atoms were held together by springs.

problem. The problem is that you can *write* the equation, you just can't *solve* it. This is due to an unsolved mathematical problem called the "three-body problem" in which we are trying to relate the positions and actions of three objects at the same time, all interconnected. Two objects are doable; three are impossible!

Nonetheless, we work with the premise that molecules can't be too terribly different from hydrogen—just being slightly larger boxes—and we start approximating. Even with the larger box size for molecules, we will still claim that virtually all atoms and molecules are in their ground state electronically, because the boxes are still very small, the mass of the electron is so small, and the resultant electronic energy levels are *sooooo* far apart.

3.6 Vibrational Energy

The actual "shivering" of molecules is what we mean by vibration. Bonds stretch and compress, bonds bend and twist—these are all vibrations. The important thing to realize is that there must be two or more atoms to have a vibration. Examples of vibrations are given in Figure 3.5.

The basic model used to analyze vibration is that the two atoms are connected by some sort of spring-like force. This force tends to pull them together as they get further apart and push them apart as they get close together. When further apart than their average distance, they receive a pull inward; when closer together than this distance, they receive a push outward. This push/pull in/out oscillation is characteristic of all vibrations. The oscillations occur with a frequency (oscillations per second) characteristic of the masses and the strength of the spring.[7] The "box" we will use for vibrational energy is called the *anharmonic oscillator potential*. The constraints here are two: The spring *force constant*, k_f, can be thought of as determining the stiffness of the "spring" holding the particles together. For molecules, k_f is related to bond strength and varies from one molecule to the next and from

---

[7] We will see that molecules in excited vibrational states are not vibrating *faster* (with different frequency ν), but instead vibrating with greater *amplitude*. Going up in energy, the atoms get further apart and compress further with each vibration. Going too high in the vibrational energy levels ends up destroying the molecule by breaking the bond, as the amplitude of vibration finally gets so large that the molecule cannot hold itself together, and the two atoms go flying apart. If you want to get an idea of how this works, hold your hands out in front of you about 1 foot apart and start moving them toward and away from each other just a few inches at some moderate tempo. Count with each vibration, "1 . . . 2 . . . 3 . . . 4 . . . ." Now, *still counting at the same tempo*, make your hands go twice as far apart and come back closer together. Keep counting and double the distance again. Double it again. Can you imagine how a bond might break?

one type of vibration to the next. The force constant k_f has units mass/time^2. For example, k_f for the vibration of the HCl molecule is 481 kg/s^2.

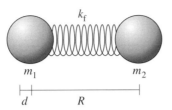

Another constraint on the system involves the masses, and is called the *reduced* mass of the system, μ (sounds like "me-you"). The reduced mass takes into account the fact that any vibration of two bodies occurs about the center of mass of the pair, and it is the lighter atom of the two that does the lion's share of the moving. Reduced mass for diatomic molecules such as H$_2$ and HCl is calculated as follows:

$$\text{by definition:} \qquad \frac{1}{\mu} = \frac{1}{m_1} + \frac{1}{m_2} \quad \text{or} \quad \mu = \frac{m_1 m_2}{m_1 + m_2} \qquad (3.10)$$

Here m_1 and m_2 are the actual masses of the two masses that are vibrating—*exact isotope* mass, not *average molar* mass. Since the exact mass of isotopes (in amu) is very close to the isotope's mass number (sum of protons and neutrons) in all cases, for our purposes we will just use mass numbers for the calculation of reduced mass and convert to units of kg (which we will need because 1 J = 1 kg·m^2/s^2) in the end. Thus, for ^{1}H^{35}Cl we have

$$\mu(^1\text{H}^{35}\text{Cl}) = \frac{(1\,\text{amu})(35\,\text{amu})}{(1\,\text{amu}) + (35\,\text{amu})} = 0.972\,\text{amu}$$

$$\times \frac{1\,\text{g}}{6.022 \times 10^{23}\,\text{amu}} \times \frac{1\,\text{kg}}{1000\,\text{g}} = 1.61 \times 10^{-27}\,\text{kg}$$

This is easier than adding and multiplying all the masses in kg. Note that the reduced mass in this case is just a little less than the mass of the lighter atom (^{1}H in this case).

Since the masses are moving back and forth when they vibrate, they are not a constant distance apart. Although we can talk about an average distance between the masses, the actual "box" is much smaller. We can think of the masses as shivering about their average position with some sort of *average displacement, d*. Thus, including both going back and forth, the box size here would be $2d$. Note that d is much smaller than the average distance between the masses, typically only on the order of 1/10 that distance. The amount of displacement that the atoms will see is a function of the force constant—the stiffer the bond (larger k_f), the smaller the average displacement, d. In addition, d is a function of the reduced mass as well (and how much energy is in the system, as we shall see).

While there is no exact equation for the energy levels due to vibration of molecules, for very simple *diatomic* molecules such as HCl and NO, the anharmonic oscillator potential will give us at least some sense of what is going on in a vibration and the order of magnitude of the separation of vibrational energy levels. The box for the anharmonic oscillator is depicted in Figure 3.6 and can be described as a potential energy "well." The x axis here is the distance between the oscillating particles.

As you can see, the energy levels get closer and closer together with increasing energy, and the probability of the two particles being further and further apart increases with increasing energy. As in the case of electronic energy, we see that the box size ($2d$, where d is the average displacement) increases with energy. What this means is that if the system is given sufficient energy, then the particles may fly out of the box (in opposite directions,

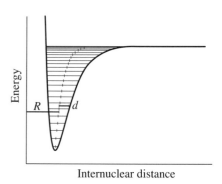

Figure 3.6
Anharmonic oscillator potential for a typical diatomic molecule. Each line represents not only the
energy, but also the general range of distance between atoms, roughly $R \pm d$, where R is the average
distance between the masses, and d is the average
displacement characteristic of that energy. Shown
here are R and d for the sixth vibrational level.

of course)! Adding energy is a great way to break a bond! Sometimes this box is referred to
as a "well" to emphasize that it has a top that particles can, given sufficient energy, achieve,
thus escaping from the box.

The energy part of the solution, at least near the bottom of the energy well, where the
two sides are nearly the same, looks like this:

$$\varepsilon_i = \left(i + \frac{1}{2}\right)^2 \frac{h^2}{4\pi^2} \left(\frac{1}{\mu d^2}\right) \quad \text{where } i = 0, 1, 2, \ldots \tag{3.11}$$

We show this equation only to emphasize how very similar vibration is to the particle-in-a-
box (Equation 3.7) and electronic energy (Equation 3.8). Note the presence once again of
both a mass term (μ this time) and a "size-of-the-box" term, d, both in the denominator.
However, this time the energy levels are much more closely spaced than in the case of
electronic energy, because μ, based as it is on the mass of the *proton* instead of the *electron*
is about 2000 times larger here.

But first, there is a slight problem with this equation. Namely, d is not a constant. It
turns out to be a function of i, μ, and k_f, the spring force constant:

$$d^2 = \left(i + \frac{1}{2}\right) \frac{h}{2\pi(\mu k_f)^{1/2}} \tag{3.12}$$

For $^1H^{35}Cl$, for example, since $k_f = 481$ kg/s^2 and $\mu = 1.61 \times 10^{-27}$ kg, for the lowest energy
level (Level 0 in the case of vibration) we get

$$d^2 = \left(i + \frac{1}{2}\right) \frac{h}{2\pi(\mu k_f)^{1/2}}$$

$$= \left(\frac{1}{2}\right) \frac{6.626 \times 10^{-34} \text{ J·s}}{2(3.14159)[(1.61 \times 10^{-27} \text{ kg})(481 \text{ kg/s}^2)]^{1/2}}$$

$$= \frac{5.99 \times 10^{-23} \text{ J·s}^2}{\text{kg}} \times \frac{1 \text{ kg·m}^2/\text{s}^2}{\text{J}} = 5.99 \times 10^{-23} \text{ m}^2$$

Taking the square root, we get $d = 7.7 \times 10^{-12}$ m, which is about 6% of the average atom–
atom distance, R, of 1.27×10^{-10} m in HCl. That means that as the H nucleus (mainly)
moves back and forth about 1.27×10^{-10} m from the Cl nucleus, it does so over a range of
about $\pm 0.08 \times 10^{-10}$ m. It really is just a little shiver!

Substituting the expression for d^2 given above into the equation for ε_i transforms that
equation into a much simpler form for the energy of the vibrational states in simple diatomic
molecules:

$$\varepsilon_i = \left(i + \frac{1}{2}\right) h\nu \quad \text{where } \nu = \frac{1}{2\pi}\sqrt{\frac{k_f}{\mu}} \text{ and } i = 0, 1, 2, \ldots \text{ (vibration)} \quad (3.13)$$

Although it does tend to hide the resemblance to the particle-in-a-box result, this form is a little better, because now everything except i is a constant. The parameter ν (nu) refers to frequency again. The idea is that a vibration has a frequency of oscillation—first compressed, then lengthened, then compressed, then lengthened, and so on. Here, again, h is Planck's constant, 6.626×10^{-34} J·s. The spring force constant, k_f, shows up along with the reduced mass, μ. For $^1H^{35}Cl$, we get $\nu = 8.70 \times 10^{13}$ s^{-1}, and for the first two levels in the vibration of this molecule we get

$$\varepsilon_1 = \frac{3}{2}h\nu = 8.65 \times 10^{-20} \text{ joules}$$

$$\varepsilon_0 = \frac{1}{2}h\nu = 2.88 \times 10^{-20} \text{ joules}$$

Calculating the ratio n_1/n_0 at 298 K to see if any molecules actually have any vibrational energy, we get

$$\frac{n_1}{n_0} = e^{-(\varepsilon_1-\varepsilon_0)/kT} = e^{\frac{-5.77\times10^{-20} \text{ J}}{(1.381\times10^{-23} \text{ J/K})298 \text{ K}}} = 8.2 \times 10^{-7}$$

This ratio is about 1×10^{-6}. Thus only about 1 molecule of $^1H^{35}Cl$ in a million is vibrationally excited at room temperature.

Notice that the energy level separation in this case, $\Delta\varepsilon$, is just $h\nu = 5.77 \times 10^{-20}$ J. We can use this difference in energy levels to determine the highest vibrational level expected for a molecule of HCl at room temperature, 298 K. In this case $\Delta\varepsilon/kT = 14.02$, consistent with the idea that virtually all of the HCl molecules are in their ground state. Borrowing from Equation 2.11 we have (Section 2.13):

$$j = \frac{\ln[6.022 \times 10^{23}(1 - e^{-14.02})]}{14.02} = 3.9$$

For a mole of HCl at room temperature, we expect to find HCl excited vibrationally only to about the 4th level.

Most importantly, you need to see that the energy level separations for vibrational levels are proportional to the square root of k_f/μ. This means systems with more closely spaced vibrational levels will be for those molecules with smaller force constants (weaker bonds), or larger reduced mass. Figure 3.7 depicts the vibrational energy systems for H_2, HCl, and Cl_2. The energy required to reach the top level in each case is referred to as the *bond dissociation energy*. Thus, given sufficient energy, any bond connection can be severed if, by chance, the molecule happens to find its way into a high energy vibrational state.

3.7 Rotational Energy

Think of rotation as *tumbling*, like a poorly thrown football. A molecule flying through space tumbles as it goes, and the faster this occurs, the higher the molecule is on the rotational energy "ladder." As with vibrations, only *molecules* can rotate, not atoms. All molecules can rotate around three axes. One of these axes is depicted in Figure 3.8.

Usually, only molecules in the liquid or gaseous state can rotate; solids are generally prevented from rotating because of weak interactions between adjacent molecules.

At least for diatomic molecules such as HCl and H_2, the box usually used to analyze molecular rotation is called the *rigid rotor model*. The constraints here are the reduced mass,

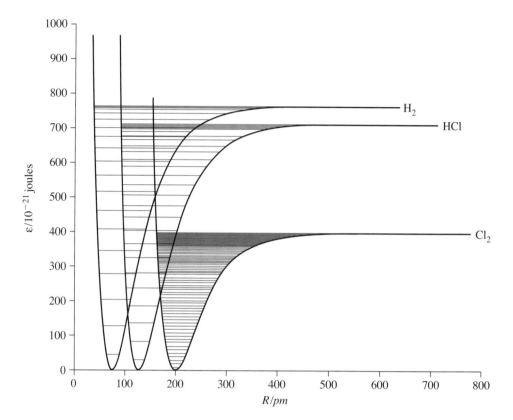

Figure 3.7
Vibrational energy levels in three different systems: H_2, with its exceptionally small reduced mass (μ) and very strong bond, has the most widely spaced levels. HCl, with a somewhat larger μ and slightly weaker bond, has more closely spaced levels. Finally, Cl_2 with much larger μ and much weaker bond has the most closely spaced levels of all.

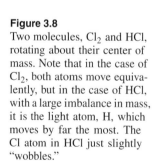

Figure 3.8
Two molecules, Cl_2 and HCl, rotating about their center of mass. Note that in the case of Cl_2, both atoms move equivalently, but in the case of HCl, with a large imbalance in mass, it is the light atom, H, which moves by far the most. The Cl atom in HCl just slightly "wobbles."

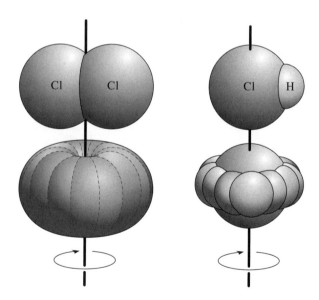

μ, once again, and the average distance, R, between the two atoms. For diatomic molecules, it turns out, we can write

$$\varepsilon_i = i(i+1)\frac{h^2}{8\pi^2}\left(\frac{1}{\mu R^2}\right) \quad \text{where } i = 0, 1, 2, \ldots \quad \text{(rotation)} \qquad (3.14)$$

where once again h is Planck's constant, 6.626×10^{-34} J·s, and μ is the reduced mass, as for vibration. (In the case of a molecule with a heavy and a light atom such as $^1H^{35}Cl$, once again it is the light atom that principally does the moving.) The parameter R here is the average atom–atom distance in the diatomic molecule, which for $^1H^{35}Cl$ is 1.27×10^{-10} m.

Note how similar this equation is to the one for vibration (Equation 3.11). The main difference here is that now we are talking about the actual bond distance, not the little back-and-forth displacement, d, involved in vibration. This R is bigger than d, and results in the energy levels for rotation being significantly more closely spaced than those for vibration. Since R is about 10 times larger than d, we expect R^2 to be around 100 times larger than d^2, and that should be reflected in the energy level spacings for rotation being about 100 times more closely spaced than those for vibration. Indeed they are. Evaluating Equation 3.14 requires including the conversion factor for 1 J = 1 kg·m^2/s^2. We get for $^1H^{35}Cl$:

$$\varepsilon_1 = 4.28 \times 10^{-22} \text{ joules}$$

$$\varepsilon_0 = 0 \text{ joules}$$

The first two *vibrational* levels for $^1H^{35}Cl$ are 5.77×10^{-20} J apart, so, compared to vibration, the levels here are indeed $4.28 \times 10^{-22}/5.77 \times 10^{-20} \approx 1/100$ as closely spaced. Once again, we use the Boltzmann law to see if these states are populated. Calculating the ratio n_1/n_0 at 298 K we get

$$\frac{n_1}{n_0} = e^{-(\varepsilon_1-\varepsilon_0)/kT} = e^{\frac{-4.28\times 10^{-22}\,J}{(1.381\times 10^{-23}\,J/K)298\,K}} = 0.901$$

So there are nearly as many particles in Level 1 as in Level 0. If we again use Equation 2.11 to at least roughly approximate the number of populated rotational levels for a mole of HCl at room temperature, we get $\Delta\varepsilon/kT = 0.104$ and

$$j = \frac{\ln[6.022 \times 10^{23}(1 - e^{-0.104})]}{0.104} = 504$$

Due to the closeness of the spacings in rotational levels, we expect to find several hundred different rotational levels populated at room temperature for a mole of HCl. What a mess!

Most importantly, the energy level separations for rotational levels are proportional to $1/(\mu R^2)$. This means that molecules with longer bonds (for example, HCl vs. H_2) or larger reduced mass (for example, HD vs. H_2) will have more closely spaced rotational energy levels. Remember, only liquids and gases can have rotational energy. Solids are generally locked from rotating by weak intermolecular forces.

Going down a group in the periodic table (for example, the series: F, Cl, Br, I), both atomic radius and mass increasees. These two effects reinforce each other so that molecules made from an element on a lower row of the periodic table *always* have rotational energy

HCl   Cl₂

Figure 3.9
Comparison of the rotational energy levels for HCl and Cl₂. The combination of much larger reduced mass and longer bond distance in Cl₂ accounts for the much closer spacings in Cl₂. Note that rotational energy levels are generally much more closely spaced than vibrational levels. The vertical scale here is about 100 times larger than the one in Figure 3.7. (Dozens of these lines would fit between the lines shown in Figure 3.7.)

levels more closely spaced than similar molecules from rows above. For example, the rotational energy levels in I_2 are about twice as closely spaced as those in Cl_2.

Figure 3.9 illustrates the effect of reduced mass and R on the rotational energy level spacings for HCl and Cl_2. Note that the energy levels in Cl_2 are much more closely spaced than those of HCl, both because the bond distance in Cl_2 is longer and because the reduced mass of the system is larger, but mainly because of the difference in reduced mass. If H_2 were put on this diagram, its first excited rotational state would show up about three times higher than the one in HCl, due mostly to the shortness of the H_2 bond (74 pm) compared to the bond in HCl (127 pm).

3.8 Translational Energy

You can think of translation as motion in a straight line. All atoms and molecules, *as long as they are in the gaseous state*, can have translational energy. We generally refer to there being three independent, perpendicular directions (x, y, and z) for translation.

The equation generally used to represent the translation of molecules through space is called the "particle-in-a-three-dimensional-box" approximation. It's just an extension of de Broglie's one-dimensional box. Due to the three dimensions of the constraining walls, we end up with three quantum numbers:

$$\varepsilon_{n_x,n_y,n_z} = (n_x^2 + n_y^2 + n_z^2)\frac{h^2}{8}\left(\frac{1}{md^2}\right) \quad \text{where } n = 1, 2, 3, \ldots \text{ (translation)} \quad (3.15)$$

Note that here we use n, not i, not because we want to be confusing, but because the lowest level is 1, not 0. Here the subscripts x, y, and z reflect the fact that there are three independent directions in which a gaseous particle can be traveling. The lowest level would be referred to as Level (1, 1, 1), and its energy would be $\varepsilon_{1,1,1}$. The next level up would involve a 3-fold *degeneracy* involving Levels (2, 1, 1), (1, 2, 1), and (1, 1, 2), all at the same energy.

Once again h is Planck's constant, 6.626×10^{-34} J·s. Parameter m is the mass of the whole atom or molecule, since the whole particle is moving through space as a unit. When calculating this, be sure to divide the molar mass taken from the periodic table by Avogadro's number to convert from *molar* mass to *atomic* or *molecular* mass. The parameter d is now the length of a side of a cube containing the gaseous atom or molecule.

Most importantly, d here is on the order of 10^{10} *larger* than the parameter R in the rotational energy equation. This results in translational energy levels being on the order of 10^{20} times as closely spaced as rotational levels (due to the $1/distance^2$ factor in these equations). For $^1H^{35}Cl$ in a 10-cm × 10-cm × 10-cm box, we have

$$\varepsilon_{2,1,1} = 5.508 \times 10^{-40} \text{ joules}$$

$$\varepsilon_{1,1,1} = 2.754 \times 10^{-40} \text{ joules}$$

This time when we use the Boltzmann law, we get quite a different result, because these levels now are *really* close in energy. For the translational energy levels of $^1H^{35}Cl$ at 298 K we get:

$$\frac{n_{2,1,1}}{n_{1,1,1}} = e^{-(\varepsilon_{2,1,1}-\varepsilon_{1,1,1})/kT} = e^{\frac{-2.754\times10^{-40} \text{ J}}{(1.381\times10^{-23} \text{ J/K})298 \text{ K}}} = 1$$

These levels are so closely spaced that for all practical purposes there is one particle per level. And if we again use Equation 2.11 to calculate the highest level j expected to be populated in a system containing a mole of HCl molecules at room temperature, we find that $\Delta\varepsilon/kT = 6.6 \times 10^{-20}$, and

$$j = \frac{\ln[6.022 \times 10^{23}(1 - e^{-6.6 \times 10^{-20}})]}{6.6 \times 10^{-20}} \approx \frac{\ln[6.022 \times 10^{23}(6.6 \times 10^{-20})]}{6.6 \times 10^{-20}} = 1.6 \times 10^{20}$$

That's a lot of levels![8]

Again, the key point here is not the details of the energy equation, but instead the relationship between energy level spacings and the various constraints. Once again, mass (this time actual, not reduced) is in the denominator, indicating that the more massive the atom or molecules, the more closely spaced the translational energy levels. Very importantly here, we see that the larger the box, the more closely spaced the energy levels will be. *Increasing the volume of a system will decrease the translational energy level spacing.*

3.9 Putting It All Together

The total energy of a particle can be represented as the sum of all of its electronic, vibrational, rotational, and translational energy. Our dissection of that energy into these four categories is somewhat artificial. Nonetheless, we do it so that we can get some sort of grasp on this very complex business. We can talk about "ground" and "excited" states for each of these types of energy. Our claim is that all molecules must be in some overall energy state, which may include various quantized amounts of electronic, vibrational, rotational, and translational energy. At room temperature, molecules are generally in their lowest electronic state, their lowest vibrational state, and (in the case of liquids and gases) in one of a few dozen lowest rotational states. This is depicted in Figure 3.10. The long lines in each well represent possible vibrational states. They vary in horizontal length from short (low energy, not much vibration) to so long that the molecule is destroyed (high-energy vibrations). A molecule vibrating to some extent can be associated with one of these lines. The short closely spaced levels represent different rotational possibilities. We think of a molecule as always being at exactly one of these energy levels. In contrast to the levels in Chapter 2, these are clearly not evenly spaced. But as far a Boltzmann or the molecule itself is concerned, one level here is just as good as another.

Levels in the upper well all represent electronically excited molecules. Levels in the lower well all represent molecules in their ground electronic state. Thus, at equilibrium, in the most probable state, all molecules will be found at one of the levels in the lower well, and no molecules will be found at any of the levels of the upper well. This is strictly a result of the large energy level separation associated with electronic excitation.

In a way, Figure 3.10 is like a locator map for molecules. Think of the two wells as being North and South America (upper and lower electronic states, respectively). The molecule is generally going to be found somewhere in South America, but it will get excited and be found in North America now and then. Beyond that, provided it is in South America, it is either in Argentina (lowest vibrational state) or Brazil (next lowest vibrational state) or Chile, or . . . . Most of the time it will be found in Argentina, where its home lies. But this molecule likes to roam, and at any given time it may be in this city or that city depending upon its energy, where the hundreds of cities in each of these countries are analogous to the hundreds of different rotational states. Finally, and this cannot be shown in Figure 3.10, once a molecule has been located in terms of electronic, vibrational, and rotational state, its translational state might be identified. What street is it on? Is it flying this way or that

---

[8] In this case the levels are not equally spaced, but there is also a lot of degeneracy. It is only a rough estimate, anyway. The approximation being used here is that for very small x, $e^{-x} \approx 1 - x$ or $1 - e^{-x} \approx x$.

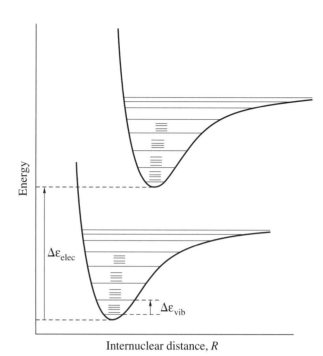

Figure 3.10
Energy level picture for the lowest two electronic states of a simple diatomic molecule such as H_2 or HCl. Long lines represent vibrational levels. Shorter lines represent rotational states.

way? Moving quickly or slowly? Essentially every molecule will be doing its own thing in this regard. Translational energy levels are incredibly closely spaced.[9]

Collisions allow molecules to transfer this energy from one to another. Thus, for example, a molecule at one moment may be tumbling along with 10 units of rotational energy and suddenly have a collision that transfers 3 of these to another molecule, starting *that* molecule tumbling. (A microwave oven works by exciting the rotational levels of water in the food. Collisions with other molecules transfer this energy into the food, warming it up.) Or a molecule might receive 1 unit of electronic energy, transfer some of that energy into vibration, transfer some into rotation, and emit light of whatever energy is left over. (This is why black materials get hot in the sun—they absorb light in the visible region, jumping up to higher electronic levels, then transfer that energy to vibration and get "hot.")

Putting this dynamic picture all together we might end up with something that looks like that depicted in Figure 3.11 for the absorption of energy by a simple diatomic molecule such as H_2 or HCl.

In this depiction, the molecules of a hypothetical yellow substance are seen as consisting of levels within levels within levels. A molecule can absorb energy in an infinite number of ways. Arrow V1 represents the absorption of an infrared photon by the molecule that is in its lowest electronic, vibrational, and rotational state, vibrating minimally. It vibrates more intensely and also starts rotating. Arrow V2 shows what happens when infrared light of slightly higher energy is absorbed (maybe the griddle was on too high!). In this case, light of sufficient energy has been absorbed that almost certainly the molecule will be destroyed on the next vibration, which will take it out of the vibrational well altogether. This leaving of the vibrational well we call *bond dissociation* or *bond breaking*.

---

[9] Try not to carry this analogy too far. It breaks down in the sense that a molecule jumping from one country (vibrational level) to another could still stay on the same street (translational level) and in the same city (rotational level). It's as if the cities in one country were exact duplicates of cities in all the other countries. If this analogy doesn't work for you, design your own!

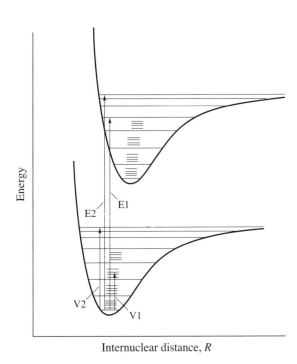

Energy

V2

V1

E2

E1

Internuclear distance, *R*

Figure 3.11
Four distinctly different types of
excitation are represented and discussed in the text.

Another possibility is represented by arrow E1. Here we imagine the molecule having
been hit by a photon in the violet region of the visible spectrum, pumping the molecule up
into its first excited electronic state. At the same time, it starts vibrating and perhaps rotating
as well. This absorption of visible light in the violet region is what gives our hypothetical
molecule its distinctive yellow color, because we see the *loss* of violet from the transmitted
light as the color yellow.

Even though this visible light is much higher energy than the light absorbed in V2, this
poses no real problem for the molecule. It simply responds by dumping that energy back
into its surroundings in smaller packets, a little at a time in a brief cascade of emissions at
lower energies (mostly in the infrared and microwave regions of the spectrum), tumbling
and vibrating its way down to the bottom of the excited electronic state's vibrational well,
then emitting a larger photon and ending up tumbling down through the levels of the ground
electronic state. Thus, one quantum of energy is absorbed, and perhaps thousands of lower-
energy quanta are released back into the environment. We have converted visible light into
heat!

Finally, the excited electronic state has the same problem as the ground state. If light
of sufficient energy is absorbed (E2), the molecule may be excited into a dangerously high
vibrational state *within* the electronic excited state. Almost certainly the molecule will be
destroyed on the next vibration, unless it can get rid of that energy *very* quickly.

The point is, all real molecules are doing this sort of thing *all* of the time. There's
simply no way to stop it. (Even in a 0 K refrigerator in total darkness a molecule would still
be vibrating.) This is the exchange of energy, the shuffling of particles from one level to
another, which must be the basis for equilibrium and all chemical reactions.

3.10 Chemical Reactions

When chemical bonds are broken, it is usually by escaping from a vibrational well, either in
the ground electronic state or via an initial excitation into a higher electronic state. Let's see
how this picture relates to chemical reactions by considering just the simplest reaction of

Figure 3.12
The energy picture for the ground electronic state of H_2. The horizontal axis represents the distance between the two nuclei in the H_2 molecule. Long horizontal lines represent vibrational levels with their increasingly large displacements around the average internuclear distance. Short lines represent rotational levels. Note how the vibrational levels become more and more closely spaced as the energy of the vibration increases. A molecule of H_2 in its ground electronic state will always be found on one of these levels. None of these levels is present for H *atoms*, as atoms cannot vibrate or rotate.

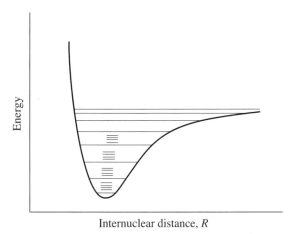

the simplest molecule: the breaking of the bond in gaseous H_2, making gaseous hydrogen atoms:

$$H_2(g) \longrightarrow 2H(g)$$

What do we know about H_2? First, all of the molecules should be in their ground electronic state. This state is depicted in Figure 3.12 where it is seen as a potential well consisting of a series of vibrational energy levels, each with its own set of rotational possibilities. Thus, gaseous H_2 molecules can vibrate, and although most will be found at the lowest vibrational level, vibrating minimally, at least a few (perhaps one in a million at room temperature) will be found in higher vibrational states.

Each molecule can also be rotating, and since rotational levels are so much more closely spaced than vibrational levels (by a factor of 10–100), many more of them will be populated at any given temperature.

Finally, since we are considering a gas, we must consider translation. All of the molecules will be moving about, and it's even fair to say that no two molecules will be in exactly the same translational state. But translation of the H_2 molecule as a whole will not assist in its dissociation.

That means that if we look closely into a box containing H_2, we will see a lot going on. What happens to this rather complex system if heat is added, and the temperature is raised? The addition of heat causes particles to move up the energy ladder. More precisely, particles are always trading up and down the energy ladder, but adding heat results in a new most-probable distribution of energy that has proportionately more particles in higher energy states. It will simply be *more likely* to find a particle in an excited rotational or vibrational state at a higher temperature.

Some of these excited states involve a fair amount of vibrational energy. As the temperature is raised, it becomes more and more likely to find particles up at these levels. Remember, vibrational excitation results in the stretching, bending, and twisting of bonds. Bond stretching, in particular, can only go so far, after which point the H_2 molecule is destroyed, breaking apart into two separate H atoms, as depicted in Figure 3.13.

Note that any given molecule may exchange energy many times along the way, rising and falling on the energy ladder due to collisions. If that molecule happens to be hit with enough energy in just the right way, then it may achieve a high energy vibrational state. At that point, the vibration may become so violent that there is simply no return. There will be a smooth transition from vibration of the molecule to translation of the atoms; the two atoms drift off on their own, with their own translational energies and *no* vibrational energy at all.

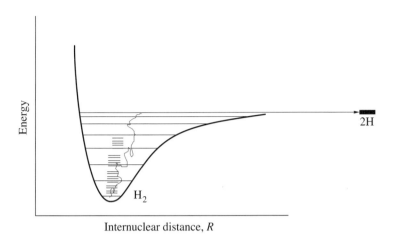

Figure 3.13
The path of destruction for a molecule of H_2 may follow many twists and turns, but it is likely to ultimately pass through a high-energy vibrational state leading to two independent H atoms separated by a large internuclear distance. The heavy line for the two H atoms represents the very closely spaced translational energy levels available to them.

We call this sort of reaction, in which a vibration of one gaseous molecule becomes the translation of two gaseous fragments, *gaseous dissociation*. In all gaseous dissociation reactions, a widely-spaced vibrational level system (with $\Delta\varepsilon$ on the order of 10^{-20} J) is replaced by an extremely closely spaced translational system (with $\Delta\varepsilon$ on the order of 10^{-40} J). In this case, in fact, since the two fragments are both atoms, the products of the reaction have no opportunity for rotation or vibration at all.

Any rotational energy the molecule had is now simply added to the translational energy of the individual atoms, and any vibrational energy it had as H_2 just adds to the "ground state" energy of the H atoms. We can state generally:

Gaseous dissociation reactions always convert vibrational energy into translational energy, leading to products with higher ground states and more closely spaced energy levels.

3.11 Chemical Equilibrium and Energy Levels

In Chapter 1 it was argued that equilibrium is the "most probable state" and that state is well characterized by the "most probable distribution" since fluctuations are going to be undetectable. In Chapter 2, where energy was introduced, it was argued that the Boltzmann law describes the most probable distribution—the equilibrium distribution—of energy in a system where energy is quantized.

Earlier in this chapter we saw how energy is, in fact, quantized in real chemical systems. Now we find that putting energy into a chemical system in the form of heat can cause chemical reactions to take place. (Or, more precisely, to be more *probable*!) But, according to Boltzmann, it does not matter at all what is represented by the levels. Vibration, rotation, H_2 molecules, H atoms, products, reactants—all are "equivalent" in terms of probability. The sole governing factor is the positions of the energy levels and the temperature!

By raising the temperature, we can shift the most probable distribution from mostly H_2 (at low temperature) to mostly H atoms (at high temperature). Note that since there will always be H_2 levels at lower energy than H levels, there will *always* be some H_2 present at equilibrium. But as the temperature rises, all Boltzmann distributions shift to particles in higher energy states, and in this case those higher energy states are called "hydrogen atoms." Thus, as the temperature rises, more and more H atoms are going to be present at equilibrium.

Basically, all we are saying is that if the H atoms released by the dissociation of H_2 are not actually removed from the system, they may very well collide with each other (or some other H atom created in the same way from another molecule somewhere) and form again a molecule of H_2. This new molecule, if it can give up just a little energy in another prompt

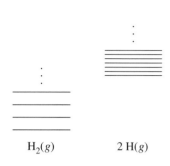

$H_2(g)$ $2\,H(g)$

Figure 3.14
A highly simplified and exaggerated depiction of the energy levels in the $H_2(g)/H(g)$ system. The idea here is to depict just the main aspects of Figure 3.13. "H_2" has a lower ground state than "2 H" because a bond must be broken to get from H_2 to 2 H. However, the reaction takes a more widely spaced vibrational system in H_2 and turns it into a very closely spaced translational system in H. This depiction is representative of all dissociation reactions.

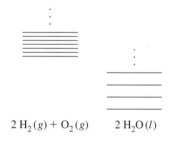

$2\,H_2(g) + O_2(g)$ $2\,H_2O(l)$

Figure 3.15
A highly simplified and exaggerated depiction of the energy levels in the $H_2(g)/O_2(g)/H_2O(l)$ system. Note that the reactants, $2\,H_2(g) + O_2(g)$, have a higher ground state as well as the more closely spaced levels.

collision or emit a photon quickly enough, can itself move down the ladder of energy and survive until it is blown apart again in a future collision. The system will, in fact, achieve equilibrium! A simplification of Figure 3.13 is shown in Figure 3.14.

Similarly, we might imagine reactions for which, overall, energy is given up (bonds are made) in the process of reaction. For example, consider the combustion of hydrogen to form liquid water:

$$2\,H_2(g) + O_2(g) \longrightarrow 2\,H_2O(l)$$

If we just consider the energy levels of reactants and products for this reaction, we might come up with Figure 3.15. Here we show the reactants with the higher ground state energy (because energy is released in the reaction) and the more closely spaced energy levels (because both are gases and can translate), while the product has the lower ground state and the less closely spaced levels (because it is a liquid). Although this is a gross simplification, with highly exaggerated scale, it will serve us well.

Considering the $H_2(g)/O_2(g)/H_2O(l)$ mixture to be one giant system, we conclude that at low temperature, liquid water is going to predominate, since it has the lower ground state. Nonetheless, as the temperature rises, or energy is put into the system, it should be possible to "split" water into its elements. (Granted, the water may boil first!) It will turn out (in Chapter 6) that the fact that the reactants have the closer energy levels is significant. It means that as the temperature increases, not only can we get the reaction to go backward and increase the amount of H_2 and O_2, but we can actually drive this reaction back far enough to make H_2 and O_2 actually *predominate* over H_2O. That is, by raising the temperature, we can actually drive this reaction so far backward that $K \ll 1$ and the compound water is broken down virtually completely into its elements hydrogen and oxygen.

We will find that simplified diagrams such as these are very helpful in our predicting the temperature effect on equilibrium, especially since there is nothing particularly difficult about drawing them. All we need is some indication of whether reactants or products have the lower ground state and whether reactants or products have the more closely spaced levels.

This discussion hints at the reasoning we will use regarding ground states: If we know that heat is *absorbed* in a chemical reaction, or we know that bonds are broken (or bonding is at least *weakened*), then we can assume that reactants have the lower ground state. This was the case in the dissociation of H_2 molecules into H atoms. On the other hand, if heat is *released*, or we know that stronger bonds are made, then the opposite will be true, and we can figure that products have the lower ground state. **Once we have an idea of which has the lower ground state, reactants or products, we also know which will predominate at low temperature as well as in what direction the equilibrium will shift as the temperature is increased.** Thus, for example, if products have the lower ground state (as depicted in Figure 3.15), then products will predominate at low temperature. However, as the temperature is raised, particles will naturally populate higher energy levels, some of which are reactant. The equilibrium will shift to the *left,* toward reactants as the temperature is increased.

The effect of energy level spacing is a much more subtle factor in equilibrium, but we will see in Chapter 6 that at high temperature whether or not $K > 1$ (that is, whether or not products predominate) depends critically upon whether or not products have the more closely spaced energy levels. This, too, will turn out to be a natural consequence of the Boltzmann law. If we want some idea as to the relative spacings of energy levels, all we really have to look at is the number of moles of gas in reactants vs. products, as it is only *gases* that have very closely spaced translational energy levels. Whichever side has more moles of gas will have the more closely spaced energy levels.

Since *all* dissociation reactions involve the conversion of vibrational energy (with relatively widely spaced energy levels) into translational energy (with extremely closely spaced levels), all molecules can be destroyed—fragmented into their component atoms—at high enough temperature. Similarly, all liquids can be completely *vaporized* (or at the very least, *destroyed*) by the application of heat, since the product gas has not only less bonding (the higher ground state) but also the more closely spaced levels.

Note that we will never actually care about the *exact* spacing of energy levels or the *exact* difference in ground state energies. Just having an idea of the *relative* spacings and *relative* ground state energies will be enough.

3.12 Color, Fluorescence, and Phosphorescence

You are probably aware that plants use energy from the Sun in the visible region of the spectrum to make virtually all of the O_2 on our planet. How do they do this? Chlorophyll is a pigment in plants that has at least two relatively low-lying excited states (Figure 3.16). Absorption of a photon (arrow A1 or A2) results in an electron in chlorophyll being excited into one of these states.

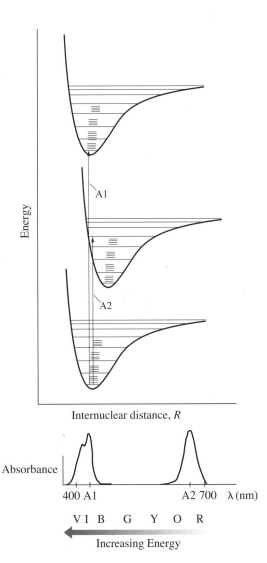

Figure 3.16
Absorption of light in the visible region of the spectrum by chlorophyll shows two main peaks, one at higher energy (A1) and one at lower energy (A2). Each corresponds to the absorption of energy into a different excited electronic state. These absorptions are quite broad, both because there are many possible vibrational and rotational levels that can be attained, and because the ground state chlorophyll molecule may be any one of hundreds of rotational states itself.

Because of the large number of available vibrational and rotational possibilities, this absorption can be effected by light across a broad range of energies.[10] But because there are two electronic possibilities, there are two maximum absorption energies, one in the red and one in the blue. This dual absorption gives plants their distinctive green color, because green is the color that is left over, or *transmitted*, when both red and blue light are absorbed.

This electronic energy is then transferred through a chain of enzyme collisions and ultimately provides the energy needed to convert CO_2 and H_2O into carbohydrates $(C \cdot H_2O)_n$:

$$n\, CO_2 + n\, H_2O \longrightarrow (C \cdot H_2O)_n + n\, O_2$$

Our precious O_2 is simply an excess unwanted byproduct for the plant.

If you look closely at Figure 3.11 or Figure 3.16, you should notice that not only are the levels *not* evenly spaced, the fact that electronic energy levels are so far apart leads to large gaps in the overall energy level system for most molecules.[11] These gaps lead to a fascinating phenomenon called *fluorescence* (Figure 3.17).

If a molecule is excited into one of its higher electronic states by absorption of a photon in the visible or ultraviolet regions of the spectrum (depicted by A1 and A2 in Figure 3.17), the Boltzmann distribution has been disturbed. That absorption of energy has put the system "out of balance" into a less probable state.

As the system starts to lose energy and reestablish an equilibrium, the particle cascades down through the energy levels, giving off photons or releasing energy through collisions. This loss of energy is depicted in Figure 3.17 by a wavy line, which just means, "I don't know exactly what it is doing, probably lots of things, to get from here to there."

All this occurs within the excited electronic state. Notice that it really doesn't matter exactly what energy of light is absorbed, as long as the molecule gets somewhere in the excited state vibrational/rotational set of levels. These all filter down to the same ultimate result: The molecule can get "stuck in the well" in one of the lowest levels of the excited electronic state.

In order to get back to the ground electronic state, the molecule must release a large amount of energy. This may take some time, especially if the arrangement of atoms in the excited state is significantly different from the arrangement in the ground electronic state. The molecule may have twisted or lost one of its double bonds in the process of electronic absorption. But lose that energy it will, by emitting a photon (E in Figure 3.17) *of lower energy than it absorbed*. This, then, is the phenomenon of fluorescence, the emitting of light at an energy lower than the energy originally absorbed.

Not only that, but while a broad range of energies can be absorbed, since they all result in the molecule filtering down to the same lower level of the excited electronic state prior to fluorescence, the fluorescence emission is generally much "cleaner" than the absorption.

---

[10] In this discussion we refer to the *energy* of the light or the *frequency* of the light synonymously, since energy and frequency of light are directly related by Planck's constant: $E = h\nu$. It is also quite common, though, to refer to the *wavelength* of light, $\lambda = c/\nu = hc/E$. In fact all instruments used to measure absorption and fluorescence are designed to report the amount of light absorbed or emitted as a function of wavelength, not frequency. When discussing data in terms of wavelength, you have to be very careful. Since wavelength and energy are *inversely* proportional, when you hear someone talking about "long wavelength" or "higher wavelength" get in the habit of immediately translating that to "low frequency" or "low energy" before you try to understand what is happening, or you are liable to get it all mixed up. Remember: large numbers for wavelength mean *small* numbers for energy. The red (low energy) region of the visible spectrum is around 700 nm wavelength, while the violet (high energy) region is around 400 nm. Be careful!

[11] In fact, not all systems have this gap, nor in all systems are the electronic levels so far apart. The types of systems we are talking about—the vast majority of substances—are electronic *insulators*. Metals, on the other hand, have huge numbers of very tightly spaced electronic levels and no gap in their energy systems. This allows the electron to travel freely throughout the metal. Electronic *semiconductors* have a set of very tightly spaced electronic levels, then a wide gap, then another set of very tightly spaced levels, giving semiconductors special properties.

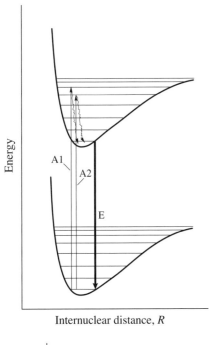

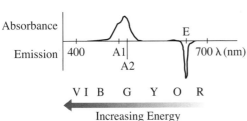

Figure 3.17

In the process of *fluorescence*, initial absorption of light at higher energy (A1 or A2) results in later emission at a lower energy (E). The emission is generally sharper and can be detected at a much higher level of sensitivity than the original absorption. The process of *phosphorescence* is similar, but the emission can be delayed up to several seconds due to difficulties in the system reestablishing the conditions necessary to regain the ground state.

There are special instruments, called *fluorimeters*, which allow sensitive testing of the system. For example, one can monitor the strength of emission at a specific energy while scanning the absorption spectrum (like trying all possible A1 and A2 in Figure 3.17). Or, one can pick an absorption energy (A1) and then scan for fluorescence spectrum looking for E. The amount of absorption and fluorescence is related directly to the amount of a substance present in solution. Since fluorescence occurs at a longer wavelength (lower energy) than the original absorption, the light used for the absorption can be filtered out, and extremely low levels of light emission can be detected. Fluorescence measurements allow concentrations in the micro- to nano-molar range to be determined easily and precisely.

All this electronic energy absorption, transfer, and emission happens in a matter of nanoseconds (10^{-9} s) in most cases. However, if somehow the system finds its way to a state where the electrons are no longer paired (remember, most molecules have all their electrons paired up), then there can be serious problems getting back to the ground state. This unpairing is called *intersystem crossing* and results in the molecule not being "allowed" to go back to its ground electronic state, because in that state the electrons must be paired. The excited state can last for seconds to *minutes*, and the final emission is called *phosphorescence*. "Glow-in-the-dark" paints contain pigments that work just this way.

3.13 Lasers and Stimulated Emission

Finally, in the process of *stimulated emission* it is possible to take advantage of the timing of absorption and fluorescence to amplify light. When that happens you have a LASER

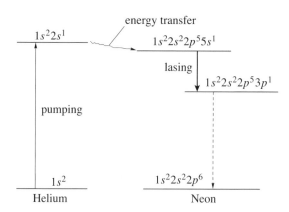

Figure 3.18
Schematic diagram of the HeNe laser. Note that there are no vibrational or rotational levels here because helium and neon are both isolated atoms, not part of larger molecules. Helium is pumped up to its electronically excited state. Collision transfers energy to neon, raising it to its excited state. Population of that state grows until stimulation causes the emission of red light. Ultimately, the neon ends up back in its ground electronic state.

(*Light Amplification by Stimulated Emission of Radiation*). The scheme for the classic HeNe (pronounced, affectionately, hee-nee) laser, which is used in all grocery store scanners, laser pointers, and most CD players, is depicted in Figure 3.18.

The HeNe laser gets its name from the two gases it contains: helium and neon, both monatomic gases. Now, the main difference between atoms and molecules is that molecules can rotate and vibrate (leading to the energy wells discussed above) while atoms can't. So in this case we have no vibrational states and no rotational states, just a few sparsely spaced electronic levels. The process of lasing goes something like this:

- First, the helium atom is placed in an electric field. This excites it to a higher energy state, which basically amounts to popping an electron out of the $1s$ level where it belongs up into the $2s$ level.
- It just so happens that this energy matches the energy it takes to bring a neon atom molecule up to *its* excited state, where one of its $2p$ electrons is excited into the $5s$ orbital to form the $1s^2 2s^2 2p^5 5s^1$ ("1-s-2-2-s-2-2-p-5-5-s-1") electronic excited state. What happens (probably!) is that there is a collision, and the result of that collision is energy transfer, bringing the helium atom down and the neon atom up at the same time.
- At this point the neon atom is in an excited electronic state, and the helium is back in its ground electronic state ready to be "pumped up" again.
- The interesting thing is that the neon atom has trouble getting back down. As in phosphorescence, it gets stuck for a brief moment in an excited electronic state. And since the helium atom can go back and get excited again, it is possible if you design the system *just right* so that photons can be "recycled," to get a relatively large number of neon atoms all stuck in the $1s^2 2s^2 2p^5 5s^1$ state at the same time.
- Now, the phenomenon of *stimulated emission* (first proposed by Albert Einstein) takes over (Figure 3.19). According to de Broglie, atoms are wave-like. That is, they can behave like waves; they can *diffract* and *interfere*. Now, when a photon is absorbed, we can think of it as an *interference* between the atom in its ground state and the photon. This interference leads to the change from the ground state to the excited state. What Einstein discovered is that the excited state, if it could be made to last long enough, has a certain probability of interfering with a photon of the same energy as well. Thus, an incoming photon can either be absorbed and pull a ground state atom up to the excited state *or* stimulate an atom already in its excited state to drop down to the ground state and emit a photon of exactly the same energy. That's *stimulated emission*. Pretty neat trick!

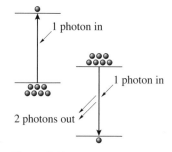

Figure 3.19
Normal absorption (above) involves a single photon and a particle moving to a higher energy state. In stimulated emission, below, when there are more particles in the upper state than in the lower state, the result is one photon in and two photons out.

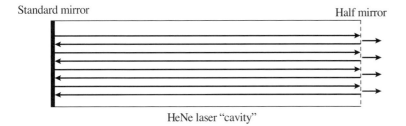

Figure 3.20
In a HeNe laser, the photons that are produced by the stimulated emission are reflected many times. One mirror (on the left) is a standard mirror, but the other (on the right) is a "half mirror" that allows some of the photons through. The photons that are lost are what we see as the red light coming out of the laser.

- It turns out that one of the many things the $1s^2 2s^2 2p^5 5s^1$ state can do is fall to the $1s^2 2s^2 2p^5 3p^1$ state, and this emission is at an energy in the red region of the spectrum. (It's the same emission that gives true neon signs their characteristic red color.) The emission is stimulated and thus produces more photons than were initially present. If it is done just right (in a box with mirrors on both ends so that the photons can be recycled) we get a tremendous rush of neon atoms in their excited electronic state all falling back to their ground electronic state at exactly the same time. We see a bright *coherent* beam of red light stream out of the laser (Figure 3.20).

3.14 Summary

This chapter introduced much information about energy levels in real chemical systems. Do not concern yourself overly much with the exact equations presented in Sections 3.5–3.8. They are there to help you get a handle on what parameters affect the energy level spacings in molecules. Focus on the trends. Focus on the overall idea of energy levels.

Note that electronic excitation is generally nonexistent at equilibrium for most substances at room temperature, because the levels are so far apart that only the lowest level is populated. However, electronic absorption of light is extremely important to understand if you ever want to understand the underlying processes involved in visible light absorption (color), fluorescence, phosphorescence, or stimulated emission.

In terms of comparing energy level spacings, we can pretty much ignore electronic states altogether, other than to say they are spaced far enough apart that unless disturbed, all molecules will be found in their ground electronic state.

Vibrational levels are about 100 times more closely spaced than electronic levels, yet for vibrations, too, more than 99.999% of the particles are generally in the lowest energy level at room temperature.

Once we get to rotations, we find that the levels are again 100 times more closely spaced, leading to significant population of several hundred levels at room temperature.

Finally, translational energy levels are incredibly closely spaced. Virtually every particle is "doing its own thing" in terms of translation. Nonetheless, by claiming that translation is quantized, we are saying that no matter how close those levels are, they could always be closer. We will find soon that changing the volume containing a gas has a direct and incredibly important relationship to the closeness of the energy levels relating to translation.

Thus, a typical molecule of gas at room temperature is in its electronic ground state, almost certainly in its vibrational ground state (vibrating gently), and may or may not be in its rotational ground state (tumbling or spinning) as it flies through space.

Beyond this, there are certain restrictions that are very important to feel comfortable with. Only molecules (not isolated atoms) can vibrate and rotate. Only liquid and gaseous molecules can rotate. Only gaseous particles can have translational energy.

In all cases, we find the following trends, which arise from the fact that all of these types of energy in molecules and atoms are quantized along the lines of de Broglie's particle in a box:

$$\varepsilon = \begin{pmatrix} \text{quantum} \\ \text{number} \\ \text{business} \end{pmatrix} \times \frac{h^2}{\begin{matrix}\text{other}\\\text{constants}\end{matrix}} \times \frac{1}{\begin{matrix}\text{mass}\\\text{constraint}\end{matrix} \times \left(\begin{matrix}\text{box}\\\text{constraint}\end{matrix}\right)^2}$$

where the constraints involve a mass-related term and a "size-of-the-box" term. The effect of these two terms is unambiguous:

- **A larger mass-related constraint** (mass of the electron in the case of electronic energy, reduced mass μ in the case of vibration and rotation, total mass in the case of translation) **always leads to more closely spaced energy levels.**
- **Relaxation of the "size-of-the-box" constraint** (larger radius in the case of electronic excitation, weaker bonds (larger displacements and smaller force constant k_f) in the case of vibration, longer interatomic distance, R, in the case of rotation, or larger volume in the case of translation) **also leads to more closely spaced energy levels**.

All of this can be interpreted in terms of probability. For example, no real movement of atoms occurs right at the instance of energy absorption. Instead, the molecule simply now finds itself in a less probable state with new constraints and forces acting upon it. Its atoms are in less probable positions in relation to their newfound constraints, and the molecule responds to this by adjusting to its new situation, shedding energy in the process or flying into pieces. The situation is not very different from Figure 1.5.

When we put all these ideas together, we start to get a new image of what a molecule "is" and what it can do. Many, many phenomena—including all chemical reactions, the appearance of color, and the emission of light by molecules under certain conditions—can be explained using the simple idea that molecules can be excited from their ground electronic or vibrational state. When that happens, the energy is going to either stay with the system, and the molecule is going to fly apart, or it is going to come back out in the form of heat or light. Only with the idea of energy quantization do these phenomena make any real sense.

Now we are ready to investigate how the distribution of energy in real systems changes due to physical and chemical changes in the state of the system (Chapters 4 and 5). We will then more fully explore the effect of temperature and why any particular equilibrium favors reactants or products (Chapter 6).

Problems

The symbol Ⓐ *indicates that the answer to the problem can be found in the "Answers to Selected Exercises" section at the back of the book.*

Note: Many of these problems involve calculation. Be sure to show your work by writing out the equation you are using first, then putting in the values **with units** unique to each particular case. You may start to feel that much of this is "busywork," but it isn't. You can make it more than that by taking the time in each case to reflect just a bit about the *magnitude* of the answer. Does the answer *feel* correct? How does the value compare to answers to other problems you have been assigned? Does this make sense in terms of the relative magnitude of electronic, vibrational, rotational, and translational energy levels? This is the sort of thinking that, during an exam, will be rewarded!

De Broglie's Box

3.1 Ⓐ Calculate the de Broglie wavelength in meters for the following objects: (a) an electron traveling at 10^6 m/s; (b) a baseball weighing 5 1/4 ounces and traveling at 90 miles per hour; (c) Superman, assuming he weighs 180 lbs., traveling at a speed where he circles the earth, circumference 40,000 km, 100 times per minute.

3.2 Calculate the de Broglie wavelength in meters for the following objects: (a) a proton traveling at 1.1×10^4 m/s; (b) a basketball weighing 600 g and traveling 1.8 m/s; (c) an aircraft weighing 800,000 lbs and traveling at 583 miles/hr.

3.3 Ⓐ Calculate the energy of the first three levels of (a) an electron confined to a one-dimensional box of length 100 pm; (b) a proton confined to the same box.

3.4 Calculate the energy of the first three levels of (a) an electron confined to a one-dimensional box of length 500 pm; (b) a proton confined to the same box.

3.5 Calculate the energy of the first three levels of (a) an electron confined to a one-dimensional box of length 2000 pm; (b) a proton confined to the same box.

Electronic Energy

3.6 Ⓐ Using Equation 3.9 and the Boltzmann equation, calculate (a) the energies of the first two electronic levels of the hydrogen atom; (b) the number of atoms in the first excited electronic energy level ($n = 2$) for a mole of hydrogen atoms at equilibrium at 5700°C (the temperature of the surface of the sun); (c) the same for a mole of hydrogen atoms at −223°C (the temperature on the surface of Pluto); (d) the temperature required for roughly 10% of the hydrogen atoms to be in the first excited electronic state (i.e., $n_2/n_1 = 1/9$).

3.7 For a helium atom the difference in energy between the first two electronic energy levels, $\Delta\varepsilon_{0,1}$, is known to be 19.82 eV. (1 eV, or *electron volt*, is 1.602×10^{-19} joules.) (a) For a mole of helium atoms at 5700°C, the temperature of the surface of the sun, how many atoms are in the first excited electronic energy level? (b) How does this number change if the mole of helium atoms is now at −223°C, the temperature on the surface

of Pluto? (c) What does the temperature need to be for 10% of helium atoms to be in the first excited electronic state (i.e. $n_2/n_1 = 1/9$)?

3.8 Ⓐ With respect to the hydrogen atom and Figure 3.4 and using Equations 3.1, 3.2, and 3.9, calculate the wavelength in nanometers associated with (a) the lowest-energy Balmer line; (b) the lowest-energy infrared emission; (c) the ionization energy.

Vibrational Energy

3.9 Draw Lewis structures and calculate μ for F_2 and N_2, and use this information to explain why the vibrational energy levels in $^{19}F_2$ are more closely spaced than those in $^{14}N_2$.

3.10 Draw Lewis structures and calculate μ for H_2 and N_2, and use this information to explain why the vibrational energy levels in $^1H_2$ are further apart in energy than those in $^{14}N_2$.

3.11 Draw Lewis structures and calculate μ for $^1H_2$ and $^2H_2$, and use this information to explain why the vibrational energy levels in $^2H_2$ are more closely spaced than those in $^1H_2$.

3.12 Ⓐ Which of the following would be the expected effect on the difference in energy of the lowest two vibrational levels ($\Delta\varepsilon_{vib}$) in a molecule HX if hydrogen is replaced by deuterium, ^2H? Explain.

    (a) $\Delta\varepsilon_{vib}$ will double.
    (b) $\Delta\varepsilon_{vib}$ will halve.
    (c) $\Delta\varepsilon_{vib}$ will go up by roughly a factor of $\sqrt{2}$.
    (d) $\Delta\varepsilon_{vib}$ will go down by roughly a factor of $\sqrt{2}$.

3.13 Argue *for* or *against* each statement based on the following data:

$^1H^{127}I$	$k_f = 291$ kg/s^2	$\mu = 0.992$ amu	$R = 161$ pm
$^1H^{79}Br$	$k_f = 408$ kg/s^2	$\mu = 0.994$ amu	$R = 141$ pm
$^{14}N_2$	$k_f = 2340$ kg/s^2	$\mu = 7.0$ amu	$R = 110$ pm
$^{12}C^{16}O$	$k_f = 1860$ kg/s^2	$\mu = 6.9$ amu	$R = 113$ pm
$^{14}N^{16}O$	$k_f = 1550$ kg/s^2	$\mu = 7.5$ amu	$R = 115$ pm
$^{16}O_2$	$k_f = 1140$ kg/s^2	$\mu = 8.0$ amu	$R = 121$ pm
$^{19}F_2$	$k_f = 450$ kg/s^2	$\mu = 9.5$ amu	$R = 141$ pm

    (a) $^{16}O_2$ has more closely spaced vibrational energy levels than $^{14}N^{16}O$.
    (b) $^{14}N_2$ has more closely spaced vibrational energy levels than $^{16}O_2$.
    (c) $^1H^{79}Br$ has more closely spaced vibrational energy levels than $^{14}N_2$.
    (d) $^1H^{127}I$ has more closely spaced vibrational energy levels than $^{14}N_2$.

3.14 Ⓐ Calculate and compare the vibrational frequency ν for the following isotopes of HCl (assume $k_f = 478$ kg/s^2): (a) $^1H^{35}Cl$ (b) $^1H^{37}Cl$ (c) $^2H^{35}Cl$ (d) $^2H^{37}Cl$.

3.15 Calculate the vibrational frequency ν for each of the following isotopes of HBr ($k_f = 290$ kg/s^2): (a) $^1H^{79}Br$ (b) $^1H^{81}Br$ (c) $^2H^{79}Br$ (d) $^2H^{81}Br$.

3.16 Ⓐ Assuming $k_f = 478$ kg/s^2, calculate and compare the fraction of molecules in the first excited vibrational state at 1000 K for (a) ^{1}H^{35}Cl (b) ^{1}H^{37}Cl (c) ^{2}H^{35}Cl (d) ^{2}H^{37}Cl.

3.17 At 1000 K, calculate the percentage of molecules in the first excited vibrational state for each of the following isotopes ($k_f = 290$ kg/s^2): (a) ^{1}H^{81}Br (b) ^{2}H^{79}Br (c) ^{2}H^{81}Br.

3.18 Ⓐ Roughly, what does the temperature need to be for 10% of ^{1}H^{79}Br molecules to be in the first excited vibrational energy level ($k_f = 290$ kg/s^2)?

3.19 Roughly what does the temperature need to be for 10% of a sample of ^{1}H^{37}Cl molecules to be in the first excited vibrational energy level (i.e. $n_1/n_0 = 1/9$)?

Rotational Energy

3.20 Ⓐ Argue *for* or *against* each statement based on the data given in Problem 3.13.

(a) ^{19}F$_2$ has more closely spaced rotational energy levels than ^{16}O$_2$.

(b) ^{14}N^{16}O has more closely spaced rotational energy levels than ^{12}C^{16}O.

(c) ^{1}H^{79}Br has more closely spaced rotational energy levels than ^{19}F$_2$.

(d) ^{1}H^{127}I has more closely spaced rotational energy levels than ^{1}H^{79}Br.

3.21 Which of the following would be the expected effect on the difference in energy of the lowest two rotational levels ($\Delta\varepsilon_{rot}$) in a molecule HX if hydrogen is replaced by deuterium, ^{2}H? Explain.

(a) $\Delta\varepsilon_{rot}$ will double.

(b) $\Delta\varepsilon_{rot}$ will go up by roughly a factor of $\sqrt{2}$.

(c) $\Delta\varepsilon_{rot}$ will halve.

(d) $\Delta\varepsilon_{rot}$ will go down by roughly a factor of $\sqrt{2}$.

3.22 Ⓐ Rotational spectroscopy can determine the energy differences between rotational energy levels. Frequently this is used to determine the bond length of the molecule, or in the case of larger molecules the bond lengths and angles. An absorption having an energy of 7.61×10^{-23} J was assigned to $\Delta\varepsilon_{0,1}$ of carbon monoxide, CO. (a) Based on this information, what is the bond length of CO? (b) What energy would you expect to observe for the absorption corresponding to $\Delta\varepsilon_{0,1}$ CO$^+$, which is proposed to have a bond length 1.3% shorter than that of carbon monoxide?

3.23 Draw an energy level picture comparing the first three rotational energy levels in a system at 25°C containing 1 mole of N$_2$ ($R = 110$ pm) vs. one containing 1 mole of Ti$_2$ ($R = 194$ pm).

3.24 Ⓐ The bond length of Na$_2$ is 307.9 pm. (a) Determine the difference in energy between the first two rotational energy levels of Na$_2$. (b) A spectrometer used to study the rotational spectroscopy of Na$_2$ can distinguish between energies that differ by 5×10^{-25} J or more. Will this spectrometer be able to distinguish between a signal due to the transition between the first and second rotational energy levels of Na$_2$ and those of Na$_2^+$, which has a bond length that is 45.1 pm longer than the bond length of Na$_2$?

3.25 The bond length of SiO is 151 pm. (a) Determine the difference in energy between the first two rotational energy levels of SiO. (b) A spectrometer used to study the rotational spectroscopy of SiO can distinguish between energies that differ by 1×10^{-25} J or more. Will this spectrometer be able to distinguish between a signal due to the transition between the first and second rotational energy levels of SiO and those of SiO$^+$, which has a bond length that is 1 pm shorter than the bond length of SiO?

Translational Energy

3.26 Ⓐ Determine the difference in energy between the first two translational energy levels, $\Delta\varepsilon_{1,1,1,2,1,1}$, of a system of He atoms confined to a cubical 15-L tank.

3.27 Determine the difference in energy between the first two translational energy levels, $\Delta\varepsilon_{1,1,1,2,1,1}$, of a system of Ar atoms confined to a cubical 15-L tank.

3.28 Argue *for* or *against*:

(a) Xe has more closely spaced translational energy levels than Kr.

(b) He has more widely spaced translational energy levels than H.

(c) SF$_6$ has more closely spaced translational energy levels than CH$_4$.

3.29 Which of the following (if any) would be the expected effect on the difference in energy of the lowest two translational levels ($\Delta\varepsilon_{trans}$) in a molecule HX if hydrogen is replaced by deuterium, ^{2}H? Explain.

(a) $\Delta\varepsilon_{trans}$ will double.

(b) $\Delta\varepsilon_{trans}$ will halve.

(c) $\Delta\varepsilon_{trans}$ will go up by roughly a factor of 4.

(d) $\Delta\varepsilon_{trans}$ will go down by roughly a factor of 4.

3.30 Ⓐ A dust mote weighing 1 μg and a hydrogen atom are trapped in a 1 cm^3 cubical vial. What is the separation between the translational energy levels of (a) the dust mote? (b) the hydrogen atom?

Putting It All Together

3.31 Complete the following table summarizing the properties of electronic, vibrational, rotational, and translational energy.

Type of Energy	Approx-imate Size of $\Delta\varepsilon$	First Quantum Number	Mass/Box Constraint Term	Present in
Electronic	10^{-18} J	1	$1/m_e a_o^2$	solids, liquids, gas, all atoms, all molecules
Vibrational				
Rotational				
Translational				

3.32 The following table is all messed up. Match the spectrum region on the left with the appropriate characteristic in each column. (The spectral regions are listed here in increasing order of energy.)

Spectral Region	Approximate Frequency Range	Approximate Wavelength Range	Typical Absorption
(a) Microwave	10^9 to 10^{11} s^{-1}	7×10^{-7} to 4×10^{-7} m	rotational
(b) Far Infrared	10^{15} to 10^{16} s^{-1}	10^{-5} to 7×10^{-7} m	electronic
(c) Infrared	10^{14} to 10^{15} s^{-1}	0.1 to 0.001 m	vibrational
(d) Visible	10^{13} to 10^{14} s^{-1}	4×10^{-7} to 10^{-8} m	electronic
(e) Ultraviolet	10^{11} to 10^{13} s^{-1}	0.001 to 10^{-5} m	rotational

3.33 ⓐ Which types of energy (electronic, vibrational, rotational, and/or translational) (a) are unique to molecules? (b) are not found in solids? (c) have the largest energy separations between levels? (d) have the smallest separations between levels?

3.34 Indicate in the blank whether the statement is true or false. Explain your reasoning.

    (a) _____ Rotational energy levels are generally more closely spaced than vibrational levels.

    (b) _____ The reduced mass of a two-body system is always less than the smaller mass.

    (c) _____ The translational energy levels in H_2 are more closely spaced than those in H_2O.

    (d) _____ For most substances, only the lowest (ground) vibrational state is populated at 298 K.

3.35 ⓐ Complete the following sentence, selecting the correct word in each case: Expansion of a gas leads to **more/less** closely spaced **translational/vibrational** levels because . . .

3.36 Complete the following sentence: At a high enough temperature every liquid will turn to gas because . . .

3.37 ⓐ Fill in the blanks with the appropriate word: electronic, rotational, translational, or vibrational. In some cases, more than one answer may be correct. Indicate *all* correct answers: (a)_____ energy levels are about 10 times more closely spaced than electronic energy levels. Consideration of reduced mass is important for calculations of (b)_____ energy. (c)_____ energy levels do not exist for molecules in the solid phase. The mass term in the (d)_____ energy equation is much smaller than the mass term for any other type of energy. (e)_____ excitation is the predominant means by which bonds are broken in ordinary chemical reactions that result from heating a substance.

3.38 Fill in each blank with one of the following words: *electronic, vibrational, rotational,* or *translational*. In some cases, more than one answer may be correct. Indicate *all* correct answers: (a)_____ excitation generally requires the least energy. Reduced mass must be used for calculating (b) _____

energies because in that case the atoms of a molecule are moving closer and further away from one another. Several hundred (c)_____ energy levels for gaseous HCl are populated at room temperature. Microwave spectroscopy generally involves (d)_____ excitation.

3.39 Fill in each blank with one of the following words: *electronic, vibrational, rotational,* or *translational*. In some cases, more than one answer may be correct. Indicate *all* correct answers: (a) _____ excitation generally requires the most energy. Ultraviolet spectroscopy generally involves (b) _____ excitation. (c)_____ excitation is generally only possible for molecular liquids and molecular gases. Only a handful of (d) _____ energy levels for gaseous HCl are populated at room temperature.

3.40 Explain (a) fluorescence (b) phosphorescence. Why might some molecules phosphoresce while others wouldn't?

3.41 For the diatomic molecule A_2, (a) sketch a partial energy level diagram showing two electronic states with their respective vibrational states. Clearly label the following: ground electronic state, excited electronic state, ground vibrational state (in both electronic states). (b) The transition between the ground vibrational state of the ground electronic state and the ground vibrational state of the first excited electronic state was observed at 488 nm. Determine the separation between the electronic energy levels (in J). (c) The excited electronic state shown is known to have a vibrational frequency of 2.42×10^{13} s^{-1}. Determine at what wavelength (in nm) the transition between the ground vibrational state of the ground electronic state and the first excited vibrational state of the first excited electronic state will be observed.

3.42 ⓐ Most current CD and DVD players, along with many other devices, use a HeNe laser that produces light at 650 nm. A proposed new standard for CDs would allow more data to be stored on a disk. This new standard uses a blue light laser that produces light at 405 nm. What is likely the major difference between the chemical systems in a HeNe laser and that of a blue light laser that results in the different wavelength of light being produced?

3.43 On the diagram shown below, indicate with a vertical arrow an absorption that might involve this molecule in its overall ground state becoming

    (a) vibrationally but not rotationally excited;

    (b) electronically but not vibrationally excited;

    (c) both electronically and vibrationally excited, but not rotationally excited.

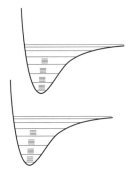

Brain Teasers

3.44   Shown below are a molecule and its infrared spectrum. The spectrum shows an absorbance of energy when the curve drops to a lower value of "% Transmittance." Note that infrared energy is absorbed more strongly at certain wavelengths than at others. Several bonds in the molecule (a–d) and areas of the spectrum (A–H) are labeled. The strong absorption **C** at about 5.8 μm is due to stretching of bond **b.** Use what you know about the effect of bond strength and atomic mass to indicate

_____   The lowest-energy absorption of **A, C,** and **E.**

_____   Of **B** and **E,** the absorption more likely to be associated with bond **a.**

_____   Of **B** and **E,** the absorption more likely to be associated with bond **c.**

_____   Of **A** and **D,** the absorption more likely associated with bond **d.**

3.45   Explain why absorbances **A** and **B** might be so much different in energy from the others. Provided vibration of bonds **a, b, c,** and **d** are seen as **A, B, C,** and **E** (but not in that order), what bonds must **A** and **B** be associated with?

3.46   What are some of the special properties that metals and semiconductors have that other substances don't have? How do you explain this using the ideas of electronic energy?

3.47   Pretend you are the electron in a hydrogen atom. Describe what you are doing (a) in the $1s$ state (b) in the $2p$ state (c) in the $1s$ state absorbing 2.04×10^{-18} J of energy.

3.48   Develop an analogy relating molecules and energy levels to people having fun at a nightclub.

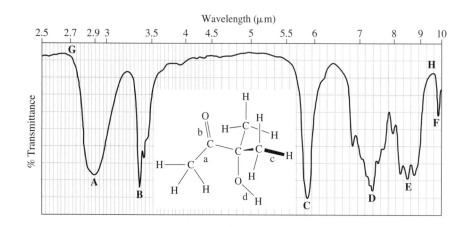

CHAPTER 4

Internal Energy (*U*) and the First Law

The energy of the universe is constant.

—*Rudolf Clausius, 1865*

4.1 The Internal Energy (*U*)

The energy we have been talking about distributing is called *internal energy* and given the symbol *U*. It is simply the sum of all of the energy of all of the particles in the system:

$$U = n_0\varepsilon_0 + n_1\varepsilon_1 + n_2\varepsilon_2 + n_3\varepsilon_3 + \cdots = \sum_{i=0}^{\infty} n_i\varepsilon_i \qquad (4.1)$$

It is the internal energy that changes when water is heated on the stove to make spaghetti. We say the water is *heated* by the stove. Whether electric or gas, a stove is a device that can transmit heat to a substance in a controlled manner, thus raising that substance's internal energy to the desired extent. The fact that the temperature rises is derivative to this internal energy change. Thus, it is not that the internal energy increases *because* the temperature rises, but instead that the temperature rises *because* the internal energy has increased.

4.2 Internal Energy (*U*) Is a State Function

It's important to realize that there are certain properties of a system that are intrinsic to that system *as it exists in its current state* and not based on *how it got there*. Clearly, the energy levels themselves in a system and the number of particles in each level in the most probable distribution are of this sort. Other such properties include the temperature, volume, and pressure of a system, and the masses and concentrations of all of a mixture's components.

These properties, also called *state functions* or *state variables,* together define the *state* of a system. If any of these properties change, then we say that the system has changed its state. Thus, for example, to describe one sample of a gas we might talk about P_1, V_1, n_1, and T_1, and to refer to another we might use P_2, V_2, n_2, and T_2.

Differences in properties between two systems in two different states are traditionally shown using a Greek Δ (delta) prior to the state function. This always means "take the second one and subtract the first one." So if $P_1 = 1.0$ bar and $P_2 = 2.5$ bar, then we would say that $\Delta P = (2.5\text{ bar} - 1.0\text{ bar}) = 1.5$ bar. The way to read this is "Delta P equals 1.5 bar."

Very often the two states we are talking about are of the same system before and after some event or disturbance (like heating, cooling, compressing, expanding, or reacting). In that case, we often read ΔX not as the *difference* in X between the two states so much as the *change* in X when the event occurs. There is a subtle difference, so perhaps an example may help. If I measure my altitude in Northfield, Minnesota, as 1000 ft above sea level, then fly to Denver, Colorado, and measure my altitude there as 5280 ft above sea level (Denver is the "mile-high city"), then I could say that ΔA (A for "altitude") for this change in state (no pun intended) is 4280 ft. Notice it doesn't matter if I flew to 30,000 ft or hovered at 500 ft off the ground all the way, ΔA would still be 4280 ft. (And if I flew the other way, from Denver to Northfield, then ΔA would be -4280 ft.)

On the other hand, "Denver" and "Northfield" are, intrinsically, at two different altitudes, whether I fly or not. We can say that the *difference* in altitude between Northfield and Denver is 4280 ft. That's also ΔA.

Thus, we could say that there is a difference in temperature between water at 30°C and water at 90°C. This difference ($\Delta T = 60$ °C) is there whether I heat 100 g of water from 30°C to 90°C or not, but if I *do* heat the water, then we will say that the *change* in temperature (also ΔT) was 60°C.

While we're on the subject of temperature, notice that this 60-degree change in temperature on the Celsius scale ($90 - 30$) is the same 60-degree change on the kelvin scale ($363 - 303$). You'll learn to switch into kelvins automatically when taking differences in temperature in degrees Celsius. If $T_1 = 20$ °C and $T_2 = 30$ °C, then $\Delta T = 10$ °C or 10 K. Units of kelvins will be preferred.

4.3 Microscopic Heat (q) and Work (w)

When you think about it, there really are only two ways to increase the internal energy of a system if the number of particles is not changed. Either we can move particles up from lower levels to higher levels, or we can move the levels *themselves* higher, as shown in Figure 4.1. That the first way, involving heat, is possible may seem obvious enough. Moving particles from lower to higher levels will certainly change the ratios n_j/n_i in the direction of increasing temperature (see Figure 2.7)—just what we would expect for "adding heat" to a system. We simply define the heat added, q, as the change in internal energy, ΔU. We say that any process for which $q > 0$, that is, for which heat is put *into* the system, is *endothermic*, and any process for which $q < 0$, that is, for which heat is generated or removed from the system, is *exothermic*.

What about this other way to change the energy, involving *work*? This is a consequence of particles having translational energy. Remember, for translation,

$$\varepsilon_{n_x,n_y,n_z} = (n_x^2 + n_y^2 + n_z^2)\frac{h^2}{8}\left(\frac{1}{md^2}\right) \quad \text{where } n = 1, 2, 3, \ldots$$

where d is a measure of the size of the box. These energies, and consequently their energy level *spacings,* are inversely proportional to d^2. If we decrease d, then the translational

Figure 4.1
Two totally different ways to add energy to a system and thus raise its temperature. In the first way, 2 units of heat are added, and a particle moves up in energy (starts moving more violently). In the second way, 1 unit of work is added, carrying the particle up to higher energy without changing its level. It is the whole level in that case that moves; the particle just goes along for the ride.

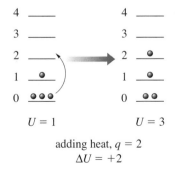

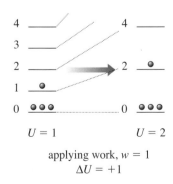

energy spacings ($\Delta \varepsilon$) in the system will increase! Recall the Boltzmann law (Equation 2.3):

$$\frac{n_j}{n_i} = e^{-(\varepsilon_j - \varepsilon_i)/kT} = e^{-\Delta \varepsilon_{ij}/kT}$$

In this case we are making no change in n_i or n_j; if $\Delta \varepsilon$ increases, *the temperature must increase as well* so that overall there is no change in the value of the exponent.

Decreasing the size of the box like this is called *compression*. Compressing a system that has translational energy (i.e., a gas) will increase its temperature! This phenomenon is really quite easy to observe. Take a bicycle pump, hook it up to a bicycle tire and start pumping. See if the tire doesn't get quite warm within seconds.

The reverse is also true: *increasing* the size of the box (*expansion*) lowers the temperature of a gas. If you've ever seen a snow-making machine, you've seen cooling due to expansion in action. Steam shot through a little hole at high pressure expands quickly, and its temperature drops so much that it freezes into snow on the fly!

Another good application of using expansion to produce cooling is a refrigerator. In a refrigerator, a gas in a sealed tube is expanded rapidly, causing the gas to cool. As it warms back up, it pulls heat out of the refrigerator compartment. The warm gas is circulated out to the back of the refrigerator, where it is compressed. This compression causes the gas to heat up, and that heat is dumped into the room. The net effect of the cycle is to draw heat out of the refrigerated compartment, transferring it into the room. An air conditioner works by the same principle. We will see how to calculate the work applied to the system, w, shortly.

4.4 "Heating" vs. "Adding Heat"

We have to be careful here. Most of us are not used to this second way of increasing the temperature of a system. We would generally say something like, "Well, if you want to heat it up, put it on the stove and add some heat." That's certainly right. Now, though, because we have a more scientific definition of heat, we have to make a distinction between "heating a system" (raising its temperature) and "adding heat."[1]

If you "heat a system," its temperature always rises. But because of the possibility of applying work, you don't have to "add heat" to "heat the system." And, to make matters worse for the uninitiated, adding heat to a system doesn't necessarily cause its temperature to rise. For example, adding heat to ice water doesn't increase its temperature as long as any ice is present. It simply melts the ice. The temperature remains constant at 0°C. Heat can be a crazy and confusing issue. No wonder it took centuries of work to sort it all out! Similarly, "cooling" means "decreasing the temperature of" and may or may not have anything to do with heat. What are we to do? Be careful! Be specific!

Here are some examples of situations that may not be obvious at first glance:

- Compressing a gas quickly raises its temperature but transfers no heat to it. We say that the gas is heated even though no heat was added.
- Adding heat to ice water does not raise its temperature; it just melts the ice. We say that adding heat to ice melts the ice. The ice is not heated, however.
- As you heat water, first its temperature rises, then, as it starts to boil, the temperature stops rising even though the heat is clearly still going in. We say at all points we are adding heat, and we are "heating it to boiling," but, technically, after it starts to boil we are no longer heating it, just adding heat.

---

[1] It is important to remember that heat is not a substance that can be added or removed from a system. When scientists refer to adding heat, they mean adding energy in a way that changes the ratios of particle numbers in the Boltzmann distribution as shown on the left in Figure 4.1.

• Food in a pressure cooker cooks faster, because it can get hotter than is possible in an open pot. We say that it is possible to heat food in a pressure cooker hotter than the normal boiling point of water.

These are all examples of situations you should be able to explain in detail with no problem very soon.

4.5 The First Law of Thermodynamics: $\Delta U = q + w$

In summary, then, there are just two ways to change the energy of a system:

1. move particles up and down the ladder of energy levels (exchange heat) or
2. move the energy levels themselves up and down (exchange work).

The statement that any change in internal energy is due to either heat or work is called the *First Law of Thermodynamics* and stated as

$$\Delta U = q + w \qquad (4.2)$$

That is, the change in internal energy is always the heat *put in* plus the *work applied*. This fundamental law was discovered long before anyone dreamed of energy levels or even molecules, for that matter. Its reinterpretation in terms of energy levels was Boltzmann's great achievement.

Interesting here is the fact that even though the internal energy U is a state function, neither heat q or work w is. Thus when the state of a system changes, there may be any number of ways of doing the change. One way may require more or less work than another. You know this intuitively. Running up ten flights of stairs followed by cooling off gets you to the same height and temperature as walking, but requires lots more work (and, consequently, the dumping of lots more heat during that cooling-off period).

So, both heat and work are always spoken of in reference to some actual *change* carried out via some specific *path*. If we are to talk sensibly about heat and work, we must be very careful to define the path taken in the process. In chemistry the way we define paths is to emphasize what is constant. For example, a reaction may be carried out at constant temperature or constant pressure or constant volume, or some combination of these. We'll see that by far the most common path used is the "constant pressure" path because that is simply the easiest to perform in the laboratory and relates well to such real systems as living organisms. But constant volume paths are quite plausible as well.

4.6 Macroscopic Heat and Heat Capacity: $q = C \Delta T$

In the laboratory there is a clear connection between heat and temperature as long as the following two conditions are met:

1. Either the volume or the overall pressure remains constant (thus defining the path), and
2. There are no chemical reactions going on (including changes of phase, such as melting, boiling, or evaporating).

If both of these conditions are met, then there should be no doubt that when we add heat to a system it will increase in temperature. In fact, there is a very simple relationship that was discovered early in the 1800s relating the heat to the temperature change in degrees Celsius or kelvins:

$$q = C \Delta T \qquad (4.3)$$

This says that the heat put into a system is directly proportional to that system's change in temperature. The proportionality constant, C, is called the *heat capacity* and has units of energy/temperature, usually joules/kelvin or sometimes calories/kelvin. It depends critically on how much material is present and being heated.[2] For example, the heat capacity of 100 g of water is 100 calories/kelvin. Thus, if a 100-g sample of water is heated from 15°C to 25°C, then ΔT is 10 K, and the heat added to the system is just

$$q = C\Delta T = \frac{100 \text{ cal}}{K} \times (10 \text{ K}) = 1000 \text{ calories}$$

(One thermodynamic calorie is 4.184 joules.) Note that it is certainly possible to take heat *out* of a system. If we had instead cooled this system from 25°C to 15°C, then ΔT would have been -10 K, and the heat added to the system would be:

$$q = C\Delta T = \frac{100 \text{ cal}}{K} \times (-10 \text{ K}) = -1000 \text{ calories}$$

Thus, $q > 0$ means heat was *added to* the system; $q < 0$ means heat was *removed from* the system.

Related to heat capacity is the *specific heat* of a substance, which has units of energy/mass-temperature. The specific heat of a substance is the amount of heat required to increase the temperature of 1 gram of the substance by 1 degree. Sometimes, too, you might see "molar heat capacity" listed for a substance. This is just what it sounds like: the amount of heat required to increase the temperature of 1 *mole* of the substance by 1 degree.

You need not worry too much about remembering all this. Just remember to consider always the *amount* of substance actually being heated and you will do fine.

For example, the specific heat of liquid water is 1.00 cal/g·K or 4.184 J/g·K. Thus, the heat capacity of 25 g of water is:

$$C = (25 \text{ g}) \times \frac{1.00 \text{ cal}}{g \cdot K} = 25 \text{ cal/K}$$

and the amount of heat in joules required to raise that 25 g of water 10°C (10 K) is

$$q = C\Delta T = \frac{25 \text{ cal}}{K} \times (10 \text{ K}) = 250 \text{ cal} \times \frac{4.184 \text{ J}}{\text{cal}} = 1050 \text{ J}$$

Just be careful with your units when working with specific heat or heat capacity, and always ask yourself, "How much material are we talking about in this case?"

4.7 Macroscopic Work: $w = -P\Delta V$

Work, like heat, can be measured experimentally. The thing to remember is that *work always requires movement*. The macroscopic definition of work is

$$\text{work} = (\text{force applied}) \times (\text{distance traveled})$$

which, because pressure is "force per area" (for example, pounds per square foot) and

---

[2] Note that heat capacity, like heat, depends upon how a change is effected, especially for gases. When gases in particular are involved—when expansion or contraction work can be done—we refer to "C_p" (C-sub-pea) as the heat capacity when the pressure is kept constant (as in a balloon) and "C_v" (C-sub-vee) when the volume is kept constant (as in a sealed chamber). Don't worry, though; at this stage in the game, you shouldn't have to worry about *which* heat capacity to use.

volume changes can be thought of as "area × distance," we have

$$\text{work} = \frac{\text{force}}{\text{area}} \times (\text{area} \times \text{distance}) \tag{4.4}$$

$$w = -P\Delta V$$

The reason for the minus sign is that we must associate a negative change in volume (compression) with an increase in internal energy U, and this is the way to do it. Notice that we are assuming here that the pressure is constant. It doesn't have to be, but if it isn't then we have to use calculus. We'll just consider cases where we have a constant pressure.

Thus, for example, if a gas is compressed using a pressure of 2.0 bar from 1.5 L to 1.0 L then ΔV is -0.5 L and $w = -(2.0 \text{ bar})(-0.5 \text{ L}) = 1.0$ L·bar. The conversion factor to remember is that $100 \text{ J} = 1 \text{ L·bar}$. Thus, in this case, we would have for the work involved

$$w = -P\Delta V = -(2.0 \text{ bar})(-0.50 \text{ L}) \left(\frac{100 \text{ J}}{\text{L·bar}} \right) = 100 \text{ J}$$

The trick in getting work, sometimes, is that neither the volume change itself nor the pressure may have been measured. We can apply the ideal-gas equation, $PV = nRT$, then, if we are just changing the temperature of a gas at constant pressure:

$$w = -P\Delta V = -nR(\Delta T) \quad \text{(constant moles)} \tag{4.5}$$

Thus, for example, if 2.5 mol of gas at a constant (perhaps even unknown) pressure is caused to increase in temperature by 10°C, then we can write that

$$w = -P\Delta V = -nR\Delta T = -2.5 \text{ mol} \times \frac{8.31451 \text{ J}}{\text{mol·K}} \times 10 \text{ K} = -208 \text{ J}$$

Note that using this trick there is no converting from L·bar to joules.

Or, perhaps we are investigating the following reaction:

$$2 \text{ Mg}(s) + \text{O}_2(g) \longrightarrow 2 \text{ MgO}(s)$$

The question is, How much work is involved if 2.5 g of Mg is combusted at 25°C and constant pressure? The key is in seeing that the number of moles of gas is *decreasing* in this reaction, since for every two moles of Mg reacting, one mole of gaseous O_2 is also consumed, and no gaseous products are produced. The temperature and pressure are constant, so we can write

$$w = -P\Delta V = -(\Delta n)RT \quad \text{(constant temperature)} \tag{4.6}$$

Thus, we calculate Δn first:

$$\Delta n = 2.5 \text{ g Mg} \times \frac{1 \text{ mol Mg}}{24.31 \text{ g Mg}} \times \frac{-1 \text{ mol gas}}{2 \text{ mol Mg}} = -0.0514 \text{ mol gas}$$

and from that we calculate w:

$$w = -P\Delta V = -\Delta nRT$$

$$= -(-0.0514 \text{ mol}) \left(\frac{8.31451 \text{ J}}{\text{mol·K}} \right) (298 \text{ K})$$

$$= 127 \text{ J}$$

Even if you are scrupulously careful with your minus signs, be sure to always ask yourself after a calculation such as this, "OK, now does this make sense? Should work be positive?" In this case, since the number of moles of gas decreased in the reaction we can think of it as a compression, and the volume would decrease as well. Anytime the volume decreases we must have $w > 0$.

4.8 In Chemical Reactions, Work Can Be Ignored

Adding heat leads to many chemical reactions (Section 3.10). We say that adding heat breaks bonds in molecules. What about work? How does work relate to chemical reactions? If we look again at the dissociation of H_2, we would write

$$H_2(g) \longrightarrow 2H(g)$$

Note that for every one mole of gas going into this reaction, two moles of gas come out. That being the case, those extra gas particles will create a harder push on the walls of the container, and the pressure will tend to increase. If the walls are fixed, as in a sealed tube, then no work will be involved, because no volume change will occur.

If, on the other hand, the greater pressure on the walls causes them to expand outward, as in a balloon or in an open flask, then as the expansion occurs, work will be transferred to the walls, and the system will lose a little bit of energy. One way to think of this is that the particles, hitting the walls, will recoil with a little less energy than they started with, having pushed the wall out just a bit.

Mathematically, we account for this energy loss by saying that the translational energy levels drop down slightly in energy due to a larger d appearing in the denominator of the translational energy equation. Due to the work, *all* of the energy levels are a bit lower.

It's important to realize that this loss in energy due to expansion work in typical chemical reactions is really very small. For example, in the case of H_2 dissociation, breaking apart a mole of H_2 molecules into H atoms requires about 432 kJ of energy. The work involved, if the walls are allowed to expand, changes with temperature, but at 298 K the work amounts to only a few kilojoules:

$$w = -P\Delta V = -\Delta nRT = -(1\ \text{mol}) \left(\frac{8.31451\ \text{J}}{\text{mol} \cdot \text{K}} \right) (298\ \text{K}) = -2500\ \text{J} = -2.5\ \text{kJ}$$

Remember, the negative sign here signifies that this is an expansion, and work is coming *out* of the system. This work is less than 1% of the total energy change of 432 kJ. Thus, even if the walls are allowed to move, we often ignore the work involved in chemical reactions, and just focus on the heat of reaction, q.

Let's look at some other typical reactions, all carried out at 25°C at a constant pressure so that expansion and compression can take place (Table 4.1). Note that at 25°C, $RT = 2.5$ kJ/mol, and that $\Delta U = q + w$. In each case, the work, w, is relatively small, and $\Delta U \approx q$. Specifically:

- In Case 1, the highly exothermic burning of magnesium, the number of moles of gas decreases as the reaction proceeds. This amounts to a "compression." We have $\Delta n = -1$ mol, and $w = -\Delta nRT = -(-1\ \text{mol})RT = 2.5$ kJ.

    This amount of work is only about 0.2% of the total energy change and can surely be ignored.

Table 4.1
A variety of chemical reactions. Work w is only of minimal importance in each case compared to heat q.

	Reaction	ΔU/kJ	q/kJ	Δn/mol	w/kJ
1	$2\,Mg(s) + O_2(g) \longrightarrow 2\,MgO(s)$	-1199	-1202	-1	$+2.5$
2	$H_2O(l) \longrightarrow H_2O(g)$	$+41$	$+44$	$+1$	-2.5
3	$H_2(g) + Cl_2(g) \longrightarrow 2\,HCl(g)$	-185	-185	0	≈ 0
4	$2\,H_2(g) + O_2(g) \longrightarrow 2\,H_2O(l)$	-564	-572	-3	$+7.5$
5	$CuO(s) + H_2O(l) \longrightarrow Cu(OH)_2(s)$	-65	-65	0	≈ 0

- In Case 2, the vaporization of a mole of water, a mole of gas is produced (Δn = +1), which causes an expansion, and $w = -2.5$ kJ. In this case, since the overall energy change is so small, work contributes as much as 6% of the total energy change—still just a minor fraction.
- In Case 3, the formation of HCl from elemental hydrogen and chlorine, $\Delta n = 0$, and thus $w = 0$ as well. Work is no issue here!
- In Case 4, the formation of liquid water from elemental hydrogen and oxygen, we have $\Delta n = -3$ mol, giving $w = 7.5$ kJ. This is the largest change in number of moles of gas of all five examples, and still the work is less than 2% of the total energy change.
- Finally, in Case 5, as in many, many reactions, no gases are involved, and $w \approx 0$.

Thus, in terms of getting an idea of how much energy change there is in a chemical reaction, we can focus on heat and safely ignore the work in these and, by extension, virtually *all* chemical reactions, whether or not the walls of the system are fixed or movable.

That is, if we measure the *heat* going into or coming out of a chemical reaction, we have either an exact measure of ΔU (if the walls are fixed) or at the very least a very good approximation of ΔU (if the reaction system is allowed to expand and contract at constant overall pressure). And once we have a measure of ΔU, we know which substance, reactant or product, will predominate at low temperature, and how the equilibrium will shift with temperature.

We will see in the next two sections that measuring the heat associated with a chemical reaction experimentally requires carrying out the reaction in a device called a *calorimeter*. The change in temperature of the calorimeter, ΔT, along with its heat capacity, C_{cal}, can then be used to determine q (and thus ΔU) for a chemical reaction.

4.9 Calorimeters Allow the Direct Determination of ΔU

A special device called a *bomb calorimeter* (depicted in Figure 4.2) allows for the direct determination of ΔU for a chemical reaction.

The reaction mixture to be studied is placed in the lower part of the chamber along with some sort of igniter, usually an electrical device. The calorimeter is closed and bolted down with several strong bolts. A very sensitive thermometer, often good to ±0.001°C,

Figure 4.2
A bomb calorimeter from the 1960s. This particular calorimeter, a Parr 1341 Plain-Jacket Oxygen Bomb Calorimeter, is specially designed to measure the heat given off by the reaction of a small amount of sample compound in an oxygen atmosphere. The reaction is carried out in the sealed ignition chamber, which is fully immersed in a water bath. Heat from the reaction is absorbed by the surroundings, including all the parts of the calorimeter and the water in the bath. Provided the heat capacity of the entire apparatus is known, the temperature change measured in the water bath can be used to calculate the internal energy change of the system. (Diagram adapted with the permission of the Parr Instrument Company, Moline, Illinois.)

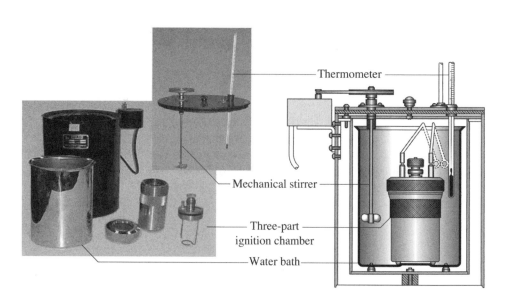

Thermometer

Mechanical stirrer

Three-part ignition chamber

Water bath

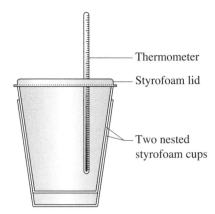

Figure 4.3
A simpler calorimeter used in introductory chemistry laboratories made from two nested styrofoam cups with a styrofoam lid is especially useful when two or more reactants are involved.

is included, and the reaction is initiated. Since the volume is fixed, the pressure changes substantially, but no work is done. Thus, $w = 0$, and $\Delta U = q$ with no approximations.

Alternatively, and more applicable to an undergraduate laboratory, one can use the simple calorimeter depicted in Figure 4.3. This calorimeter is simply two nested styrofoam cups covered by a styrofoam lid. In this case, the reaction must take place in water, but the idea is the same. One reactant is placed in the calorimeter and the temperature is measured. Then the other reactant, either solid or dissolved in water already, is added. The lid is closed and the temperature is monitored. The main difference between this calorimeter and a bomb calorimeter is that here the volume is not truly fixed, even when no gases are produced. The volume is not exactly constant because when two solutions are mixed, the total volume is not necessarily the sum of the two initial volumes. Nonetheless, the approximation that $w = 0$ is very good because ΔV is so very small.

4.10 Don't Forget the Surroundings!

A schematic of the calorimeter in Figure 4.3 is shown in Figure 4.4. One of the biggest mistakes beginners make in this business is to not make the distinction between the *system* (the thing we're interested in) and everything else, which we call the *surroundings*. When the temperature changes, and we arrive at ΔT, what are we really measuring the temperature of? The system? The surroundings? Both? To be sure, always consider the following general points:

- We almost never measure the temperature of the system itself, especially if that "system" is a chemical reaction.

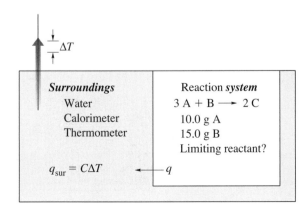

Figure 4.4
Schematic diagram of the calorimeter in 4.3. The thermometer is part of the *surroundings*, so $C\Delta T$ measures q_{sur}, not q. In this case the reaction is exothermic: $q < 0$.

- When a chemical reaction occurs in solution, such as the reaction between H^+ and OH^- to make H_2O, even if "H_2O" is in the chemical equation, our thermometer is always in the *solution*, containing moles and moles of water, not in the "reaction."
- Invariably when a reaction occurs—even if it is just the melting of ice—that reaction is contained in some sort of vessel. The vessel could be as simple as a beaker or a styrofoam cup, or as fancy as a bomb calorimeter. Invariably that vessel heats or cools along with the solution it contains, so it must be considered part of the surroundings.
- The heat capacity is the sum of all the individual heat capacities of everything being heated. In the case of reaction in solution, the heat capacity of the solution itself is often approximated as just being the heat capacity of an equivalent amount of pure water.
- The heat going into the system is coming out of the surroundings (and vice-versa):

$$q_{sur} = -q$$

This is just another way of saying that no heat is ever lost, $q + q_{sur} = 0$.

Note carefully that a *chemical equation* is different from a *chemical reaction*. While it is common in chemistry to use these two words synonymously, here we need to make a distinction. We never measure the heat given off by a whole mole of reactants. In this book, if we have the equation

$$3\,A + B \longrightarrow 2\,C$$

for which we want to specify "the change in internal energy," we will use the international standard notation $\Delta_r U$, with a little subscript "r" after the delta sign.[3] In the *equation* we have 3 moles of A; in the *reaction* we might have only 0.050 mol of A, and it may not all react, because B may be the limiting reactant. We will reserve the notation ΔU for the actual change in internal energy for the *chemical reaction*.

For example, say that we used something like the equipment listed in Figure 4.3, and we found that when 0.050 mol of A was reacted with 0.020 mol of B in a 50-mL solution, the temperature rose 1.5°C. Furthermore, say that we had determined previously that the heat capacity of the calorimeter was 14 J/K. Then we would write:

$$\Delta U = q + w \approx q = -q_{sur} = -C\Delta T_{sur}$$
$$= -(C_{soln} + C_{cal})\Delta T_{sur}$$

Now, the heat capacity of 50 mL of the solution is approximated as being that of 50 g of pure water, for which the specific heat is 4.184 J/g·K:

$$C_{soln} \approx (50\text{ g }H_2O) \times \frac{4.184\text{ J/K}}{1\text{ g }H_2O} = 209\text{ J/K}$$

Adding this to our 14 J/K for C_{cal} and multiplying by ΔT, we get

$$\Delta U \approx -q_{sur} = -\left(\frac{223\text{ J}}{K} \times 1.5\text{ K}\right) = -335\text{ J}$$

So the reaction involved a change in internal energy of -335 J.

---

[3] The technical term for $\Delta_r U$ is *reaction internal energy*. In this book we will also see $\Delta_r S$ (reaction entropy), $\Delta_r H$ (reaction enthalpy), and $\Delta_r G$ (reaction Gibbs energy). Despite the word "reaction" here, try to remember that these terms critically depend upon exactly how the chemical *equation* for the reaction is written.

Now the question is how to write this in terms of the chemical equation, which we would be free to write as either

$$3\,A + B \longrightarrow 2\,C$$

or

$$1.5\,A + 0.5\,B \longrightarrow C$$

or, for that matter, any combination of coefficients that are in a 3:1:2 ratio. In any case we could carry out a quick limiting reactant calculation:

$$0.050 \text{ mol A} \times \frac{2 \text{ mol C}}{3 \text{ mol A}} = 0.033 \text{ mol C}$$

$$0.020 \text{ mol B} \times \frac{2 \text{ mol C}}{1 \text{ mol B}} = 0.040 \text{ mol C}$$

Thus, even though there is more A than B present, because of the way they react, A is the limiting reactant. So we would calculate

$$\Delta U \approx \frac{-335 \text{ J}}{0.050 \text{ mol A}} \times \frac{1 \text{ kJ}}{1000 \text{ J}} = -6.7 \text{ kJ/mol A}$$

This is a perfectly unambiguous statement. For every mole of A reacting, 6.7 kJ of energy will be released. If we choose to write our chemical equation for this reaction as

$$3\,A + B \longrightarrow 2\,C$$

we would multiply ΔU by the factor "3 moles of A per mole of reaction" to get $\Delta_r U$ for this particular writing of the chemical equation:

$$\Delta_r U = \Delta U \times \frac{3 \text{ mol A}}{\text{mol of reaction}} = -20 \text{ kJ/mol of reaction}$$

We write

$$3\,A + B \longrightarrow 2\,C \qquad \Delta_r U = -20 \text{ kJ/mol}$$

It does not matter which substance we initially choose to write the energy as "per mole of." For example, we could also have calculated

$$\Delta U \approx \frac{-335 \text{ J}}{0.033 \text{ mol C}} \times \frac{1 \text{ kJ}}{1000 \text{ J}} = -10 \text{ kJ/mol C}$$

Now, since there are two moles of C in the equation, we would multiply this time by "2 moles of C per mole of reaction" and still conclude

$$3\,A + B \longrightarrow 2\,C \qquad \Delta_r U = -20 \text{ kJ/mol}$$

How exactly we write the chemical equation is not particularly important *as long as we always specify it when we indicate the value for* $\Delta_r U$. So, alternatively, we could have written:

$$1.5\,A + 0.5\,B \longrightarrow C$$

Then we would have calculated for $\Delta_r U$:

$$\Delta_r U = \Delta U \times \frac{1.5 \text{ mol A}}{\text{mol of reaction}} = -10 \text{ kJ/mol of reaction}$$

and we would have written:

$$1.5\,A + 0.5\,B \longrightarrow C \qquad \Delta_r U = -10\,kJ/mol$$

Note that if we were given either chemical equation with its associated value for $\Delta_r U$ and then asked, "How many joules of energy would be released if 0.033 mol of C were produced?" then we would get the same answer:

$$\Delta U = 0.033\,mol\,C \times \frac{1\,mol\,of\,reaction}{2\,mol\,C} \times \frac{-20\,kJ}{mol\,of\,reaction} = -0.33\,kJ$$

$$\Delta U = 0.033\,mol\,C \times \frac{1\,mol\,of\,reaction}{1\,mol\,C} \times \frac{-10\,kJ}{mol\,of\,reaction} = -0.33\,kJ$$

The question asks for the amount of energy *released*. This is our clue that we should express our answer as a positive number: 0.33 kJ.

4.11 Engines: Converting Heat into Work

An engine is simply any device that turns heat into work. Thus, imagine the hypothetical sequence of events of Figure 4.5. Given any sequence of energy diagrams, the easiest way to figure out what is happening to ΔU, q, and w is to first calculate U for each state by adding up the energies of all the particles shown. From that, figure ΔU between each state. Then, if the populations change but the positions of the levels don't, you can say $q = \Delta U$ and $w = 0$. If, on the other hand, the *populations* don't change but the levels do change position, then you can say $w = \Delta U$ and $q = 0$. *If both change, then you cannot know exactly how much work and heat were involved without knowing more about exactly how the change was carried out.*

Notice that the state at the beginning of this particular sequence is exactly the same as the state at the end. This is called a *cycle* of events. The total change in energy of a system around a cycle is zero. Thus, ΔU overall is $4 + 9 - 7 - 6 = 0$. But wait! Look at q and w:

$$total\ q = 0 + 9 + 0 - 6 = 3$$
$$total\ w = 4 + 0 - 7 + 0 = -3$$

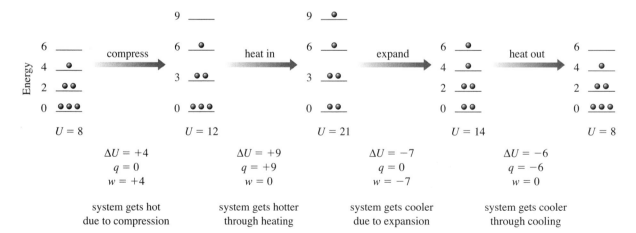

Figure 4.5
A highly simplified view of the four cycles in a standard engine: compression, heating, expansion, and cooling (exhaust).

Thus, over the whole cycle, three units of heat (input) were turned into three units of work (output). *All* engines work this way, transforming heat into work, although not all are perfect cycles.

Invariably, heat is generated by some sort of chemical reaction such as the burning of gasoline. In a four-stroke automobile engine, for example, the first down stroke draws in the fuel/air mixture. A second compression stroke compresses the fuel mixture in the cylinder. The fuel is ignited by a spark (generally) or just by the heat caused by the compression itself (in the case of diesel engines). The production of voluminous amounts of gaseous CO_2 and H_2O then drives the piston back on the power stroke, doing work ($w < 0$). Finally, the gas is cooled (partially at least) as it exits the cylinder in one last stroke.

4.12 Biological and Other Forms of Work

One thing you might consider a bit odd is that work *always* involves movement. Yet, just try holding a heavy book out at arm's length for even just 10 minutes. It's hard! This takes a lot of work. Yet nothing appears to have moved. What's wrong here?

The problem is simply that our muscles are not designed to stay rigid. Instead, they are designed to move rapidly. There are two kinds of muscles in your body: *striated* and *smooth*. Striated muscle is what you have in your arms. Smooth muscle, on the other hand, is what your intestines and blood vessels are made of (and what clams have). Smooth muscles are made to hold a position for a long time and move very slowly. The problem is, when you hold your arm in the air, you are using your striated muscle, not your smooth muscle.

Your striated muscle acts by contracting every time a nerve pulse comes along *and then immediately relaxing again* (so that it can move again to some other position very quickly). Thus, holding that book out at arm's length requires innumerable "firings" of the muscle, which requires work to be expended, and your muscles really are moving. A clam, on the other hand, using its smooth muscle, can hold its shell open indefinitely expending no work at all.

By the way, that work ultimately ends up as heat transferred out into the surroundings (your body), which you remove by sweating. The evaporation of water from your skin is Case 2, in Section 4.8. Note that $q > 0$, which means that heat is being taken *into* the system (the water liquid/gas reaction system) when evaporation occurs. That heat has to come from somewhere: Your body cools.

Our definition of work as $P \Delta V$ applies only when the pressure is constant. When it is not, then a more precise definition of work involves using derivatives:[4]

$$dw = P_{ext}dV$$

Here P_{ext} is the external pressure and is not necessarily constant. dw means "an infinitesimally small amount of work" and dV means "an infinitesimally small change in volume."

This use of derivatives allows a more precise explanation of the other kinds of work besides this sort of P–V work. Fundamentally, work is any change in internal energy of a system that is *not* attributable to the transfer of heat. Several important definitions of work include those shown in Table 4.2. Thus, lifting this book off your table takes energy. The book doesn't get hot, yet relative to where it was, its energy has increased. We say that work has been done *on* the book against the force of gravity: $w_{book} > 0$. Basically, in types 2–5, you can think of the work as just lifting the *whole energy level system* to higher energy

---

[4] If we were to use calculus at this point, we would integrate to get w:

$$w = \int P_{ext}dV$$

Table 4.2

A variety of chemical reactions. Work w is only of minimal importance in each case compared to heat q.

	Work against . . .	Based on . . .	Changes . . .
1	an external pressure, P_{ext}	$dw = P_{ext}dV$	volume, V
2	the force of gravity, g	$dw = mgdh$	height, h
3	a torque, τ	$dw = \tau d\theta$	angle, θ
4	a tension, τ	$dw = \tau dl$	length, l
5	a surface tension, γ	$dw = \gamma dA$	surface area, A
6	an electrical potential, E	$dw = Edq$	charge, q
7	a chemical potential, μ	$dw = \mu dn$	moles, n

without having anything at all to do with the populations of the levels *or* the energy level spacings. In a sense, that's a third fundamental way of increasing the energy of a system.

Sometimes it is less obvious to see what is moving when work is done. Obviously when you lift a weight against the force of gravity or turn a screw against a torque, there is movement. But what about those last two forms, electrical and chemical?

In the electrical case, that "charge" is really an electron moving in a wire. So there really is movement of a sort. We'll get to electrochemistry in Chapter 13. The last type, involving "chemical potential," is a matter of interpretation. For now we will ignore it and show that we don't need it to understand what is fundamentally involved in chemical reactions. "Chemical potential" turns out to be related to "Gibbs energy," which we'll get to in Chapter 9.

The point of this discussion was not to fully understand work, rather to argue that pressure–volume work in particular can pretty much be ignored by chemists.

4.13 Summary

We've seen in this chapter how energy is conserved. Namely, there are only two ways to increase the energy of a system: by adding heat and by applying work. The total internal energy change of a system is the sum of these two energy sources:

$$\Delta U = q + w$$

The sort of work we will concern ourselves with here is pressure–volume work, which is calculated as

$$w = -P\Delta V$$

When no chemical reactions of any sort are occurring in the system (including phase changes), we can measure the amount of heat involved in a change in state as

$$q = C\Delta T$$

where C is the *heat capacity* of whatever is changing temperature.

When chemical reactions are involved, it gets a little tricky. For our purposes we can safely ignore the small amount of work involved in chemical reactions. Thus, ΔU can be approximated well as just q. For reactions that do not involve gases or do not involve changes in the overall number of moles of gas, this is a particularly good approximation.

Thus, if we want to determine ΔU for a chemical reaction, we can carry out that reaction either in a simple calorimeter open to the atmosphere, or if it is really important to determine ΔU exactly, in a bomb calorimeter. In either case we will measure the temperature change in the *surroundings*, which include all of the water and gadgetry of the calorimeter itself. Thus, the heat capacity of the calorimeter must be known so that q_{sur} can be determined from $(C\Delta T)_{sur}$. We then determine q from

$$q = -q_{sur}$$

In Chapter 5 we'll see a way of predicting the change in internal energy for a reaction using tables of bond dissociation energies. Later, in Chapter 9, we'll see still another way of predicting the energy change of reactions that is much more accurate.

Problems

The symbol $\boxed{A}$ *indicates that the answer to the problem can be found in the "Answers to Selected Exercises" section at the back of the book.*

Microscopic Heat and Work

4.1 $\boxed{A}$ Explain why state functions, such as U and T, are frequently referred to as *path-independent* functions while q and w are referred to as *path-dependent* functions. Give a real-world example of a change that can occur by two different paths, leading to different values of q in each case.

4.2 Consider each of the four changes, A–D, below. Note: aJ = attojoule = 10^{-18} J. (a) State for each change whether compression, expansion, addition of heat, or removal of heat has occurred. (b) Calculate U for each system before and after the change. (c) Calculate ΔU, q, and w for each change. (d) Indicate in each case whether the temperature increases or decreases.

4.3 1.00 mol of nitrogen gas, N_2, is at 735 torr and 298 K. (a) Calculate the volume of the gas and d, the length of the side of a cube containing this volume. (b) Using your value for d, calculate the energies of the two lowest translational energies for the system, $\varepsilon_{1,1,1}$, $\varepsilon_{2,1,1}$, and their difference, $\Delta\varepsilon$. (c) If the volume of gas is doubled, and $n_{2,1,1}/n_{1,1,1}$ stays constant (that is, no heat is added or removed), what is the new temperature of the gas?

4.4 $\boxed{A}$ A 15.0-L tank containing 1.0 mol of helium gas is at 298 K. (a) Calculate the initial pressure of the gas and d, the length of the side of a cube containing this volume. (b) Using your value for d, calculate the energies of the two lowest translational energies for the system, $\varepsilon_{1,1,1}$, $\varepsilon_{2,1,1}$, and their difference, $\Delta\varepsilon$. (c) The tank is opened, and the gas is allowed to expand rapidly into a 2000-L room such that there is no heat exchange with the surroundings. Assuming that $n_{2,1,1}/n_{1,1,1}$ remains constant, determine the new temperature of the gas. (d) The tank is opened and the gas is allowed to expand rapidly to a final pressure of 1.00 bar. Again assume that $n_{2,1,1}/n_{1,1,1}$ remains constant. Determine the new temperature of the gas.

4.5 A 15 L tank containing 1.0 mol of argon gas is at 298 K. (a) Calculate the initial pressure of the gas and d, the length of the side of a cube containing this volume. (b) Using your value for d, calculate the energies of the two lowest translational energies for the system, $\varepsilon_{1,1,1}$, $\varepsilon_{2,1,1}$, and their difference, $\Delta\varepsilon$. (c) The tank is opened and the gas is allowed to expand rapidly into a 2000-L room such that there is no heat exchange with the surroundings. Assuming that $n_{2,1,1}/n_{1,1,1}$ remains constant, determine the new

temperature of the gas. (d) The tank is opened and the gas is allowed to expand rapidly to a final pressure of 1.00 bar. Again assume that $n_{2,1,1}/n_{1,1,1}$ remains constant. Determine the new temperature of the gas.

Measuring Heat and Work

4.6 $\boxed{A}$ A sample of 4.00 moles of oxygen gas, O_2, is initially at 1.00 bar and 300 K. Calculate (a) V_1, V_2, and w when the volume of the container is doubled against 1 bar of external pressure; (b) V_1, V_2, and w when the volume is decreased when the external pressure is increased to 4 bar; (c) w when the temperature is raised to 400 K at a constant pressure of 1 bar; (d) w when the temperature is raised to 900 K while keeping the volume constant.

4.7 A sample of 2.0 moles of N_2 gas is initially at 1.0 bar and 298 K. Calculate (a) V_1, V_2, and w when the volume of the container is doubled against 1.0 bar of external pressure; (b) V_1, V_2, and w when the volume of the container is decreased when the external pressure is raised to 2.0 bar; (c) w when the temperature is raised to 350 K at constant pressure of 1.0 bar; (d) w when the temperature is raised to 1000 K while keeping the volume constant.

4.8 A sample of 5.0 moles of helium gas is initially at 2.0 bar and 298 K. Calculate (a) V_1, V_2, and w when the volume of the container is doubled against 1.0 bar of external pressure; (b) V_1, V_2, and w when the volume of the container is decreased when the external pressure is raised to 6.0 bar; (c) w when the temperature is raised to 350 K at constant pressure of 2.0 bar; (d) w when the temperature is raised to 1000 K while keeping the volume constant.

4.9 Give two specific real-world examples of each of the following: (a) a situation where putting heat into a system raises its temperature, (b) a situation where putting heat into a system does *not* raise its temperature, and (c) a situation where a system's temperature is changed even though no heat is added or removed.

4.10 $\boxed{A}$ In handbooks, heat capacities are sometimes given as *molar heat capacities*, with units of J/mol·K, and sometimes as *specific heats*, with units of J/g·K. (a) Complete the table that follows by converting molar heat capacities to specific heats for each substance. (b) Which substance can absorb the most heat with the smallest temperature increase on a mole per

$$4d \quad \frac{\Delta\varepsilon_1}{\Delta\varepsilon_2} = \frac{T_1}{T_2} = \frac{3h^2/8md_1^2}{3h^2/8md_2^2} = \frac{d_2^3}{d_1^3} \qquad d_2 = \sqrt[5]{\frac{P_1}{P_2}} \, d_1^5$$

basis? (c) Which substance can absorb the most heat with the smallest temperature increase on a gram per gram basis?

Liquid	Molar Heat Capacity (J/mol·K)	Specific Heat (J/g·K)
acetone, $(CH_3)_2CO$	124.7	
benzene, C_6H_6	136.5	
ethanol, CH_3CH_2OH	112.1	
methanol, CH_3OH	80	
water, H_2O	75.4	

4.11 Ⓐ Assuming all substances remain liquids in the temperature range under consideration, use the information in the previous problem to calculate the amount of heat needed to raise the temperature of (a) 100.0 g of acetone by 1°C, (b) 100.0 g of ethanol by 10 K, (c) 10.0 g of water to boiling from room temperature (298 K), (d) a 50.0-g mixture consisting of 10% by weight of ethanol in water by 5°C.

4.12 Heat capacities are sometimes given as *molar heat capacities*, with units of J/mol·K, and sometimes as *specific heats*, with units of J/g·K. (a) Complete the table by converting molar heat capacities to specific heats for each substance. (b) Which substance can absorb the most heat with the smallest temperature increase on a mole per mole basis? (c) Which substance can absorb the most heat with the smallest temperature increase on a gram per gram basis?

Solid	Molar Heat Capacity (J/mol·K)	Specific Heat (J/g·K)
aluminum	24.25	
sodium chloride	50.21	
hexachlorobenzene, C_6Cl_6	201.29	
gold	25.41	
naphthalene, $C_{10}H_8$	196.06	

4.13 Assuming all substances remain solids in the temperature range under consideration, use the information in the previous problem to calculate the amount of heat needed to raise the temperature of (a) 100.0 g of aluminum by 1°C; (b) 100.0 g of naphthalene by 10 K; (c) 10.0 g of hexachlorobenzene to its melting temperature of 505°C from room temperature (298 K); (d) a 75.0-g mixture consisting of 20% aluminum and 80% gold (by weight) by 20°C.

4.14 Why is w necessarily zero for a reaction that proceeds in a bomb calorimeter?

4.15 Ⓐ Bomb calorimeters are calibrated by combusting a substance that gives off a known quantity of heat (called the *heat of combustion* for that substance). An example of such a substance is benzoic acid, $C_7H_6O_2$. The heat of combustion of benzoic acid is 26.38 kJ/g. If a 2.451-g sample of benzoic acid is combusted in a bomb calorimeter, and the temperature of the calorimeter rises 8.241°C, what is the heat capacity of the calorimeter?

4.16 A bomb calorimeter with a heat capacity of 8.00 kJ/K was used to combust 10.0 g of liquid hydrazine, N_2H_4. The temperature in the calorimeter rose 24.1°C. (a) Determine q, w, and ΔU in this case. (b) Fill in the value for $\Delta_r U$ when the equation for this reaction is written:

$$N_2H_4(l) + O_2(g) \longrightarrow N_2(g) + 2\,H_2O(g)$$
$$\Delta_r U = \underline{\quad\quad} \text{ kJ/mol}$$

(c) Fill in the value for $\Delta_r U$ when the equation for this reaction is written:

$$\tfrac{1}{2}N_2H_4(l) + \tfrac{1}{2}O_2(g) \longrightarrow \tfrac{1}{2}N_2(g) + H_2O(g)$$
$$\Delta_r U = \underline{\quad\quad} \text{ kJ/mol}$$

4.17 Ⓐ When 1.00 g of glucose, $C_6H_{12}O_6$, is combusted, 15.57 kJ of heat is released. A 2.500-g sample of glucose is combusted in a bomb calorimeter, raising the temperature from 20.55°C to 23.25°C. (a) What is the heat capacity of the calorimeter in kJ/K? (b) If the same calorimeter is then used to combust a 1.250-g sample of octane, C_8H_{18}, and the temperature in the calorimeter rises 5.15°C, determine $\Delta_r U$ for the reaction as written:

$$2\,C_8H_{18}(l) + 25\,O_2(g) \longrightarrow 16\,CO_2 + 18\,H_2O$$

(c) Determine $\Delta_r U$ for the reaction as written:

$$4\,C_8H_{18}(l) + 50\,O_2(g) \longrightarrow 32\,CO_2 + 36\,H_2O$$

CHAPTER 5

Bonding and Internal Energy

We may, I believe, anticipate that the chemist of the future who is interested in the structure of proteins, nucleic acids, polysaccharides, and other complex substances with high molecular weight will come to rely upon a new structural chemistry, involving precise geometrical relationships among the atoms in the molecules and the rigorous application of the new structural principles, and that great progress will be made, through this technique, in the attack, by chemical methods, on the problems of biology and medicine.

—*Linus Pauling, 1954 Nobel Prize Address*

5.1 The Chemical Bond

You are no doubt aware of the idea that bringing together atoms to make molecules involves making bonds. We have single, double, triple, and (yes, in some coordination compounds) even *quadruple* bonds. Whether you think of bonds in terms of localized pairs of electrons or probability functions called *orbitals*, you know that bonds are a key part of chemistry and, if nothing else, a very handy convention.

Most chemical reactions make and break bonds. In the reaction

$$2\,H_2(g) + O_2(g) \longrightarrow 2\,H_2O(g)$$

we are breaking two H−H bonds and one O=O bond and making four H−O bonds. In this reaction, which often takes place with a loud *BANG!* and a flash of light, a tremendous amount of heat is released. In fact, almost 500,000 joules of heat is released per two moles of H_2, and one mole of O_2 consumed. That's enough energy in 4 g of hydrogen to take almost *7 liters* of water from 25°C to its boiling point. Talk about exothermic!

Or consider the burning of natural gas (methane):

$$CH_4(g) + 2\,O_2(g) \longrightarrow CO_2(g) + 2\,H_2O(g)$$

Here we still make four H−O bonds, but we also make two C=O bonds. We're breaking twice as many bonds as in the case of H_2: four C−H bonds and two O=O bonds. One mole of O_2 in this reaction can generate about 400,000 joules of heat—not quite as much as for the combustion of H_2, but if you think about it in terms of energy per mole of *fuel* (H_2 or CH_4), then the methane reaction comes out ahead (400,000 J vs. 250,000 J in the case of H_2). Then again, if you were going into space and needed to save weight, comparing these per gram of fuel makes the H_2 reaction come out way, way ahead, with 125,000 J/g H_2 vs. only about 25,000 J/g for CH_4.

In terms of q, we would say that q in the first reaction is −500 kJ/mol O_2, and q in the second reaction is −400 kJ/mol O_2. (Notice that we've added minus signs here, because the energy was referred to as heat *released*, not heat *added*. Our q is heat *added*.) Since work is such a small component of chemical changes, we can assume that these values for q are pretty good estimates for ΔU in each case.

5.2 Hess's Law

The fact that internal energy, U, is a thermodynamic state function (Section 4.2) is important. It means that, like other state functions, such as temperature, pressure, and volume, it behaves according to *Hess's law:*

If a change in state from A to B involves more than one step, then the overall change in a state function X, ΔX_{AB}, is the sum of the changes for each of the steps leading from A to B. For example, since temperature is a state function, it makes no difference if a system is heated from 100 K to 300 K directly ($\Delta T = +200$ K), or heated from 100 K way up to 500 K ($\Delta T_1 = +400$ K) and then cooled back down to 300 K ($\Delta T_2 = -200$ K). The result is the same:

$$\Delta T = \Delta T_1 + \Delta T_2 = (400 \text{ K}) + (-200 \text{ K}) = +200 \text{ K}$$

(The "+" sign here isn't necessary, but we use it anyway just to remind ourselves that this is a *change* in temperature, not the actual temperature "200 kelvins.") Hess's law is simply another way of saying that a change in a state function is independent of the path (or however many paths) taken. A state function is a property of the *state* of a system, not a property of the *path* to that state. Remember, heat and work are properties of paths, but internal energy, pressure, volume, temperature, and number of moles of each substance in a system are all properties of states.

Hess's law becomes important when we want to know the overall change going from State A to State B, and we can't figure out how to measure that change directly. The key is that if there is a standard reference point from which both State A and State B can be compared, then that point serves as a means for calculating the overall change from A to B.

An excellent example of a state function is altitude. What is the change in altitude in going from Northfield, Minnesota, to Chicago, Illinois? Has anyone ever measured it directly? With a ruler? Probably not. But there is a standard reference called "mean sea level" to which both Northfield and Chicago can be compared. The situation is illustrated in Figure 5.1.

Since Northfield is 1000 ft above sea level ($A_N = 1000$ ft) and Chicago is 595 ft above sea level ($A_C = 595$ ft), the path from Northfield to Chicago can be imagined as carried out

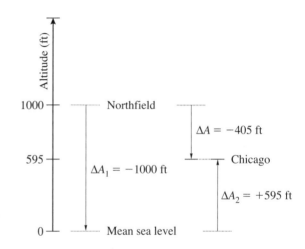

Figure 5.1
The change in altitude in going from Northfield to Chicago is −405 ft based on the two-step process of going through "mean sea level."

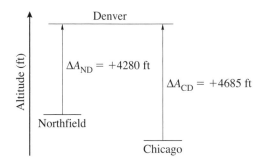

Figure 5.2
Data referencing the altitudes of both North-field and Chicago to the altitude of Denver, Colorado. These data provide an alternative method of determining the altitude change upon going from Northfield to Chicago without reference to actual altitude or mean sea level.

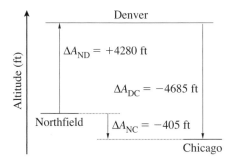

Figure 5.3
The altitude change on going from Northfield to Chicago is determined by Hess's law from data referencing both of their altitudes to the altitude of Denver, Colorado.

in two steps. First we go from Northfield to sea level ($\Delta A_1 = -1000$ ft), then we go from sea level up to Chicago ($\Delta A_2 = +595$ ft). The overall change in altitude is simply

$$\Delta A = \Delta A_1 + \Delta A_2 = (-1000\ \text{ft}) + (595\ \text{ft}) = -405\ \text{ft}$$

If you say, "Look, there's an easier way. Just subtract Northfield's altitude from Chicago's altitude," you would be correct: 595 ft − 1000 ft = −405 ft. But that's possible only because you happen to *know* the altitudes of both Northfield and Chicago already. What if you didn't know either altitude? You might think the problem is impossible, but it isn't! Say you didn't know any altitudes, but what you did know was the *change* in altitude in going from any city to a third *reference* city, such as Denver, Colorado (Figure 5.2). Denver simply becomes our reference instead of sea level. This is all we need to determine the altitude change in going between any two cities for which we have data.

Realize that Hess's law says that a path "Northfield $\longrightarrow$ Denver $\longrightarrow$ Chicago" should give the same result as the direct path "Northfield $\longrightarrow$ Chicago." The data in Figure 5.2 indicate that the change in altitude in going from Northfield to Denver, ΔA_{ND}, is +4280 ft. If the change in altitude going from Chicago to Denver, ΔA_{CD}, is +4685 ft, then it stands to reason that the altitude change going the other way, from Denver to Chicago, ΔA_{DC}, is −4685 ft.

Hess's law then tells us that the change in altitude going from Northfield to Chicago can be calculated as the sum of the two individual steps "Northfield $\longrightarrow$ Denver" and "Denver $\longrightarrow$ Chicago" (Figure 5.3):

$$\Delta A_{\mathrm{NC}} = \Delta A_{\mathrm{ND}} + \Delta A_{\mathrm{DC}} = (4280\ \text{ft}) + (-4685\ \text{ft}) = -405\ \text{ft}$$

Note that we have had to switch the sign of the Chicago/Denver reference because our path takes us in the opposite direction as that used as a reference.

Thus, it is possible to determine the change in altitude (or any other state function) without knowing *any* actual altitude (or values of that state function) at *any* place along the path.

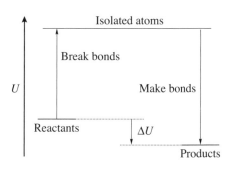

Figure 5.4
Hess's law and the fact that U is a state function allow us to calculate ΔU using the state of "isolated atoms" as a reference and using the ideas of breaking and making bonds.

5.3 The Reference Point for Changes in Internal Energy Is "Isolated Atoms"

The extension of these ideas to chemistry and internal energy is based on the fact that we can never actually calculate the internal energy of any real system. Another way of saying this is that internal energy is a relative thing. It's not the actual internal energy of a system that is important. What is important is the internal energy of a state *relative* to the internal energy of another state. For example, consider the reaction again between hydrogen and oxygen to form water, for which we have the balanced chemical equation

$$2\,H_2(g) + O_2(g) \longrightarrow 2\,H_2O(g)$$

We would like to at least estimate the expected change in internal energy for this reaction. The first state is "two moles of H_2 gas and one mole of O_2 gas." The second state is "two moles of water vapor." Basically, since work is such a minor part of chemical reactions, the question is this: Is the reaction going to release heat or isn't it? And if so, how much heat should we expect to be released? The key lies in defining a suitable reference point and building a database of information.

The reference point used for internal energy is the state of *isolated atoms*, where there are no molecules and no bonds, just atoms (Figure 5.4). This reference point is higher in internal energy than any molecule, because it *always* takes energy to dissociate molecules into individual atoms. Breaking bonds always requires energy. The energy required to break all the bonds in a molecule is called the *total bond dissociation energy* (ΣBDE) of a molecule, and in the reaction of interest we have:

reactants:	$2\,H_2(g) + O_2(g) \longrightarrow 4\,H + 2\,O$	$\Delta_r U = \Sigma BDE_{\text{reactants}}$
products:	$2\,H_2O(g) \longrightarrow 4\,H + 2\,O$	$\Delta_r U = \Sigma BDE_{\text{products}}$

Note that the state of isolated atoms, $4\,H + 2\,O$, provides the needed reference point for determining $\Delta_r U$ for our reaction.

However, we have to be careful! In order to use Hess's law, as for altitude, we will have to reverse this second step. The overall change in internal energy (the *reaction internal energy*, $\Delta_r U$, based on the above equation) is the sum of all the energy it takes to dissociate all the bonds in the reactants ($\Sigma BDE_{\text{reactants}}$) plus all the energy *that would be released* when all the bonds in the products are made ($-\Sigma BDE_{\text{products}}$):

$$\Delta_r U = \Sigma BDE_{\text{reactants}} + (-\Sigma BDE_{\text{products}}) \tag{5.1}$$

The only question is, how can we estimate these reference values? The key is to focus on what "dissociation into individual atoms" primarily involves: the breaking of bonds.

5.4 Two Corollaries of Hess's Law

Hess's law (1) and two corollaries (2 and 3) can now be stated:

1. $\Delta_r U$ for A $\longrightarrow$ C is the sum of $\Delta_r U$ for A $\longrightarrow$ B and $\Delta_r U$ for B $\longrightarrow$ C.
2. $\Delta_r U$ for B $\longrightarrow$ A is the opposite of that for A $\longrightarrow$ B.
3. $\Delta_r U$ for nA $\longrightarrow$ nB is n times that for A $\longrightarrow$ B.

These three statements are basic truths relating to any state function. Thus, for example, if we somehow have determined (or estimated) that

$$2\,H_2(g) + O_2(g) \longrightarrow 2\,H_2O(g) \qquad \Delta_r U = -482\,\text{kJ/mol}$$

then we would also know all of the following:

$$2\,H_2O(g) \longrightarrow 2\,H_2(g) + O_2(g) \qquad \Delta_r U = +482\,\text{kJ/mol}$$
$$H_2(g) + {}^1\!/_2\,O_2(g) \longrightarrow H_2O(g) \qquad \Delta_r U = -241\,\text{kJ/mol}$$
$$H_2O(g) \longrightarrow H_2(g) + {}^1\!/_2\,O_2(g) \qquad \Delta_r U = +241\,\text{kJ/mol}$$

Sometimes this is useful if we want to use some data we have already to derive a change in internal energy or any other state function for a reaction of interest.

For example, say we wanted to know the change in internal energy for the combustion of methane gas based on the following equation:

$$CH_4(g) + 2\,O_2(g) \longrightarrow CO_2(g) + 2\,H_2O(l)$$

This reaction is slightly different from the one used in Section 5.1, because here the product water is liquid. However, say we know the following information:

$$CH_4(g) + 2\,O_2(g) \longrightarrow CO_2(g) + 2\,H_2O(g) \qquad \Delta_r U = -803\,\text{kJ/mol}$$
$$H_2O(l) \longrightarrow H_2O(g) \qquad \Delta_r U = +41\,\text{kJ/mol}$$

(This second piece of information is the energy required to evaporate a mole of liquid water and was introduced in Section 4.8.) Then we could write

$$CH_4(g) + 2\,O_2(g) \longrightarrow CO_2(g) + 2\,H_2O(g) \qquad \Delta_r U = -803\,\text{kJ/mol}$$
$$2\,H_2O(g) \longrightarrow 2\,H_2O(l) \qquad \Delta_r U = -82\,\text{kJ/mol}$$
$$\overline{CH_4(g) + 2\,O_2(g) \longrightarrow CO_2(g) + 2\,H_2O(l) \qquad \Delta_r U = -885\,\text{kJ/mol}}$$

Note that "-82 kJ/mol" comes from reversing and doubling the value "$+41$ kJ/mol" given in the data. Thus, Hess's law has many applications, all based on the idea that an overall equation for a reaction can be thought of as the sum of two or more steps.

5.5 Mean Bond Dissociation Energies and Internal Energy

During the mid part of the 20th century chemists used bomb calorimeters (Section 4.11) to determine the change in internal energy, ΔU, for thousands upon thousands of chemical reactions. Call it the "human genome project" for chemists. Many, many chemists from all around the world were involved. They realized that if enough data could be collected, then it might be possible to sort out what was going on in reactions in terms of those little lines we call bonds in Lewis structures.

Imagine for a moment a gigantic set of simultaneous equations where the variables look like (H−H), (O=O), (C−O), (C=O), and (O−H). The equations would then look like

$$2(\text{H}-\text{H}) + (\text{O}=\text{O}) - 4(\text{O}-\text{H}) = -481 \text{ kJ}$$
$$4(\text{C}-\text{H}) + 2(\text{O}=\text{O}) - 2(\text{C}=\text{O}) - 4(\text{O}-\text{H}) = -803 \text{ kJ}$$

These variables would represent how much energy each type of bond is worth—how many kJ of energy it would take to break each bond. The numbers would be the experimentally determined changes in energy for actual reactions.

Clearly, though, to get values for all the important types of bonds, one would have to carry out a lot of reactions, and one would need to find some way to solve the huge set of simultaneous equations that would define all of those reactions. We already have five variables and just two equations. Imagine what it would be like to get all the possible bonds this way! Complicating the issue, there really isn't any guarantee that a C−H bond in CH_4 is worth the same as a C−H bond in, say CH_3OH.

Some of these bond strength values could be determined directly. For example, it wasn't too difficult to determine the strength of the Br−Br bond in Br_2. One simply had to prepare some bromine atoms and see how much heat was released when they come back together:

$$\text{Br}(g) + \text{Br}(g) \longrightarrow \text{Br}_2(g) \quad q = -193 \text{ kJ}$$

This gave a Br–Br bond strength of 193 kJ/mol.

So it turned out to be an interesting job of finding the "best fit" to the data, and the result was the average or *mean* values that seem to give the best results when applied generally. It took a lot of money, a lot of data, and a lot of time to do the job, but they did it! A partial synopsis of literally *thousands* of person-years of work is summarized in the table of *Mean Bond Dissociation Energies* shown in Table 5.1.

Note that the "strongest" single bond on this table is H−F (565 kJ/mol), followed by C−F (485 kJ/mol). The strength of carbon−fluorine bonds has a lot to do with why Teflon (empirical formula CF_2) is so slick and why Freon, CF_2Cl_2, makes it all the way up to the

Table 5.1
Mean bond dissociation energies (kJ/mol)

Single Bonds								Multiple Bonds	
H−H	432	N−H	391	F−F	154	S−H	347	C=C	614
H−F	565	N−N	160	F−Cl	253	S−C	259	C=N	615
H−Cl	427	N−O	201	F−Br	237	S−F	327	C=O [a]	799
H−Br	363	N−F	272			S−Cl	253		
H−I	295	N−Cl	200	Cl−Cl	239	S−Br	218	C≡C	839
		N−Br	243	Cl−Br	218	S−S	266	C≡N	891
				Cl−I	208			C≡O	1072
C−H	413					Si−H	393		
C−C	347	O−H	467			Si−C	360	N=N	418
C−N	305	O−O	146	Br−Br	193	Si−O	452	N=O	607
C−O	358	O−F	190	Br−I	175	Si−Si	226		
C−F	485	O−Cl	203						
C−Cl	339	O−I	234	I−I	149			N≡N	941
C−Br	276								
C−I	240							O=O	495

[a] This value is for the C=O bond in CO_2.

stratosphere to cause its damage there to the ozone layer. (Most other molecules are broken down or dissolved in rain long before they diffuse up so high in the atmosphere.)

The O−H bond is also very strong (467 kJ/mol), and this explains to large measure the great stability of water and its great abundance on our planet.

For a given pair of atoms, single bonds are easier to break than double bonds, and double bonds are easier to break than triple bonds. For example, we have N−N (160 kJ/mol), N=N (418 kJ/mol), and N≡N (941 kJ/mol). (Notice, in fact, that the N≡N bond is a full *six times* stronger than the N–N bond!)

All of the multiple bonds to carbon are very strong as well. The C=O bond in CO_2 (799 kJ/mol) is stronger than twice the average C−O bond (347 kJ/mol). Notably weak among multiple bonds is the O=O bond (in O_2, 495 kJ/mol). The combination of weak bonding in O_2 and very strong bonding in H_2O and CO_2 probably explains why O_2 is such a good reactant in the burning process. Not only are reactant bonds weak, but the products, CO_2 and H_2O, are two of the most stable compounds known.

The fact that the N≡N bond (in N_2, 941 kJ/mol) is very strong compared to other bonds to nitrogen explains why the N_2 in air is so unreactive (and, thus, probably why it is so plentiful on our planet).

5.6 Estimating $\Delta_r U$ for Chemical Reactions Using Bond Dissociation Energies

The basic idea is that we should be able to use mean bond dissociation energies along with Hess's law to at least estimate the change in internal energy expected for a chemical reaction based on a specific writing of the balanced chemical equation:

$$\Delta_r U \approx \Sigma BDE_{reactants} + (-\Sigma BDE_{products})$$

Let's see how this works for the two reactions mentioned above. Remember, these values are *averages* taken over a vast set of reactions. We shouldn't expect perfection here. Consider the reaction of hydrogen with oxygen to form water:

$$2\,H_2(g) + O_2(g) \longrightarrow 2\,H_2O(g)$$

$$\Sigma BDE_{reactants} = 2(H-H) + (O=O) = 2(432) + 495 = 1359 \text{ kJ/mol}$$
$$\Sigma BDE_{products} = 4(O-H) = 4(467) = 1868 \text{ kJ/mol}$$
$$\Delta_r U \approx 1359 + (-1868) = -509 \text{ kJ/mol}$$

The experimental value is actually −481 kJ/mol. Not bad, for a quick calculation! We would expect the energy diagram for this system to look something like that shown below.

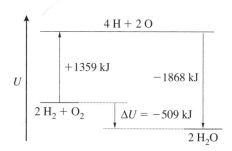

Similarly, for

$$CH_4(g) + 2\,O_2(g) \longrightarrow CO_2(g) + 2\,H_2O(g)$$

we get

$$\Sigma BDE_{\text{reactants}} = 4(C-H) + 2(O=O) = 4(413) + 2(495) = 2642 \text{ kJ/mol}$$
$$\Sigma BDE_{\text{products}} = 2(C=O) + 4(O-H) = 2(799) + 4(467) = 3466 \text{ kJ/mol}$$
$$\Delta_r U \approx 2642 + (-3466) = -824 \text{ kJ/mol}$$

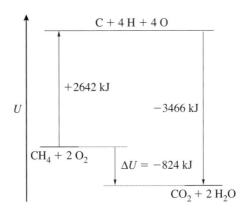

The experimental value in this case is -803 kJ/mol. So, we're off by about 3 to 5% in both cases—not really very bad at all for a rough calculation.

Notice that if we write a slightly different equation for the very same reaction:

$$\tfrac{1}{2}CH_4(g) + O_2(g) \longrightarrow \tfrac{1}{2}CO_2(g) + H_2O(g)$$

we get a different number:

$$\Sigma BDE_{\text{reactants}} = \text{½}\,[4(C-H)] + (O=O) = \text{½}\,[4(413)] + (495) = 1321 \text{ kJ/mol}$$
$$\Sigma BDE_{\text{products}} = \text{½}\,[2(C=O)] + 2(O-H) = (799) + 2(467) = 1733 \text{ kJ/mol}$$
$$\Delta_r U \approx 1321 + (-1733) = -412 \text{ kJ/mol}$$

This is just half as much as we got for the equation involving a whole mole of CH_4, which should be no surprise. The point is, when someone tells you, "ΔU for this reaction is -412 kJ," without showing you a chemical equation, be sure to ask them what exact *balanced chemical equation* they are using to describe that reaction.

5.7 Using Bond Dissociation Energies to Understand Chemical Reactions

We can use bond dissociation energies to start to understand why certain chemical reactions occur and others don't. The exothermic nature of the reactions given as examples in this chapter is common to virtually all "spontaneous" reactions. (We will have to wait until Chapter 9 to see why this is true.) Still, why are these reactions exothermic?

The answer is that they all involve O_2 as a reactant and either H_2O or CO_2 as a product. O_2, with its two unpaired electrons in antibonding orbitals, has *especially* weak bonds for a double-bonded molecule. In addition, the $O-H$ bond in H_2O is an especially strong combination. Note that the cost of breaking *four* bonds in CH_4 is very large, but the payoff in also generating the very stable CO_2 molecule makes up for it.

Contrast this to what we get for the hypothetical "combustion of the atmosphere" involving the burning of N_2 to form NO:

$$N_2(g) + O_2(g) \longrightarrow 2\,NO(g)$$

for which we have

$$\Sigma BDE_{\text{reactants}} = (N\equiv N) + (O=O) = 941 + 495 = 1436 \text{ kJ/mol}$$
$$\Sigma BDE_{\text{products}} = 2(N=O) = 2(607) = 1214 \text{ kJ/mol}$$
$$\Delta_r U \approx 1436 + (-1214) = +222 \text{ kJ/mol}$$

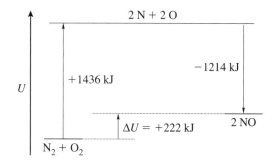

What a difference between this and the reaction of hydrogen with oxygen! It isn't that breaking one $N\equiv N$ bond (941 kJ/mol) is so difficult in comparison to breaking two $H-H$ bonds (864 kJ/mol), but there's certainly no payoff when that NO is made! We can be thankful that NO is so relatively unstable and that the combustion of N_2 is not generally spontaneous. Otherwise, there would be no life-sustaining atmosphere on Earth!

5.8 The "High-Energy Phosphate Bond" and Other Anomalies

If you ever take biology or biochemistry, you are going to hear a lot about the "high-energy phosphate bond." The principle idea is that ATP (adenosine triphosphate) can react to form ADP (adenosine diphosphate) and P_i (inorganic phosphate). The structures of ATP, ADP, and P_i in their fully protonated forms are shown in Figure 5.5. However, it should be noted that in reality, at biological pH, all of these substances, being weak polyprotic acids, are really mixtures of several partially deprotonated species. In this case, it's probably most honest to leave them just as their acronyms, allowing the acronyms to be somewhat vague.

The important thing for us to consider here is that ATP is seen as the storehouse of energy for the body, and that the breaking of the high-energy phosphate bond in ATP

Figure 5.5
The structures of ATP, ADP, and inorganic phosphate in their fully protonated forms. At pH 7, ATP has lost three of its four phosphate OH hydrogens completely and one to the extent of about 50%. Similarly for ADP, two of its phosphate OH hydrogens are lost, and only one remains partially dissociated. For H_3PO_4, one of the three hydrogens is lost completely and another is about 50% dissociated.

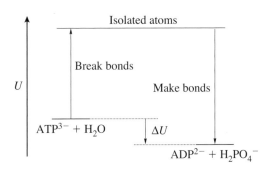

Figure 5.6
The "high-energy phosphate bond" in ATP stores energy in the sense that its bonds are *weaker* than those in ADP and inorganic phosphate. Breaking of the phosphate bond in ATP does, indeed, require energy, but even more energy is released when a new, stronger, phosphorus–oxygen bond is made. This reaction can be understood only when water is added to the overall equation.

provides that energy on demand. There is absolutely nothing wrong with this statement (although we will see later that the energy referred to here is not really internal energy, but, rather, *free* energy, to be introduced in Chapter 10). Thus, you will see something like this:

$$ATP \longrightarrow ADP + P_i + \textbf{energy}$$

Now, the problem from our perspective is that it sure looks from this that breaking a bond can release energy. How can this be? Didn't we just learn that it always *takes* energy to break a bond?

Part of the answer lies in understanding that the writer is using acronyms for complex species, not for actual substances. In chemistry we try to be very specific and use balanced chemical equations indicating actual *pure substances* as much as possible. In fact, a possibly better chemical, albeit just as inaccurate, way of writing this equation would be to include charges and a mole of water:

$$ATP^{3-} + H_2O \longrightarrow ADP^{2-} + H_2PO_4^- + \textbf{energy}$$

Although we have *broken* a bond in ATP, we have actually *made* a stronger bond in the inorganic phosphate, as seen in Figure 5.6. So there really isn't any problem. Breaking a bond *always* requires energy. In fact, the only problem is that some students have this as one of their first introductions to bonding and get the wrong impression that bonds have energy to be gotten by breaking them. Well, yes, in breaking a bond you can get energy, but in order to do that, you *always* have to make another bond!

Another good example of the problem with abbreviating chemical equations is closer to home in the "dissociation" of an acid. We write

$$HCl \longrightarrow H^+ + Cl^-$$

We know from experience that this reaction goes to completion and releases a lot of energy. After all, HCl is a strong acid, and its reaction with water is dangerously exothermic. But how can that be? Aren't we breaking a bond? Once again, the problem lies in our not writing the real equation for this reaction:

$$HCl + H_2O \longrightarrow H_3O^+ + Cl^-$$

Ah, now we have no problem. Sure this is a "dissociation," but the dissociation of an acid is really just the *transfer* of a proton from it to water. A bond is broken in dissociating HCl, but another is made in forming H_3O^+. So once again, bonds are broken and made in this dissociation reaction, and it is perfectly reasonable that the reaction releases heat. In fact, *all* dissociation reactions occurring in aqueous solution are much more than simple bond breaking. The solvent must be included to make real sense of what is going on.

5.9 Computational Chemistry and the Modern View of Bonding

The quantum ideas of Planck, Einstein, de Broglie, Schrödinger, and others revolutionized the world of atomic physics in the early 20th century. The atom was no longer the "pea soup" or "currant pudding" of the 19th century. Now there was a massive nucleus containing protons and neutrons and bearing all the positive charge. Now there were the electrons, 1/2000 the mass of the proton, moving at enormous speeds and having wave-like properties themselves. Energy and mass now became two aspects of the same thing, related by the conversion factor c^2. The hydrogen atom was fully described. The nuclear age was born.

The seemingly intractable problem of solving the Schrödinger equation for anything more complex than hydrogen spawned many creative approximations. Our current ideas of the "covalent bond" we write as $-$, $=$, and $\equiv$, "orbital," "sigma," "pi," "bonding," "antibonding," "resonance," and "delocalization" all arose during this most creative period. These concepts have served chemists well, because they allow chemists to predict the direction and extent of many chemical reactions. However, they have serious limitations and are by no means the end of the road.

With the discovery of the transistor (which itself requires quantum ideas to understand) and the widespread use of computers came a revolution in the way the mathematical problem of molecules could be treated. This revolution is far from over, and it is safe to say that we are not yet done understanding the chemistry of molecules. However, a method of thinking about and working with the mathematics of molecules has been developed which is very powerful and not really very hard to understand. It's a thought experiment that goes like this:

1. First you need an energy-o-meter. This is a device that reads out the energy of a system.
2. Start with a set of atoms, perhaps 4 H atoms and 2 O atoms, all far, far apart from one another in a neutral state, with the same number of protons as electrons. Call this the "isolated atoms" reference state and "tare" your energy-o-meter (set it to read 0).
3. Now separate all the nuclei (which are positively charged) from all the electrons (which are negatively charged). Pull everything "infinitely far" apart. The energy-o-meter reading will go way up, because opposite charges attract, and you are pulling negatively charged electrons away from positively charged protons.
4. Now bring the nuclei back into the vicinity of each other and arrange them any way you want. (On a computer, you simply drag them in with the mouse and drop them where you want them.) No electrons here, just nuclei, remember. Since positive charges repel each other, the energy-o-meter will read even higher.
5. Now bring in the electrons, and confine them to "boxes" called *orbitals*. Orbitals are simply definitions that limit the electron movement in the sense that they say, "Electrons, if you're in this box, then you have to stay over here more of the time than over there, and if you're in *this* box, then you have to spend more of your time over *here* and less of your time over *there*."

    You can design your orbitals any way you want. If it helps, you can assign some electrons to "core" orbitals to say they should stay right around one particular atom, or "lone pairs" that will stay *pretty much* around one atom, or "bonds" which will allow them to spend time around two atoms. This process of confining the electrons to these three small box types is called *localization*. The reading on the energy-o-meter will initially go way down, becoming negative.
6. OK, here's the trick: Monitor your energy-o-meter and fiddle with the *orbitals*. Make them bigger. Make them smaller. Combine them any way you like, all the time checking to see if your energy-o-meter reading goes up or down.

7. Finally, monitor your energy-o-meter while you fiddle with the *nuclear positions*. Pull the nuclei apart a bit, push them together, move them around. You might even just want to randomly jiggle the positions to see what happens. Each time you move the nuclei, repeat step 6. You are looking for the absolute minimum reading on the energy-o-meter.
8. When you can't get the energy-o-meter reading to go down any further, call it quits. You've got a molecule!

Of course, the real challenge is two-fold. First we must design, build, and test that energy-o-meter, and then we have to design the boxes and the way we're going to fiddle with them. Many people have worked very hard on these problems, and much success has been achieved. New methods that are simpler and faster continue to come out all the time.

The interesting thing for us, though, is that with the introduction of very fast desktop computers, you and I can carry out this sort of computation using a variety of box types and energy-o-meters in just seconds or minutes.

Thus, *computational chemistry*, also called *molecular modeling*, provides an alternative to going to tables of bond energies to figure out ΔU for a chemical reaction. Of course, you need a good fast computer to do this, but every method has its limitations. Recent advances have allowed the rapid modeling of reactions between drugs and enzymes, and between molecules and metal surfaces.

In addition, computational chemistry can give us much more than just ΔU. Because we get detailed information about the boxes, we end up with the capability of estimating not only the total energy, but also the entire "picture" of the energy levels of a system. By doing this we can predict many of the properties of the substances we are studying, such as heat capacity, color, and stability. If you get a chance, you should play around a little with one of these computational systems. It's fun!

5.10 Beyond Covalent Bonding

Tables of mean bond dissociation energies are useful, but caution is recommended for two reasons. First, mean bond dissociations are by definition only approximations and are bound to be off by a few percent. Second, they can't account for differences when the reactants or products are not all the same phase. For example, calculating $\Delta_r U$ for the evaporation of liquid water, $H_2O(l) \longrightarrow H_2O(g)$, would just give 0 kJ/mol. But the real value is 41 kJ/mol. Why?

It is important to realize that there is more to bonding than just the "bonds" we write in Lewis dot structures. Any interaction between molecules that lowers the internal energy of a system can be classified as a "bond" in a certain sense, but generally referred to as a *van der Waals* force, which includes hydrogen bonding and several other related interactions all based on the fact that the electrons in one molecule are also slightly attracted to the nuclei in *other* molecules. These forces result in all solid substances being of lower internal energy than their liquid counterparts and in all liquid substances being of lower internal energy than their associated vapors.

Water is a particularly interesting substance in that it makes strong hydrogen bonds with itself and other polar molecules. In order to vaporize water, these weak interactions between molecules must be overcome.

Most importantly, we will use internal energy, which is really the *sum* of all of the energies of all of the particles in a system, along with its associated connection with bonding, as a handle on the relative ground states of reactants and products in a chemical reaction. The argument, as introduced in Section 3.11, is that at least at low temperature, when the system

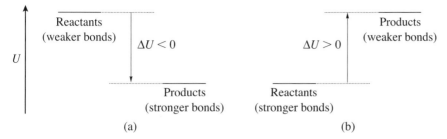

Figure 5.7
In situation (a) the internal energy is lower and the bonding is stronger for the products than for the reactants. Products will be favored at low temperature. In (b) the reverse is true. Reactants have lower internal energy and stronger bonding. In this case, reactants will be favored at low temperature. As the temperature increases, though, in each case the reaction will shift away from the lower-energy state as particles are excited out of the lowest energy levels of the system into higher energy levels.

is mostly in its ground state, the side of the equation representing the lower ground state will be favored at low temperature, and as the temperature rises, the equilibrium position will shift away from this lower-energy state. This is illustrated in Figure 5.7.

5.11 Summary

Hess's law simply states that when going from State A to State B, as long as we are talking about a thermodynamic state function (in particular, internal energy), the change is independent of the path taken. If more than one step is involved, we simply (carefully!) add up all the changes along the way to determine the final change.

Realize that Hess's law does not require that you actually *take* the path used in the calculation. That's the best part! We didn't have to actually go through Denver to get from Northfield to Chicago. Rather, Denver, like "mean sea level," is simply a convenient reference for the calculation of altitude.

In addition, Hess's law has broader use than just being the basis for estimating the change in internal energy for a chemical reaction. You will see Hess's law used over and over again in this book, because it lets us determine the change in *any* state function based on breaking the reaction down into a sequence of steps, each with its own balanced chemical equation. In all cases, we can visualize the state function along the y axis of a diagram. In that case, vertical arrows will represent changes in the state function. Arrows pointing up will be positive changes; arrows pointing down will be negative changes. The overall change is seen as a sum of these arrows, which ultimately may be positive, negative, or zero.

It's quite easy to estimate the change in internal energy, $\Delta_r U$, for a chemical reaction if we can write the Lewis structures for the reactants and products and we have in mind a particular writing of the balanced chemical equation for that reaction. Using Hess's law and the idea of a mean bond dissociation energy we can estimate the internal energy change in a chemical reaction. The idea is that going from reactants to products we imagine first breaking all of the bonds in reactants ($\Delta_r U_1 = \Sigma BDE_{reactants}$) followed by making all of the bonds in the products ($\Delta_r U_2 = -\Sigma BDE_{products}$) so that

$$\Delta_r U \approx \Delta_r U_1 + \Delta_r U_2 = \Sigma BDE_{reactants} + (-\Sigma BDE_{products})$$

We do this trickery because we simply have no access to actual internal energy values for either reactants or products. (If we had that information, just as for altitude, we could simply calculate $\Delta_r U = U_{products} - U_{reactants}$. But we don't have that information.)

It is *very* important to remember that the numbers listed in tables sometimes titled "bond energies" are really bond *dissociation* energies.[1] **It always takes energy to break a bond. Stronger bonding always correlates with a lower internal energy as well as a lower ground state.** This is a good start, and now we are ready (in Chapter 6) to consider more fully the equilibria of some real chemical reactions and how these equilibria depend upon temperature.

---

[1] Many authors and most chemists drop the word *dissociation* here and call these "bond energies," but please, don't do that! Many, many students then get the wrong idea that a strong bond has more energy than a weak bond. No! A strong bond has *less* energy than a weak bond, because it is "further down on the energy ladder" and requires more energy to break. If it helps, draw a little picture with the vertical arrows showing the numbers as you do your calculation.

Problems

The symbol Ⓐ *indicates that the answer to the problem can be found in the "Answers to Selected Exercises" section at the back of the book.*

Hess's Law

5.1 Ⓐ Use Hess's law and the fact that going from New York City to Denver involves an increase in altitude of 5230 ft to determine the change in altitude in going from Chicago to New York City. Express your answer as in Figure 5.3.

5.2 Use Hess's law and the fact that going from Des Moines to Northfield involves an increase in altitude of 50 ft to determine the change in altitude in going from Denver to Des Moines. Express your answer in the form of a diagram similar to Figure 5.3.

5.3 Use Hess's law and the fact that going from Bangor, Maine, to Chicago involves an increase in altitude of 403 ft to determine the change in altitude in going from Denver to Bangor. Express your answer in the form of a diagram similar to Figure 5.3.

5.4 Use Hess's law and the fact that going from Santa Fe to Northfield involves a decrease in altitude of 5300 ft to determine the change in altitude in going from Denver to Santa Fe. Express your answer in the form of a diagram similar to Figure 5.3.

5.5 Draw diagrams similar to Figure 5.4 for the following reactions, as written. (No need to do any calculations here, just focus on relative ground states.)

(a) $2\,C(s) + 2\,H_2(g) \longrightarrow C_2H_4(g)$ $\Delta_r U = 227$ kJ/mol
(b) $C(s) + 2\,H_2(g) \longrightarrow CH_4(g)$ $\Delta_r U = -75$ kJ/mol
(c) $Br_2(l) \longrightarrow Br_2(g)$ $\Delta_r U = 28$ kJ/mol
(d) $C_2H_5OH(l) + 3\,O_2(g) \longrightarrow$ $\Delta_r U = -1367$ kJ/mol
    $2\,CO_2(g) + 3\,H_2O(l)$

5.6 Draw diagrams similar to Figure 5.4 for the following reactions, as written. (No need to do any calculations, just focus on relative ground states.)

(a) $CO_2 + CCl_4 \longrightarrow 2\,CCl_2O$ $\Delta_r U = 70$ kJ/mol
(b) $2\,CF_2O \longrightarrow CO_2 + CF_4$ $\Delta_r U = -112$ kJ/mol
(c) $CBr_2O + H_2O \longrightarrow 2\,HBr + CO_2$ $\Delta_r U = -205$ kJ/mol
(d) $CCl_2O + H_2O \longrightarrow CO_2 + 2\,HCl$ $\Delta_r U = -72$ kJ/mol

5.7 Draw diagrams similar to Figure 5.4 for the following reactions, as written. (No need to do any calculations, just focus on relative ground states.)

(a) $C_2H_6Zn \longrightarrow CH_3Zn + CH_4$ $\Delta_r U = 270$ kJ/mol
(b) $C_2H_6Cd \longrightarrow CH_3Cd + CH_4$ $\Delta_r U = 240$ kJ/mol
(c) $C_4H_{12}Ge \longrightarrow C_3H_9Ge + CH_4$ $\Delta_r U = 340$ kJ/mol
(d) $H_2 + 2\,CH_3I \longrightarrow 2\,CH_4 + I_2$ $\Delta_r U = -126$ kJ/mol

5.8 Draw diagrams similar to Figure 5.4 for the following reactions, as written. (No need to do any calculations, just focus on relative ground states.)

(a) $H_2 + C_3H_6O \longrightarrow C_3H_8O$ $\Delta_r U = -55$ kJ/mol
(b) $2\,H_2 + C_2H_3Cl \longrightarrow C_2H_6 + HCl$ $\Delta_r U = -214$ kJ/mol
(c) $CO_2 + 4\,HF \longrightarrow CF_4 + 2\,H_2O$ $\Delta_r U = 174$ kJ/mol
(d) $C_3H_8 + HF \longrightarrow H_2 + C_3H_7F$ $\Delta_r U = 92$ kJ/mol

5.9 Ⓐ Based on the following information:

$$2\,SO_2 + O_2 \longrightarrow 2\,SO_3 \quad \Delta_r U = -200 \text{ kJ/mol}$$
$$2\,S + 3\,O_2 \longrightarrow 2\,SO_3 \quad \Delta_r U = -793 \text{ kJ/mol}$$

calculate the change in internal energy (ΔU, in kJ) for the oxidation of a mole of sulfur to give a mole of sulfur dioxide.

5.10 Based on the following information:

$$2\,SO_2 + O_2 \longrightarrow 2\,SO_3 \quad \Delta_r U = -200 \text{ kJ/mol}$$

calculate $\Delta_r U$ for each of the following reactions, as written:

(a) $SO_2 + \frac{1}{2}\,O_2 \longrightarrow SO_3$
(b) $2\,SO_3 \longrightarrow O_2 + 2\,SO_2$
(c) $SO_3 \longrightarrow SO_2 + \frac{1}{2}\,O_2$

5.11 Ⓐ Based on the following information:

$$2\,H_2 + C_2H_3Cl \longrightarrow C_2H_6 + HCl \quad \Delta_r U = -214 \text{ kJ/mol}$$

calculate the internal energy change for each of the following reactions, as written:

(a) $H_2 + \frac{1}{2}\,C_2H_3Cl \longrightarrow \frac{1}{2}\,C_2H_6 + \frac{1}{2}\,HCl$
(b) $6\,H_2 + 3\,C_2H_3Cl \longrightarrow 3\,C_2H_6 + 3\,HCl$
(c) $4\,C_2H_6 + 4\,HCl \longrightarrow 8\,H_2 + 4\,C_2H_3Cl$

5.12 Based on the following information:

$$CO_2 + CF_4 \longrightarrow 2\,CF_2O \quad \Delta_r U = 112 \text{ kJ/mol}$$

calculate the internal energy change for each of the following reactions, as written:

(a) $\frac{1}{2}\,CO_2 + \frac{1}{2}\,CF_4 \longrightarrow CF_2O$
(b) $10\,CF_2O \longrightarrow 5\,CO_2 + 5\,CF_4$
(c) $4\,CF_2O \longrightarrow 2\,CO_2 + 2\,CF_4$

Mean Bond Dissociation Energies

5.13 In all our work, "Δ" means "products minus reactants" or "final minus initial." Explain why $\Delta_r U$ is "$\Sigma BDE_{reactants} - \Sigma BDE_{products}$" instead of "$\Sigma BDE_{products} - \Sigma BDE_{reactants}$."

5.14 Ⓐ Estimate $\Delta_r U$ for each of these reactions, as written, using values for mean bond dissociation energies from Table 5.1.

(a) $CH_4 + 4\,Cl_2 \longrightarrow CCl_4 + 4\,HCl$
(b) $2\,HI \longrightarrow H_2 + I_2$
(c) $H + Br_2 \longrightarrow HBr + Br$

5.15 Estimate $\Delta_r U$ for the reactions, as written, using mean bond dissociation energies from Table 5.1.

(a) $S + 3\,BrF_5 \longrightarrow SF_6 + 3\,BrF_3$
(b) $CH_4 + Br_2 \longrightarrow HBr + CH_3Br$
(c) $CH_4 + CH_2I_2 \longrightarrow 2\,CH_3I$

5.16 Estimate $\Delta_r U$ for the reactions, as written, using mean bond dissociation energies from Table 5.1.

(a) $CF_4 + 4\,HF \longrightarrow CH_4 + 4\,F_2$
(b) $2\,H_2 + C_2H_3Br \longrightarrow HBr + C_2H_6$
(c) $CH_2F_2 + Br_2 \longrightarrow HBr + CHBrF_2$

5.17 Draw diagrams for the reactions in Problem 5.14 similar to those in Section 5.6, including in your diagram the values for $\Sigma BDE_{reactants}$ and $\Sigma BDE_{products}$.

5.18 Ⓐ Draw out Lewis dot structures for reactants and products in the following equations and estimate $\Delta_r U$ for each using values for mean bond dissociation energies from Table 5.1.

(a) $2\,H_2O + 2\,F_2 \longrightarrow 4\,HF + O_2$
(b) $4\,HCl + O_2 \longrightarrow 2\,Cl_2 + 2\,H_2O$
(c) $N_2H_4 + O_2 \longrightarrow N_2 + 2\,H_2O$
(d) $N_2 + 3\,F_2 \longrightarrow 2\,NF_3$

5.19 The Lewis dot structure of ozone, O_3, must be written as a pair of resonance structures. Draw these. Given

$$3\,O_2 \longrightarrow 2\,O_3 \quad \Delta_r U = +145\,\text{kJ/mol}$$

and the mean bond dissociation energy for O_2 from Table 5.1, determine the mean bond dissociation energy for the oxygen–oxygen bond in O_3. Compare this to the average of a single and a double oxygen–oxygen bond. Argue the case that electron delocalization or "resonance" leads to a lowering of energy in a system and illustrate your case using a diagram.

5.20 Using mean bond dissociation energies from Table 5.1, estimate $\Delta_r U$ for the following reaction, as written:

$$2\,NX_3 \longrightarrow N_2 + 3\,X_2$$

where X is (a) F, (b) Cl, and (c) Br. Use this information to argue the case that anyone handling NCl_3 or NBr_3 should be prepared for an explosion.

5.21 Ⓐ In 1962, the compound xenon tetrafluoride was synthesized for the first time by Neil Bartlett. (a) Write a balanced chemical reaction for the reaction between xenon and fluorine gases to form xenon tetrafluoride. (b) Predict the value of $\Delta_r U$ for this reaction given the mean bond dissociation energy Xe−F 130 kJ/mol.

5.22 The Lewis dot structure of benzene, C_6H_6, can be written as a pair of resonance structures. The experimentally determined $\Delta_r U$ for the reaction of benzene with hydrogen gas to give *cyclohexane*, C_6H_{12}, as written below, is -199 kJ/mol. (a) Compare this number with the value you would get using the mean bond dissociation energies of H−H, C−H, C−H, and C=C bonds given in Table 5.1. (b) Explain how this comparison supports the statement that resonance makes a compound more stable in terms of overall energy.

benzene   cyclohexane

Brain Teasers

5.23 Using Hess's law and the following information:

$C(\text{diamond}) \longrightarrow C(g)$	$\Delta_r U = +713\,\text{kJ/mol}$
$CO(g) + \frac{1}{2}\,O_2(g) \longrightarrow CO_2(g)$	$\Delta_r U = -282\,\text{kJ/mol}$
$O_2(g) \longrightarrow 2\,O(g)$	$\Delta_r U = +496\,\text{kJ/mol}$
$CO(g) \longrightarrow C(g) + O(g)$	$\Delta_r U = +1075\,\text{kJ/mol}$

answer the following question: Your plane, carrying a cargo of diamonds, has crashed in Antarctica. While trying to stay warm waiting for rescue, you have burned all available paper, rubber, wood, and plastic—everything you can think of that will burn. Should you burn the diamonds? Hint: Determine $\Delta_r U$ for the following reaction from these data:

$$C(\text{diamond}) + O_2(g) \longrightarrow CO_2(g)$$

5.24 Draw a diagram illustrating the states and their relative energies involved in using computational chemistry to estimate the structure and energy of CH_4. Consider both possibilities: planar and tetrahedral. The vertical axis on your diagram should be "energy-o-meter reading."

CHAPTER 6

The Effect of Temperature on Equilibrium

Although mechanical energy is indestructible, there is a universal tendency to its dissipation, which produces throughout the system a gradual augmentation and diffusion of heat, cessation of motion and exhaustion of the potential energy of the material Universe.

—*Sir William Thompson (Lord Kelvin)*

6.1 Chemical Reactions as Single Systems: Isomerizations

Consider the simple reaction of one substance becoming another:

$$A \rightleftharpoons B$$

If the equation is balanced, there will be just as many atoms, just as many electrons, protons, and neutrons in A as in B. Thus, B can be seen as just another form of A. We say that A and B are *isomers*. For example, the substances N≡C−O−H (cyanic acid) and H−N=C=O (isocyanic acid) are isomers. Both contain one atom each of H, N, C, and O, but in cyanic acid the H is on the O, while in isocyanic acid, the H is on the N. In the gas phase at room temperature, the equilibrium

$$\begin{array}{ccc} \text{N}\equiv\text{C}-\text{O}-\text{H} & \rightleftharpoons & \text{H}-\text{N}=\text{C}=\text{O} \\ \text{cyanic acid} & & \text{isocyanic acid} \end{array}$$

lies far to the right, while in aqueous solution both isomers are present.

Although isomers A and B may be thought of as having two completely different energy systems, the fact that A and B are in equilibrium with each other means that it is possible for a particle in one system to "jump" to the other system. Thus, really we have one giant system that encompasses both isomers. We really have two independent energy systems, one for substance A and one for substance B, both involving electronic, vibrational, rotational, and translational energy. However, it is possible to consider these two as really being one and the same energy system.

Imagine that for some reason isomer A has a lower ground-state energy than isomer B. Maybe the arrangement of atoms in A is slightly more stable than that in B. Perhaps the bonds in A are a bit stronger. Imagine also that the energy level spacings in B are twice as closely spaced as those in A. If there were 50 million molecules total, we might then have a situation like that shown in Figure 6.1, which is for a temperature of 200 K.

In this particular distribution, five levels of A and nine levels of B are populated. Interestingly, even with twice as many levels in B as in A, at this temperature there are clearly many more molecules of A than of B. That appears to be due to the fact that A has

J/10^{-21} J	A	B
50	0	0
45		3
40	20	20
35		127
30	783	783
25		4,782
20	29,241	29,241
15		178,806
10	1,093,378	1,093,378
5		6,685,891
0	40,883,547	

Figure 6.1
The most probable distribution of energy at 200 K for a system containing 50,000,000 molecules that exist in two distinct isomers A and B, each of which has the energy level spacings indicated. Isomer A has the lower ground state by 5×10^{-21} J; isomer B has the more closely spaced levels by a factor of 2. Note the degeneracies! At this temperature the total number of molecules of isomer A, n_A, is 42,006,969; the total number of molecules of isomer B, n_B, is 7,993,031. Thus, at 200 K, the system is 84% A and 16% B, and the ratio $n_B/n_A = 0.19$.

the lower ground state, which itself represents more than 80% of all of the molecules in the system!

Since both A and B must be in the same container in order to be at equilibrium, we can write

$$K = \frac{[B]}{[A]} = \frac{n_B}{n_A} = 0.19$$

Let's take a close look at what is going on here. The idea is that this system is behaving simply as a large Boltzmann distribution. At any given time there may be a collision between two molecules that results in the trading of energy, as always. But in this case it is possible for a molecule of A to gain just the right amount of energy to become a particle of B, and vice versa. So we can think of this as one large system for which the lowest level happens to be labeled "A."

6.2 The Temperature Effect on Isomerizations

At the extreme of absolute 0 temperature, only the very lowest level will have any particles in it, because only the lowest level of any Boltzmann distribution is populated at absolute zero. All of the material will be in the form of A, and there will be no B at all. So at the limit of low temperature in this case, $K = 0$:

$$K = \frac{[B]}{[A]} = \frac{0}{n} = 0 \quad \text{when } T \to 0 \text{ and lowest level is of A}$$

As the temperature increases, more and more levels of both A and B will become populated. Since now there will be some B and there will be less A, K in this case should increase with increasing temperature. Compare now the situations in Figure 6.2, at 200 K, with the situation in Figure 6.3, for the same system at the higher temperature of 600 K. Starting with the energy levels (top center panel), we see that at the higher temperature of 600 K both A and B do indeed have more levels populated. In fact, all told, there are 15 levels of A and 29 levels of B populated at 600 K. (Fewer than these are shown, because only levels with at least 25,000 particles are shown.) In addition, looking at the top right panel in Figure 6.3, we see that at this higher temperature there is almost as much B as A, and

$$K = \frac{n_B}{n_A} = \frac{0.458}{0.542} = 0.846$$

The value of K has increased substantially in going from 200 K to 600 K.

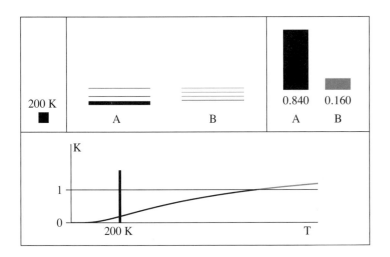

Figure 6.2
Isomerization equilibrium involving two substances, A and B, at 200 K depicting the system described in Figure 6.1. The first panel is the temperature. In the top center panel is a depiction of the energy levels in A and B. The thickness of line in this center panel indicates the relative number of particles in each level containing at least 25,000 particles. The two vertical bars in the right-hand panel indicate relative amounts of A and B at this particular temperature. The lower graph shows the effect of temperature on the value of K. At this particular temperature, 200 K, $K = n_B/n_A = 0.160/0.840 = 0.19$.

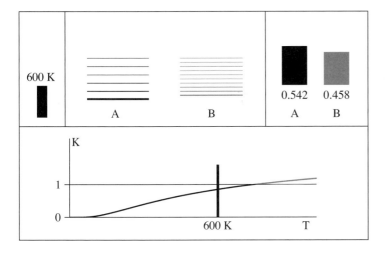

Figure 6.3
The same equilibrium as in Figures 6.1 and 6.2, this time at the higher temperature of 600 K. At this higher temperature, more levels of both A and B are populated, and $K = n_B/n_A = 0.458/0.542 = 0.846$.

The question remains, why did K increase? A simple argument can be made that ***K increases with temperature in this case simply because the ground state of B is higher than the ground state of A***. Believe it or not, it has nothing to do with the relative spacings of the levels in A and B. Here's how the argument goes in this case:

- At very low temperature there are no particles of B, only A. Thus, at the limit of absolute zero we have $K = 0$. This is because for *all* Boltzmann distributions, this one included, all of the particles must be in the ground state at 0 K, and the ground state in this case is labeled "A," not "B."
- As the temperature rises, as for *all* Boltzmann distributions, this one included, more and more levels become populated. The fact that some of those levels happen to be labeled "A" or "B" makes no difference whatsoever.
- Since there are only so many particles in the first place, if *any* particles take on the identity "B" then there will be more B and less A, and the equilibrium constant must increase. That is exactly what happens: Levels we call "B" start to become more populated as the temperature increases, and this must be *at the expense* of particles we call "A." $K = $ [B]/[A] increases simply because as the temperature increases, we are guaranteed to have more B and less A as long as B has the higher ground state.

Notice that nothing has changed between Figure 6.2 and Figure 6.3 with respect to the levels themselves. In both cases we have exactly the same relationships among the

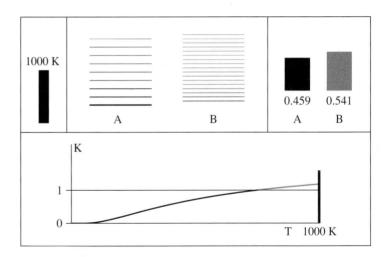

Figure 6.4
The same system as depicted in Figures 6.1, 6.2, and 6.3, now at the much higher temperature of 1000 kelvins. At this temperature there is actually more B than A, and $K = n_B/n_A = 0.541/0.459 = 1.18$.

energies of all the levels as in Figure 6.1. Think of the energy level system itself as being completely independent of the temperature. You may remember from Chapter 3 that none of our equations for translational, rotational, vibrational or electronic energy have T in them. (See the Appendix A entry for ε for a summary.) It is only the *populations* of those levels that vary with temperature.

Now consider Figure 6.4, where we have heated the system up to 1000 K. At this very high temperature the system has finally "crossed the line," and there is more B present than A—we now have $K = 1.18$. The fact that at this very high temperature there is more B than A present is, in fact, finally the result of B having more closely spaced levels than A. If the temperature is increased further and further, the equilibrium constant will continue to rise, approaching a high temperature limit of exactly 2.0. We will see how this arises in the next section.

6.3 *K* vs. *T* for Evenly Spaced Systems

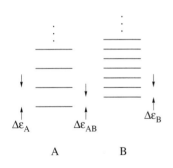

Figure 6.5
The three parameters defining a hypothetical system containing two different substances with evenly spaced energy levels. In this example, the energy levels in B are twice as closely spaced as those in A, so $\Delta\varepsilon_A/\Delta\varepsilon_B = 2$. In addition, the difference in energy between the two ground states is $\Delta\varepsilon_{AB}$, which in this case is positive, since B has the higher ground state. If B had had the lower ground state, then $\Delta\varepsilon_{AB}$ would have been negative.

For an equilibrium such as shown in Figures 6.1–6.4, involving just two substances, both with evenly spaced energy levels, there are only three parameters (Figure 6.5):

$\Delta\varepsilon_A$ the energy level separations for substance A
$\Delta\varepsilon_B$ the energy level separations for substance B
$\Delta\varepsilon_{AB}$ the difference in energy between the lowest two levels in A and B

These three parameters along with temperature completely define the system and allow us to define K explicitly:

$$K = \frac{1 - e^{-\Delta\varepsilon_A/kT}}{1 - e^{-\Delta\varepsilon_B/kT}} \, e^{-\Delta\varepsilon_{AB}/kT} \qquad (6.1)$$

The derivation of this rather complicated looking equation is not terribly difficult and serves as a good exercise in handling exponential functions. After showing it, we will see that it indeed predicts just what we have seen in Figures 6.1–6.4 about the behavior of K at low and high temperature.

The key to deriving this equation is to apply a finding from Section 2.12, which relates the number of particles in the lowest energy level, n_0, to the total number of particles, n, for evenly spaced systems:

$$n_0 = n(1 - e^{-\Delta\varepsilon/kT})$$

In this case, we relate the lowest-level populations in A and B, n_{0A} and n_{0B}, respectively, with the energy level spacing for each isomer:

$$n_{0A} = n_A(1 - e^{-\Delta\varepsilon_A/kT})$$

$$n_{0B} = n_B(1 - e^{-\Delta\varepsilon_B/kT})$$

In addition, we know that the molecules in the lowest level of A are in equilibrium with the particles in the lowest level of B, and thus their populations are related to their energy level separation, $\Delta\varepsilon_{AB}$:

$$\frac{n_{0B}}{n_{0A}} = e^{-\Delta\varepsilon_{AB}/kT}$$

Substituting the expressions for the ground state populations n_{0A} and n_{0B} into this equation, we get

$$\frac{n_B(1 - e^{-\Delta\varepsilon_B/kT})}{n_A(1 - e^{-\Delta\varepsilon_A/kT})} = e^{-\Delta\varepsilon_{AB}/kT}$$

Noting that $K = n_B/n_A$, a little rearrangement gives Equation 6.1, and we find that K for evenly spaced energy systems is indeed just a function of the three parameters $\Delta\varepsilon_A$, $\Delta\varepsilon_B$, and $\Delta\varepsilon_{AB}$, along with the temperature, T.

In order to understand how this function behaves, we have to apply some mathematics of exponentials which are summarized in Table 6.1.

Both the numerator and the denominator of the fraction in Equation 6.1 are the sort of function listed on the last line of this table. At very low temperature, both of these terms approach the number 1, and Equation 6.1 can be approximated as just the final exponential:

$$K_{\text{low } T} \approx e^{-\Delta\varepsilon_{AB}/kT} \tag{6.2}$$

which is just to say that at low temperature, the position of the equilibrium is governed by the difference in *ground state energies* of A and B. If the ground state of B is higher in energy than the ground state of A, as is the case in Figures 6.1–6.5, then $\Delta\varepsilon_{AB} > 0$ and $K_{\text{low } T} < 1$.

However, if the ground state of B is below the ground state of A, as in Figure 6.6, then $\Delta\varepsilon_{AB} < 0$, and we have a function that is of the form $e^{a/T}$. We expect K to approach ∞ as T approaches 0. This can also be seen by considering the effect of having all B and no A at 0 K:

$$K = \frac{n_B}{n_A} = \frac{n}{0} = \infty \quad \text{(when } T \to 0 \text{ and lowest level is of B)}$$

Table 6.1
Some useful relationships. **All results presume** $a > 0$. The last result for $T \to \infty$ is based on the approximation for $e^{-x} = 1 - x + x^2/2! - x^3/3! + \cdots \approx 1 - x$ when x is very small.

as $T \to 0$	as $T \to \infty$
$1/T \to \infty$	$1/T \to 0$
$e^{a/T} \to \infty$	$e^{a/T} \to 1$
$e^{-a/T} \to 0$	$e^{-a/T} \to 1$
$1 - e^{-a/T} \to 1$	$1 - e^{-a/T} \to a/T$

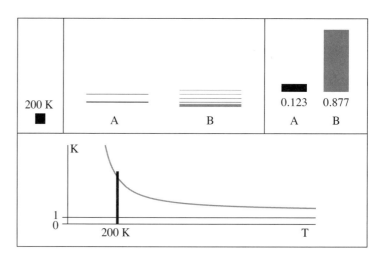

Figure 6.6
This system now consists of two isomers A and B where the ground state of B is lower than the ground state of A. The value of K is infinite at 0 kelvins and gets smaller with increasing temperature, reaching a limit ultimately of $K = 2$ at very high temperature.

The high temperature extreme, where K is governed only by the relative spacings of the energy levels in A and B, is a little trickier. At high temperature all of the exponential terms approach the number 1. However, we can't just say that K is $(1-1)/(1-1)$, because that is undefined. Instead we apply the trick that for very small x:

$$1 - e^{-x} \approx x$$

(Check this out for yourself using your calculator.) So, in this case, we have

$$K_{\text{high } T} = \frac{1 - e^{-\Delta\varepsilon_A/kT}}{1 - e^{-\Delta\varepsilon_B/kT}} e^{-\Delta\varepsilon_{AB}/kT} \approx \frac{\Delta\varepsilon_A/kT}{\Delta\varepsilon_B/kT}(1) = \frac{\Delta\varepsilon_A}{\Delta\varepsilon_B} \qquad (6.3)$$

and we see that K at high temperature approaches a limit depending solely upon the energy level spacings in A and B. Thus, for the systems shown above, where $\Delta\varepsilon_A/\Delta\varepsilon_B = 2$, we have that K approaches the value 2 either from below (Figures 6.1–6.5) or from above (Figure 6.6).

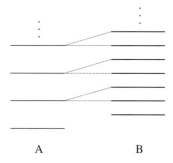

Figure 6.7
Each level of A corresponds to two levels of B. At the limit of high T, the value of K will approach 2.0.

Note that the system with the *closer* energy levels has the *smaller* $\Delta\varepsilon$. A simple way to think about the high temperature limit for a system of evenly spaced energy levels is to consider the *density* of states (Figure 6.7). In this figure is shown once more exactly the same system as in Figures 6.1–6.5. The idea is to count the number of levels in A vs. B, *disregarding the difference in ground state*. We see that for every level in A there are two levels in B, and we can rewrite Equation 6.3 in terms of number of levels:

$$K_{\text{high } T} = \frac{(\text{number of levels of B})}{(\text{number of levels of A})} \qquad (6.4)$$

In effect, we are saying that at very high temperature there are going to be particles spread out fairly evenly among all the levels in the system, and so it is just as easy to count the levels as it is to count the particles! This is the case, however, only at very high temperature or for systems with extremely closely spaced energy levels (such as in the case of the translational energy levels of gases). We come to the same conclusion we found in Section 3.11:

At the limit of high temperature, the isomer with the more closely spaced energy levels will be favored.

6.4 Experimental Data Can Reveal Energy Level Information

It is important to realize that regardless of the relative spacing of levels in A vs. B, as the temperature increases, more and more of the substance with the higher ground state will be populated, as demonstrated in all of these examples. This is *guaranteed* by the Boltzmann law.

Thus, just a little experimental data about K in relation to T can tell us quickly which substance has the lower ground state. If the ratio n_B/n_A increases with increasing T, then we have the situation depicted in Figures 6.1–6.5, with the ground state of B being higher in energy than the ground state in A. If the opposite is true, if the ratio n_B/n_A *decreases* with increasing T, as in Figure 6.6, then B must have the lower ground state.

Which substance, A or B, has the more closely spaced levels is trickier to discern from limited data—at least from what we know so far. Unless we actually have data that cross the line $K = 1$, we would not know at this point if the energy level spacings were closer in A than in B. For example, consider the two data sets in Table 6.2, which are for two unrelated equilibria, which we simply write as

$$reactants \rightleftharpoons products$$

For Reaction 1, we see that K is increasing with increasing temperature, indicating that adding heat shifts the equilibrium to the right, and products have the higher ground state. Exactly the opposite is true for Reaction 2. In that case, the value of K decreases with increasing temperature, shifting the equilibrium toward reactants and indicating that it is the reactants which have the higher ground state.

What about the relative spacings of the energy levels in reactants vs. products in each case? We see in Reaction 1 that the line $K = 1$ has been crossed (from below) and it must be true that as the temperature increases, the products will be favored more and more. We may conclude that in Reaction 1 the products have the closer energy levels.

As for Reaction 2, note that the line $K = 1$ has not been crossed. We might be able to project what will happen to K as the temperature increases further (it turns out, for example, that a plot of $\ln K$ vs. $1/T$ is roughly linear) but doing anything that fancy is not the goal here. At this point we will simply say that we do not know anything about the relative spacings of the energy levels between reactants and products in Reaction 2 because we lack sufficient data. We would need data showing that K drops below the value of 1, and we just don't have it. (As it turns out, the high temperature limit for *both* of these reactions is $K = 2.0$.)

Table 6.2
Hypothetical laboratory results showing equilibrium constants for two related reactions determined at five different temperatures.

$T/\,°C$	K_{rxn1}	K_{rxn2}
0	0.34	85
100	0.53	33
200	0.68	20
400	0.92	12
600	1.10	9.4

6.5 Application to Real Chemical Reactions

The general ideas presented above need not be limited to isomerization reactions or even to systems with evenly spaced energy levels. Although a full treatment of more complex reactions will have to wait, at this point it should be interesting to extend what we have to real systems anyway. Our ultimate interest is in the prediction or rationalization of the direction of a reaction based on writing just the equation for that reaction, with the help of a periodic table or a table of bond dissociation energies.

We need to know two things before we can make predictions about whether a reaction is going to be favorable or not at low temperature and what the effect of increasing the temperature will be:

- We need to know something about the relative positions of ground states. This involves estimating $\Delta_r U$ from tables of mean bond dissociation energies (Table 5.1) or determining it in the lab using a calorimeter (Section 4.11).
- We need to know something about the relative spacings of the energy levels. At this point, this has to come from taking a close look at what we have in the

reaction and seeing if we can guesstimate the relative spacings. We'll see soon how this, too, can be done using tables (Section 7.7) or based on experimental data (Section 11.4).

We will look at three examples to illustrate these points. In each case we will consider first the relative ground states based on what we know about bonding from our experience or from Chapter 5. Then we will consider the relative energy level spacings based on what we know from Chapter 3. Thus, we will rely on the fact that gases, which can translate, have more closely spaced levels than either liquids or solids, and that only liquids and gases have rotational levels, while solids can only vibrate.

Case 1: $H_2(g) + F_2(g) \rightleftharpoons 2\,HF(g)$

Relative ground state energies. Using a table of bond dissociation energies (Table 5.1) and Hess's law we know that

$$\Delta_r U \approx \Sigma BDE_{reactants} + (-\Sigma BDE_{products})$$

We can use $\Delta_r U$ as an estimate of ground state difference between reactant and product:

$$\Delta_r U \approx (H\text{–}H) + (F\text{–}F) - 2(H\text{–}F) = 432 + 154 - 2(565)$$
$$= -544\,kJ/mol$$

Clearly this reaction is exothermic, and the ground state of products must be *way* below the ground state of reactants.

Relative energy level spacings. There is nothing particularly notable here—same number of gas moles on the left as on the right.

$$H_2 + F_2 \qquad\qquad 2\,HF$$

We expect to see something like that shown above. There will be a preponderance of product HF at low temperature, because it has the lower ground state. We expect K, which equals $[HF]^2/[H_2][F_2]$ in this case, to be much larger than 1 at low temperature and then decrease as the temperature is raised. Whether or not K ever drops below 1 is an unanswerable question at this point. (See Figure 2.1 for a hint.)

Case 2: $N_2(g) + O_2(g) \rightleftharpoons 2\,NO(g)$

Relative ground state energies. Using a table of bond dissociation energies, we have

$$\Delta_r U \approx (N\equiv N) + (O=O) - 2(N=O) = 941 + 495 - 2(607)$$
$$= +222\,kJ/mol$$

In this case we have an endothermic reaction, and it is the *reactants* that therefore have the lower ground state and should be favored at low temperature.

Relative energy level spacings. Once again, it is too difficult to tell. There are no more moles of gas on one side than on the other, and the molecules are all very similar. We won't see much of an effect here.

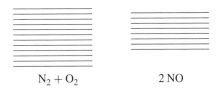

$$N_2 + O_2 \qquad\qquad 2\,NO$$

We expect to see something like that depicted above. Since reactants N_2 and O_2 have the lower ground state, they should be favored at low temperature. In fact, the ground state of reactants is *way* below the ground state of products (by over 200 kJ). At low temperature there will be very little NO, and $K \ll 1$. Once again, the side with the higher ground state (products here) will start to get populated as the temperature increases, so we expect K to increase a bit with temperature, but we cannot predict with any certainty whether product NO will ever be favored. A graph of K vs. T for this reaction is also shown in Figure 2.1, where even there it is unclear what will be the case at high temperature.

Case 3: $H_2O(l) \rightleftharpoons H_2O(g)$

Relative ground state energies. Bond dissociation energy tables won't help us here, because as far as covalent bonds are concerned, there is no difference between reactants and products in this case. We must instead consider phases. The liquid phase is characterized by weak intermolecular bonding (hydrogen bonding in this case, Section 5.8), which is not present in the gas. Thus, we expect the liquid to have the lower ground state (more bonding) than the gas, and liquid water should be more prevalent at low temperature.

Relative energy level spacings. In terms of energy level spacings, here we have an increase in the number of gas molecules as the reaction proceeds from reactants to products: $\Delta n_{gas} = +1$ mole. Thus, the product (water vapor) has many more translational energy levels than the reactant (liquid water) and should have energy levels much more closely spaced. We expect water vapor to be favored over liquid water at high temperature. Sounds just like water!

$$H_2O(l) \qquad\qquad H_2O(g)$$

Overall, for the equilibrium between liquid water and water vapor we envision an energy level picture something like that shown above. Liquid water, with its stronger bonding and lower ground state, should be favored at low temperature. But as the temperature increases, the relative amount of water vapor *in equilibrium* with the liquid (for example, in a closed container) should increase. Indeed, if you heat a can of beans without opening it (don't!), the increasing pressure of water vapor in the can could be enough to create an explosion.

6.6 The Solid/Liquid Problem

Now consider a related reaction, the melting of ice:

$$H_2O(s) \rightleftharpoons H_2O(l)$$

The solid phase has stronger intermolecular interactions than the liquid phase, so we rightly expect it to be favored at low temperature. On the other hand, since the liquid has the potential to rotate, it should have the more closely spaced energy levels and should be favored at high temperature. Right again.

But what about K? There really isn't any equilibrium here, is there? (Perhaps you have learned to "ignore solids and liquids" when writing equilibrium expressions.) The answer is, yes, there is an equilibrium, but there is no equilibrium *constant*! Equilibrium in this case happens only at one single temperature: 0°C. And when the system is at 0°C, any amount of solid or liquid water may be present!

Below 0°C, solid water is *so* favored that every last trace of liquid water will disappear, and above that temperature every last trace of ice will melt. We know that's true from experience. But how do we explain it within this model?

The answer to *that* question is simple: We can't. We're missing an important part of the puzzle called the "surroundings," which will be sorted out in Chapter 8. But first we need to quantify this "relative-energy-level-spacing" business. We need a real measure of it—a state function we can use in a way similar to the way we can use U as a measure of relative ground states. That state function is called *entropy* and is introduced in Chapter 7. In fact, we will see that this whole "lower ground state favored at low temperature" idea really comes from the fact that in going to that state a system must deliver heat to its surroundings, and it is what happens *there*, in the surroundings, that drives the reaction in one direction or another.

6.7 Summary

In this chapter we have brought together several threads which have been hinted at since Chapter 1. These include energy, heat, temperature, and probability. The only new idea here is the extension of the idea of the Boltzmann distribution to describe the most probable distribution in more complex systems. This was done by identifying some levels of the system as belonging to "reactants" and other levels as belonging to "products."

The end result is two-fold:

1. Substances on the side of a balanced chemical equation having the lower ground state will be favored at low temperature.
2. Substances on the side of a balanced chemical equation having the more closely spaced energy levels will be favored at high temperature.

Both of these findings derive from the way energy is distributed randomly in chemical systems where only discrete "quanta" of energy are possible. The Boltzmann law governs all such cases, and, at least for evenly spaced systems, we get Equation 6.1:

$$K = \frac{1 - e^{-\Delta\varepsilon_A/kT}}{1 - e^{-\Delta\varepsilon_B/kT}} \, e^{-\Delta\varepsilon_{AB}/kT}$$

In principle, we could use this equation to predict exactly how evenly spaced systems would behave as a function of temperature. To simplify matters at this stage, though, we just look at the extremes of low and high temperature.

At the extreme of low temperature, this equation approximates to Equation 6.2:

$$K_{\text{low } T} \approx e^{-\Delta\varepsilon_{AB}/kT}$$

This supports the idea that at low temperature the factor that determines the position of equilibrium is the relative ground state energies of reactants and products, $\Delta\varepsilon_{AB}$. If this number is positive (if reactants have the lower ground state) then at low temperature it is the reactants which will be favored. However, increasing the temperature will shift the equilibrium to the right, because in increasing the temperature particles will start to populate higher energy levels in accordance with the Boltzmann law, and some of these levels will be of product. It is for this reason and for this reason alone that water evaporates more at high temperature than at low temperature. (We say that the "vapor pressure" of boiling water is higher than the vapor pressure of ice water.) Water requires heat to evaporate. Any reaction requiring heat as a reactant will tend to shift toward product as the temperature is raised.

If the relative ground state energies of reactants and products, $\Delta\varepsilon_{AB}$, is negative, then we see the opposite: Products, having the lower ground state, will be favored at low temperature, and as the temperature rises, the equilibrium will shift to the left in the direction of reactants.

However, even though the equilibrium may shift toward products, for example, nothing about bonding or relative ground states says *anything* about whether products will ever be *favored* (that is, $K > 1$) at some temperature. Whether or not that happens will depend upon the relative energy level spacings in reactants vs. products. Equation 6.1 boils down to the following at high temperature (Equation 6.4):

$$K_{\text{high } T} = \frac{\text{(number of levels of B)}}{\text{(number of levels of A)}}$$

The substance having the more closely spaced (proportionately more numerous) energy levels will be favored at high temperature. Notice that "bonding" (which involves differences in ground state energies) has nothing to do with it! We propose a radical statement: *Whether or not reactants or products are favored at high temperature has nothing to do with bonding!*

It is true that breaking bonds requires energy, and increasing the temperature of a system by heating is a good way to add energy to a system. But a system will never go over to the state of "mainly products" at high temperature unless those products have more closely spaced energy levels. Most dissociation reactions involve both the breaking of bonds *and* the creation of species with more closely spaced energy levels.

We generally associate spontaneity with the release of energy, but this is not always the case. Sometimes spontaneous reactions absorb energy from their surroundings. There are plenty of reactions that occur despite the fact that bonds are broken (energy is required) in going from reactants to products. A classic example is the dissolving of salt in ice water. Try this at dinner: Test the temperature of ice water using your finger in a glass filled with ice and just a little liquid water. Then add salt. (It takes several grams to get a good effect.) Test the temperature of the water again, and you will find that the water is much colder than it was before you added the salt. This indicates that the reacting system has absorbed heat (from the ice and water in the glass). The system has *spontaneously* gone from a lower ground state (solid NaCl and liquid water) to a higher ground state (Na^+ and Cl^- ions dissolved in water)! The reason this is possible is that the energy levels are more closely spaced in the dissolved state than in the solid state. The system absorbs energy from the surroundings simply because it leads to a more probable distribution of energy overall.

We now have more than half of the puzzle. Boltzmann rules. Probability rules. Systems always tend toward the most probable distribution of energy. We have a suitable measure related to ground states in U; what is now needed is a suitable measure of the closeness of energy levels that will work for any system, regardless of whether that system has evenly spaced energy levels. Chapters 7 and 8 introduce precisely this measure, which we call "entropy."

Problems

The symbol Ⓐ *indicates that the answer to the problem can be found in the "Answers to Selected Exercises" section at the back of the book.*

6.1 Ⓐ The following energy level diagrams are for four different cases where the reactants (A) and products (B) are in equilibrium as A ⇌ B. You may presume that the levels continue to higher and higher energy as evenly spaced energy systems.

Case 1 A B	Case 2 A B	Case 3 A B	Case 4 A B
─ ─	─ ─	─	─ ─
─ ─	─	─ ─	─ ─
─ ─	─	─	─ ─
─ ─	─	─ ─	─ ─
─ ─	─ ─	─ ─	─ ─
─	─ ─	─ ─	─ ─
─	─ ─	─ ─	─ ─

In which cases are (a) products favored at low temperature? (b) reactants favored at low temperature? (c) neither reactants nor products favored at low temperature?

6.2 For the cases in Problem 6.1, in which cases are (a) products favored at high temperature? (b) reactants favored at high temperature? (c) neither reactants nor products favored at high temperature?

6.3 Ⓐ (a) Determine the limit of K at low temperature for the four cases in Problem 6.1. (b) Determine the limit of K as T approaches infinity for the four cases in Problem 6.1.

6.4 Sketch a graph of K vs. T for the four cases in Problem 6.1. Be sure to indicate the correct behavior of K with increasing T, including the precise values at the limiting cases of $T = 0$ K and $T = \infty$ K.

6.5 Ⓐ What factors determine which side of a reaction is favored at low temperature? What factors determine whether K will increase or decrease with increasing temperature? What factors determine the limit of K at very high temperature?

6.6 Sketch the expected energy level picture, similar to the depictions given in Problem 6.1, for each of the following equilibria:

(a) $\qquad\qquad CO_2(s) \rightleftharpoons CO_2(g)$
(b) $\quad N_2H_4(l) + O_2(g) \rightleftharpoons N_2(g) + 2\,H_2O(g)$
(c) $\qquad\quad 2\,HI(g) \rightleftharpoons H_2(g) + I_2(g)$

6.7 Ⓐ Which of the reactions in Problem 6.6, if any, (a) favor products at low temperature? (b) favor reactants at low temperature? (c) will be expected to have a K that increases with temperature? (d) will be expected to have a K that decreases with temperature?

6.8 Sketch the expected energy level picture, similar to the depictions given in Problem 6.1, for each of the following equilibria:

(a) $\qquad\qquad H_2O(l) \rightleftharpoons H_2O(g)$
(b) $\quad 2\,N_2H_2(l) + 5\,O_2(g) \rightleftharpoons 4\,NO_2(g) + 2\,H_2O(g)$
(c) $\quad Ba^{2+}(aq) + SO_4^{2-}(aq) \rightleftharpoons BaSO_4(s)$

6.9 Ⓐ For the previous problem, sketch K vs. T graphs. Be sure to indicate the correct behavior of K with increasing T, including whether the values at the limiting cases of $T = 0$ K and T approaching infinity are above or below $K = 1$.

6.10 Sketch the expected energy level picture, similar to the depictions given in Problem 6.1, for each of the following equilibria:

(a) $\quad C_5O_5W(g) + C_2H_6(g) \rightleftharpoons C_7H_6O_5W(g)$
(b) $\qquad C_2H_6(g) \rightleftharpoons 2\,H_2(g) + C_2H_2(g)$
(c) $\quad 2\,HCl(g) + C_4H_{10}AlCl(l) \rightleftharpoons AlCl_3(s) + 2\,C_2H_6(g)$

6.11 For the previous problem, sketch K vs. T graphs. Be sure to indicate the correct behavior of K with increasing T, including whether the values at the limiting cases of $T = 0$ K and T approaching infinity are above or below $K = 1$.

6.12 Sketch the expected energy level picture, similar to the depictions given in Problem 6.1, for each of the following equilibria:

(a) $\qquad CH_3NO(g) \rightleftharpoons CO(g) + NH_3(g)$
(b) $\quad Mg(s) + \frac{1}{2}\,O_2(g) \rightleftharpoons MgO(s)$
(c) $\quad S(s) + 3\,BrF_5(l) \rightleftharpoons SF_6(g) + 3\,BrF_3(l)$

6.13 For the previous problem, sketch K vs. T graphs. Be sure to indicate the correct behavior of K with increasing T, including whether the values at the limiting cases of $T = 0$ K and T approaching infinity are above or below $K = 1$.

CHAPTER 7

Entropy (*S*) and the Second Law

The law that entropy always increases—the second law of thermodynamics—holds I think, the supreme position among the laws of Nature. If someone points out to you that your pet theory of the universe is in disagreement with Maxwell's equations—then so much worse for Maxwell's equations. If it is found to be contradicted by observation—well these experimentalists do bungle things sometimes. But if your theory is found to be against the second law of Thermodynamics, I can give you no hope; there is nothing for it but to collapse in deepest humiliation.

—*"Sir Arthur Stanley Eddington," in* The Nature of the Physical World, *Maxmillan, New York, 1948, p. 74.*

7.1 Energy Does Not Rule

The 19th century was a revolutionary time for the scientific understanding of energy, heat, and work. In 1850, Rudolf Clausius argued successfully that heat was not some physical fluid that flowed from one substance to another, but was instead a transfer of pure energy. In many ways, in fact, it was this scientific revolution that led to the one in industry, as scientists and engineers started to understand how to control chemical and physical processes on a huge scale.

The discovery that the transfer of energy involves only heat and work was so important that it became the First Law of Thermodynamics (Section 4.5):

$$\Delta U = q + w$$

The idea that energy is conserved, that overall the change in energy of the "universe" is constant:

$$\Delta U_{\text{universe}} = 0$$

is a direct corollary of the First Law. It arises when we consider the idea that the "universe" includes not just our system but also the surroundings (Section 4.10 and Figure 7.1).

Thus, we can write

$$q_{\text{sur}} = -q$$
$$w_{\text{sur}} = -w$$
$$q_{\text{sur}} + w_{\text{sur}} = -(q + w)$$
$$\Delta U_{\text{sur}} = -\Delta U$$
$$\Delta U + \Delta U_{\text{sur}} = 0$$
$$\Delta U_{\text{universe}} = 0$$

Figure 7.1

The universe according to thermodynamics consists of the *system* and the *surroundings*. Energy may be transferred from one to another in only two ways: through heat (q) or work (w, shown here as movement of the wall). Both heat and work are defined relative to the system, but we could just as well define them in terms of the surroundings. Thus, $q_{sur} = -q$ and $w_{sur} = -w$.

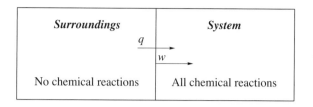

and we have that energy is conserved.[1]

The problem with the conservation of energy is that if the net result of chemical reactions is no overall change in energy (of the universe), what could possibly explain their "happening"? Most chemical reactions are exothermic, releasing heat and thus dropping to lower energy *of the system* as they proceed. But certainly the release of energy in a chemical reaction is not required. Have you ever seen one of those "instant ice" packs that you twist and it gets cold? How does that work? What about evaporation? You sweat to stay cool. As the water in sweat evaporates, energy in the form of heat is absorbed by the H_2O molecules and carried away from your body by air currents. Neither of these phenomena can be explained in terms of the lowering of energy being something that "drives" the process.

So the mystery of what drives chemical reactions and determines the position of equilibrium must involve more than energy. After successfully showing that heat was pure energy and not some real fluid, Clausius's second great achievement, published in 1865, was the discovery of a new state function he referred to as *entropy* (S). Clausius demonstrated that entropy rules the world, and that it is related to heat and temperature. His finding became the Second Law of Thermodynamics:

$$\Delta S_{universe} = \Delta S + \Delta S_{sur} > 0 \qquad (7.1)$$

That is, for a reaction to proceed, for any change to be observed, the entropy of the universe, which includes both the system and the surroundings, must increase. Simple. Effective. But what is *entropy*? The truth is, Clausius didn't know! He had a definition involving heat and work, but fundamentally he had no idea what entropy could be. In fact, Clausius's definition referred only to the *change* in entropy for states *not undergoing chemical reactions* (such as the surroundings):

$$\Delta S_{sur} = \frac{q_{sur}}{T} \qquad (7.2)$$

Believe it or not, Clausius was never able to actually define S itself. He only showed that its *change* (at least in the surroundings, where no chemical reactions are occurring) could be measured by measuring the amount of heat flowing into the surroundings (using heat capacities, Section 4.6).

7.2 The Definition of Entropy: $S = k \ln W$

The fundamental understanding that entropy is related to the number of ways energy can be distributed was Boltzmann's second great achievement. Boltzmann knew very well of Clausius's discovery of entropy—it was published just four years before Boltzmann's own landmark paper of 1869 detailing what is now known as the "Boltzmann law." He knew that

---

[1] Notice that in this and all equations in this book, the subscript "sur" means "of the surroundings." We could add the subscript "sys" to mean "of the system," but generally we don't. The convention is that if no subscript is present, we are referring to the system.

his own work had to be consistent with Clausius's, and in 1877, Boltzmann published an extension of his theory that redefined entropy in his own terms as

$$S = k \ln W \qquad (7.3)$$

Here W is the thermodynamic probability (number of all the ways) for the most probable distribution of energy in a system, and, of course, k is again Boltzmann's constant. In many ways this statement was Boltzmann's greatest contribution to science.

Thus, "entropy" was given a strict mathematical definition which related it to probability. Remember (Section 2.5):

$$W = \frac{n!}{n_0! \, n_1! \, n_2! \cdots}$$

Entropy S gets larger when W gets larger and smaller when W gets smaller. If we want to understand what entropy is, we must understand how various changes in the system affect W. What do we know about W?

W increases when the number of particles in a system increases. This is largely due to the $n!$ term in the numerator in the definition of W, and has been true in every system we have looked at in this book. The number of ways four cards can be distributed is easily counted, but the distribution of more than just a few cards starts to involve an incredible number of ways. Thus, we should find it no surprise that **entropy increases as the number of particles in a system increases.**

W increases with increasing temperature (Section 2.9 and Figure 7.2). As particles move up the energy ladder, n_0 becomes smaller, causing W to increase. In the example in Figure 7.2, note that W increases immensely (by a factor of about 5×10^7), while due to the natural logarithm in the definition of entropy, S only doubles. Thus, we expect that **as the temperature of a system is increased, the entropy of that system will increase as well.**

Finally, consider the effect of energy level spacing on W and S (Figure 7.3). If the energy level spacings of a system are made closer through expansion, the system cools

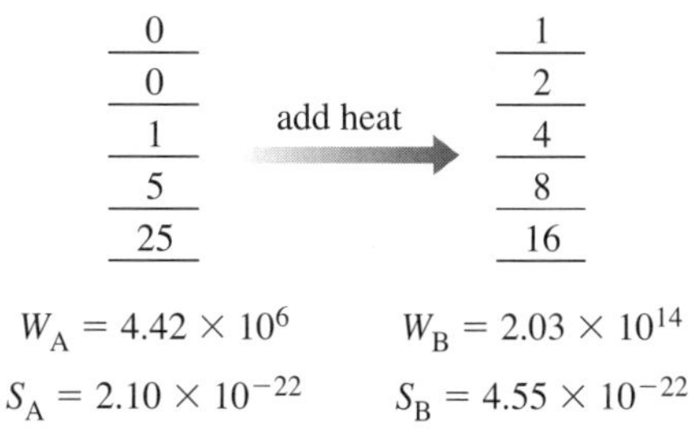

Figure 7.2
Thermodynamic probability W and entropy S both increase when heat is added to raise the temperature of a system. In this case, the entropy more than doubles when the temperature is increased. The unit of S is J/K.

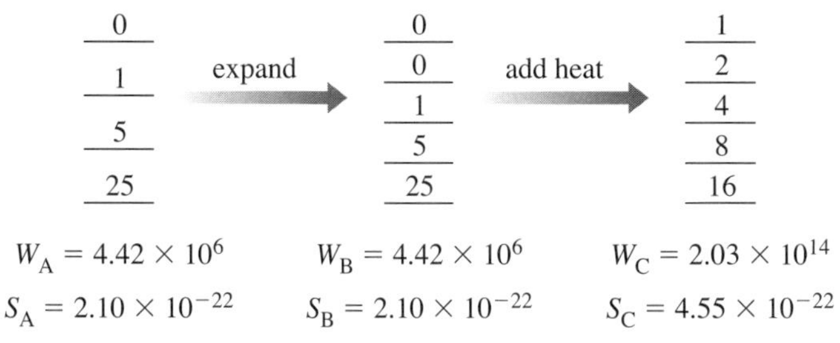

Figure 7.3
Thermodynamic probability W and entropy S both increase when the energy levels of a system are made to be more closely spaced *while at the same time keeping the overall total energy constant.* (The system cools due to the expansion, but it is brought back up to its original temperature by the addition of heat.) The unit of S is J/K.

but no change in either W or S occurs, since the number of particles in each level remain constant. However, bringing the system back to its original temperature requires heating and an increase in W and S for the overall "expansion at constant temperature." By Hess's law (Section 5.2), which states that the change in a state function is independent of path, we can infer that for two systems compared *at the same temperature* (A and C in Figure 7.3) **the system with the more closely spaced energy levels will have the higher entropy.**

The connection sometimes made between entropy and disorder is based on how a system appears after an increase in W. Basically, the more ways there are to distribute energy, the messier a system appears. More particles doing more different things with the energy allotted to them could be thought of as being more disordered. For all systems, at very low temperature only a few levels are populated—there are few ways the system can distribute its meager internal energy. As more particles are added to the system, or the temperature is raised, or the energy levels are brought closer together, we are simply going to see more going on in the system. Still, entropy is not "disorder."

7.3 Changes in Entropy: $\Delta S = k \ln(W_2/W_1)$

The primary reason for the natural logarithm in the definition of entropy is to give S the very nice property that, when you go from "State 1" with W_1 to "State 2" with W_2, the change in entropy is related to the relative probability of the two states, W_2/W_1:

$$\Delta S = S_2 - S_1 = k \ln W_2 - k \ln W_1 = k \ln \frac{W_2}{W_1} \qquad (7.4)$$

The mathematical trick used here is that $\ln x - \ln y = \ln(x/y)$. The amazing thing was that the value of the proportionality constant k wasn't actually known at the time Boltzmann developed all this. Boltzmann just knew it was needed and realized it had to be greater than zero!

7.4 The Second Law of Thermodynamics: $\Delta S_{universe} > 0$

Clearly if the final state *of the universe,* "State 2," is the *most probable state* then W_2 is larger than any other thermodynamic probability, including W_1, and certainly it must be true that $W_2 / W_1 > 1$. So, in that case, since $\ln(1) = 0$, Clausius's Second Law immediately follows:

$$\Delta S_{universe} = k \ln \frac{W_2}{W_1} > k \ln 1 = 0 \qquad (7.5)$$

In addition, since we can consider the system and the surroundings as two independent structures, we can say that their probabilities multiply to give the overall probability of the universe:

$$W_{universe} = W \times W_{sur}$$

And, since $\ln(xy) = \ln(x) + \ln(y)$, we have that the entropy of the universe is the *sum* of the entropies of the system and the surroundings:

$$\ln W_{universe} = \ln W + \ln W_{sur}$$
$$k \ln W_{universe} = k \ln W + k \ln W_{sur}$$
$$S_{universe} = S + S_{sur}$$

And in that case, we can write that the *change* in entropy of the universe (which must be positive according to Clausius) is the sum of the *changes* in entropy for the system and its surroundings (Equation 7.1):

$$\Delta S_{universe} = \Delta S + \Delta S_{sur} > 0$$

Thus, using the definition of entropy discovered by Boltzmann, the Second Law is seen as a natural outcome of the fact that the universe obeys the laws of probability.

7.5 Heat and Entropy Changes in the Surroundings: $\Delta S_{sur} = q_{sur}/T$

The second requirement Boltzmann needed to show in order to be taken seriously was that Clausius's definition of entropy change in terms of heat and temperature could be derived from probability. The proof of this is quite complex, but for a simple system of evenly spaced energy levels, it is rather easy to derive. The idea is similar to our demonstration that the Boltzmann distribution really is the most probable distribution in the first place (Section 2.11).[2]

Our "energy level system" in this case is really the surroundings, the place where no chemical reactions occur. This is important, because all of what is done here presumes that the energy levels of the system are fixed in their separations. When chemistry is going on, the situation is a bit more complicated. But in the surroundings, since no chemistry is happening, we just consider the effect of adding heat. The "surroundings" really are just more molecules—styrofoam, beakers, thermometers, water, and so on—so we see the surroundings as just another (rather large and complex) energy level system.

We start by imagining just two levels of this huge energy level system we call the surroundings. Consider the two states A and B, *both* of which are most probable states, before (A) and after (B) adding just the tiniest amount of heat (Figure 7.4). The temperature of B is perhaps infinitesimally higher than the temperature of A. One single particle has made the jump from Level i to the next level up, Level j, leaving Level i with one less particle and Level j with one more particle. The heat required, q, is just the difference in energy of the two levels, $\Delta \varepsilon$.

Notice that the two thermodynamic probabilities are almost exactly the same:

$$W_A = \frac{n!}{n_0! \, n_1! \, n_2! \cdots n_i! \, n_j! \cdots}$$

$$W_B = \frac{n!}{n_0! \, n_1! \, n_2! \cdots (n_i - 1)! \, (n_j + 1)! \cdots}$$

The only differences are two factorial terms in the denominator. In fact, we can rewrite W_B in terms of W_A by using the following two relationships of factorials:

$$(n - 1)! = \frac{n_i!}{n_i} \quad \text{and} \quad (n_j + 1)! = n_j! \, (n_j + 1)$$

Thus, we have

$$W_B = \frac{n!}{n_0! \, n_1! \, n_2! \cdots (n_i - 1)! \, (n_j + 1)! \cdots} = \frac{n!}{n_0! \, n_1! \, n_2! \cdots \frac{n_i!}{n_i} \, n_j! \, (n_j + 1) \cdots}$$

$$= \frac{n!}{n_0! \, n_1! \, n_2! \cdots n_i! \, n_j! \cdots} \left(\frac{n_i}{n_j + 1} \right) = W_A \left(\frac{n_i}{n_j + 1} \right)$$

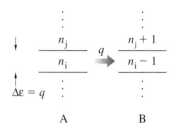

Figure 7.4
Adding heat (q) to a system changes the number of particles with specific amounts of energy.

---

[2] The mathematics of this section may be skipped, but the conclusion is going to be critically important to an understanding of how energy changes in a chemical reaction relate to probability, so at least try to understand that the conclusion is *reasonable* based on the definition of W.

The change in entropy, ΔS_{sur}, comes now from $k \ln(W_B / W_A)$, where we have

$$\frac{W_B}{W_A} = \frac{n_i}{n_j + 1} \approx \frac{n_i}{n_j}$$

We drop the "+1" in the denominator since we are dealing, after all, with chemical substances, which can be presumed to involve a huge number of particles. (This is the same trick we used in Chapter 1 when we were dealing with probability involving large populations.)

Note that this ratio, n_i / n_j, is the ratio of the population of the *lower* state over the population of the *upper* state in Distribution A, which is just the reverse of what we would write using the Boltzmann law:

$$\frac{n_j}{n_i} = e^{-\Delta \varepsilon / kT}$$

Thus, using the fact that $1/(e^{-x}) = e^x$ and recognizing that $q_{sur} = \Delta \varepsilon$, we have

$$\frac{W_B}{W_A} = \frac{n_i}{n_j} = \frac{1}{e^{-\Delta \varepsilon / kT}} = e^{\Delta \varepsilon / kT} = e^{q_{sur} / kT}$$

Finally, using the fact that $\ln(e^x) = x$, we have

$$\Delta S_{sur} = k \ln \frac{W_B}{W_A} = k \ln \left(e^{q_{sur}/kT} \right) = k \frac{q_{sur}}{kT} = \frac{q_{sur}}{T}$$

just as Clausius required!

This analysis was for the surroundings, but it works for any case where chemical reactions are not occurring. Boltzmann's proof, of course, is a bit more rigorous, but surely you get the point. He was successful in demonstrating that *his* definition of probability as (ways of x)/(ways total) and *his* definition of entropy as $k \ln W$ work out to give entropy the very same relationship to heat and temperature as discovered by Clausius.

7.6 Measuring Entropy Changes

The really interesting thing is that these theories can be checked out experimentally, although that is a tricky operation, to be sure. Using Clausius's definition, it's not too hard to imagine that one might be able to measure q_{sur}/T for some process. The trick, of course, is to make sure that as energy is added or removed, the temperature doesn't change too much. It's even possible to choose systems such as the melting of ice, where adding energy is possible with no temperature change. (Adding energy in the form of heat to ice water merely melts the ice; it doesn't increase the temperature of the mixture.)

Notice that ΔS_{sur} is proportional to how much material there is in the surroundings, because the heat absorbed is going to be related to a heat capacity (C), and the heat capacity of a material is proportional to how much material we have (Sections 4.6 and 4.12).

So measuring entropy changes generally involves the same sort of tools as measuring heat: thermometers, calorimeters, and the like. For example, say we melt 10.0 g of ice at 0°C. For now we won't worry about the entropy change of the system; let's just look at the entropy change *of the surroundings*. We need to know first how much heat is required to get this job done. This reaction absorbs heat to the tune of 6010 J/mol. Therefore, we have:

$$q = 10.0 \text{ g } H_2O \times \frac{1 \text{ mol } H_2O}{18.0 \text{ g } H_2O} \times \frac{6010 \text{ J}}{\text{mol } H_2O} = 3340 \text{ J}$$

This is the heat put into the system, coming out of the surroundings. So $q_{sur} = -3340$ J, and

$$\Delta S_{sur} = \frac{q_{sur}}{T} = \frac{-3340\,J}{273\,K} = -12.2\,J/K$$

We conclude that in melting that ice, the entropy of the surroundings decreases by 12.2 J/K. The only problem with all this is that in order to do this calculation we have to know the temperature *and it must be constant*. Of course, most of the time when we add energy in the form of heat to a substance, its temperature increases. We can get around this difficulty by making sure that we have so much substance present that adding heat to it doesn't change its temperature very much.[3] For example, if 1000 J of heat were dumped into the surroundings and the temperature increased from 273.4 K to 273.6 K, we can just say the temperature was 273.5 K, and

$$\Delta S_{sur} = \frac{1000\,J}{273.5\,K} = 3.66\,J/K$$

7.7 Standard Molar Entropy: S°

Using calorimetry it has been possible to measure the amount of heat required to raise the temperature of many substances slowly, leading to "standard" values of entropy for those substances. However, one must be careful here. Since expansions and contractions of a gas will change its energy level separations, the amount of heat required to raise the temperature of a gas depends upon its pressure. We need a "standard state" for these "standard" values. In addition, since the heat capacity of a substance is tied to the amount of substance being heated, in order to make valid comparisons, we need to specify how much substance is being tested. The state chosen to be "standard" is **1 mol of substance at 1 bar pressure.** (We'll see how entropy changes with pressure in Chapter 8, but for now let's just assume the pressure is 1 bar.)

We designate the standard state with a degree sign: S°. We'll see this "$^\circ$" applied several more times. Just remember, whenever you see the superscript "$^\circ$" it always means "for one mole at 1 bar pressure."

In addition, it's handy to compare entropies for different substances. To do that, it's important to compare them at the same temperature as well. Otherwise the comparisons are meaningless. To indicate the temperature, which is *not* part of the standard state, since that just involves pressure, you might see the temperature added as a subscript: S°_{298} is read, "the standard entropy at 298 kelvins," and means, "the entropy of 1 mole at 1 bar pressure and 298 kelvins." Values of S°_{298} determined for several substances are listed in Table 7.1. Many more are given in Appendix D. You can find even more at the National Institutes of Standards and Technology web site *(http://webbook.nist.gov/chemistry)*.

7.8 Entropy Comparisons Are Informative

It's one thing to be able to read a table; it's another thing to understand it. Let's take a closer look at the numbers in Table 7.1 and see what generalizations we can make from it. Remember, the entropy of a system will increase (a) if the number of particles increases, (b) if the temperature increases, or (c) if the energy levels are somehow made closer together. Of these, only (c) is important to this discussion. All of the numbers on this table are for a mole of substance, and they are also all at 298 K. The only difference between values

---

[3] Alternatively, we can use calculus. For simple heating, $q = C\Delta T$, which in the language of calculus becomes $dq_{sur} = C_{sur}dT$. We then write that $dS_{sur} = dq_{sur}/T = (C_{sur}dT)/T$ and

$$\Delta S_{sur} = \int dS_{sur} = \int \frac{C_{sur}}{T}dT = C_{sur}\ln\frac{T_2}{T_1}$$

Table 7.1
Standard molar entropies (J/mol·K) at 298 K (S°_{298}).

Solids		Liquids		Gases			
C(d)	2.4	$H_2O(l)$	70	He(*g*)	126	HCN(*g*)	202
C(gr)	5.7	Hg(*l*)	76	$H_2(g)$	131	$F_2(g)$	203
P(red)	23	$CH_3OH(l)$	127	HD(*g*)	143	$O_2(g)$	205
P(black)	23	$Br_2(l)$	152	$D_2(g)$	145	$PH_3(g)$	210
Fe(*s*)	27	$HNO_3(l)$	156	Ne(*g*)	146	NO(*g*)	211
Mn(*s*)	32	$H_2SO_4(l)$	157	Ar(*g*)	155	$CO_2(g)$	214
Mg(*s*)	33	$N_2O_4(l)$	209	Xe(*g*)	170	$Cl_2(g)$	223
LiF(*s*)	36	$CCl_4(l)$	216	HF(*g*)	174	$C_2H_6(g)$	230
P(white)	41			Hg(*g*)	175	$O_3(g)$	239
SiO_2(q)	42			$CH_4(g)$	186	$NO_2(g)$	240
Sn(gray)	44			HCl(*g*)	187	$Br_2(g)$	245
Sn(white)	52			$H_2O(g)$	189	$I_2(g)$	261
LiCl(*s*)	58			$N_2(g)$	192	$C_5H_{10}(g)$	293
NaCl(*s*)	72			$NH_3(g)$	193	$N_2O_4(g)$	304
KCl(*s*)	83			CO(*g*)	198	$PCl_3(g)$	312
KI(*s*)	106			HBr(*g*)	199	$PCl_5(g)$	361
$I_2(s)$	116						

must be due specifically to differences in substance, and based on Chapter 3 we know that different substances do indeed have differently spaced energy levels.

An important aspect of entropy is that just as we could separate molecular energies into different *independent* sets of energy distributions—electronic, vibrational, rotational, and translational—we can talk about the "electronic entropy" or the "vibrational entropy" in a system. This is because, since the different energy distributions are independent in this model, the thermodynamic probabilities (ways) multiply, just as for the system and the surroundings:

$$W_{\text{total}} = W_{\text{elec}} \times W_{\text{vib}} \times W_{\text{rot}} \times W_{\text{trans}} \tag{7.6}$$

And, since $\ln(xy) = \ln(x) + \ln(y)$, we have

$$S_{\text{total}} = k \ln W_{\text{total}} = k \ln W_{\text{elec}} + k \ln W_{\text{vib}} + k \ln W_{\text{rot}} + k \ln W_{\text{trans}}$$
$$S_{\text{total}} = S_{\text{elec}} + S_{\text{vib}} + S_{\text{rot}} + S_{\text{trans}} \tag{7.7}$$

This is true for any substance, be it part of the system or part of the surroundings. We'll see that this is useful in comparing one substance's entropy to another's, because it allows us to focus on the part of the entropy that is really important. For example, in going from liquid to gas we expect the entropy to increase, because the translational levels in the gas are so closely spaced, leading to an increase in both W_{trans} and S_{trans}. Several generalizations regarding entropy can now be made based on the data in Table 7.1:

Standard molar entropy increases in the series solid → liquid → gas. Although there is some overlap, from Table 7.1 we see that generally solids have less molar entropy than liquids, and liquids have less molar entropy than gases. This is a direct consequence of liquids and solids not being able to translate, and solids not being able to rotate either. Increasing the degrees of freedom of movement in a system in going from solid to liquid to gas amounts to loosening the constraints, leading to more closely spaced energy levels. Remember, *translational* energy levels are very closely spaced, but available only to gases.

Likewise, rotational levels are relatively closely spaced, but are available only to gases and liquids. Solids are restricted to vibrate only. This constraint to their motion shows up as their lower standard molar entropies.

For gases, standard molar entropy generally increases with increasing mass. For example, S_{298}° for the series He, Ne, Ar, Xe increases in that order, just as does mass. Why? Because mass m appears in the denominator of the translational energy equation:

$$\varepsilon_{n_x,n_y,n_z} = (n_x^2 + n_y^2 + n_z^2)\frac{h^2}{8}\left(\frac{1}{md^2}\right) \quad \text{(translation)}$$

Increasing m reduces every ε, thus also decreasing every $\Delta\varepsilon$, making the levels more closely spaced for more massive gases.

The atomic (noble) gases seem to have inordinately low standard molar entropies. For example, although argon (Ar) has an atomic mass of 39.95 amu and H_2O has a molecular mass of 18.016 amu, in the gaseous state, argon's entropy (155 J/K) is lower than water's (189 J/K). This has to do with the fact that even though argon is more massive, water has many more *ways* of distributing its energy. Water is more *complex* than argon. $H_2O(g)$ can vibrate, rotate, and translate, while $Ar(g)$, being monatomic, can only translate. Thus, $H_2O(g)$ ends up with a higher standard molar entropy than $Ar(g)$, despite its lower molecular mass.

Diamond has the lowest standard molar entropy. The structure of diamond is an infinite lattice of very strong C–C bonds in a tetrahedral arrangement. No translation and no rotation is allowed here. Only vibrations are allowed, and since the bonding is so strong, these vibrations have huge force constants. Remember, the force constant acts as a constraint, widening the levels:

$$\varepsilon_i = h\nu\left(i + \frac{1}{2}\right) \quad \text{where } \nu = \frac{1}{2\pi}\sqrt{\frac{k_f}{\mu}} \text{ and } i = 0, 1, 2, \ldots \quad \text{(vibration)}$$

where k_f, the force constant, is in the numerator. Diamond has few types of energy level distributions, and those that it has (vibrational) are spaced relatively far apart. Thus, diamond's standard molar entropy is especially low.

It is very important to realize that temperature is *not* a part of the standard state. (The "standard state" specifies only pressure and number of moles.) Thus, the numbers on Table 7.1 are simply the values obtained at 298 K. We know that since W increases with temperature, so, too, must entropy. In fact, a graph of S° vs. T for a variety of substances all shown together would look something like Figure 7.5. The values for S_{298}° given on our tables are simply the values for where these curves cross the vertical line at 298 K.

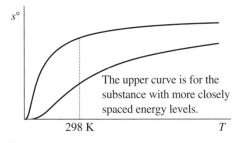

The upper curve is for the substance with more closely spaced energy levels.

298 K

T

Figure 7.5
The entropy of all substances increases with temperature, but substances with more closely spaced levels increase in entropy faster, at least at low temperature. The values given in Table 7.1 are the entropies specifically at 298 K.

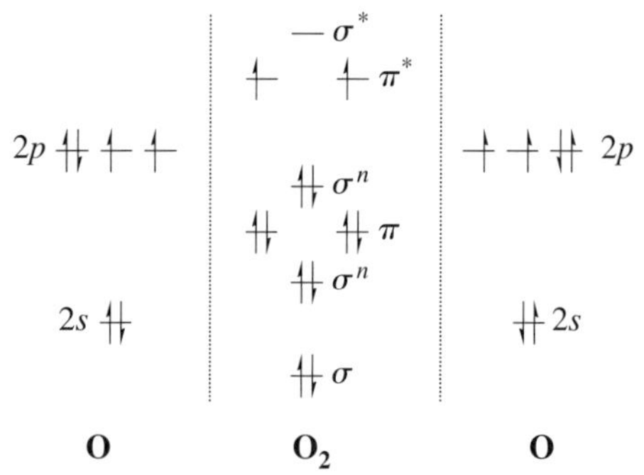

Figure 7.6
The electronic ground state of O_2 with its 12 valence electrons, two of which are in $\pi*$ orbitals and unpaired. Three equivalent-energy ground states are possible based on the possible electron spins of these two electrons: both up, both down, and one up/one down. This degeneracy leads to an increase in molar entropy for O_2.

7.9 The Effect of Ground State Electronic Degeneracy on Molar Entropy

There seems to be something wrong going down the series:

$N_2(g)$	192 J/mol·K
$O_2(g)$	205 J/mol·K
$F_2(g)$	202 J/mol·K

Shouldn't these be in order? Either something is strange about F_2 or something is strange about O_2. What's the problem? The problem is that O_2 is a very special sort of molecule. Specifically, O_2 is one of the rare nonmetals that has unpaired electrons in its ground electronic state.

You might recall that if you make an orbital diagram for O_2, which has 12 valence electrons, it will end up with two degenerate $\pi*$ (antibonding) orbitals, each containing just one unpaired electron (Figure 7.6). The central part of this diagram is a detailed view of what we refer to as the O_2 electronic ground state. In that state there is not one but *three* possible equal-energy configurations of the two unpaired electrons: both electrons spin up, both spin down, and one up and one down. (We say that O_2 is a *ground-state triplet*.)

Thus, the presence of two unpaired electrons results in a three-fold *degeneracy* (Section 2.8), where there are three distinct quantum states at the same energy. The effect on the electronic part of the entropy for O_2, S_{elec}, due to this three-fold ground state degeneracy is an *extra* $R \ln 3$ (about 9 J/mol·K) in entropy. Had O_2 *not* been a ground state triplet, its entropy would have been about 9 J/mol·K lower (196 J/mol·K), placing its entropy perfectly between N_2 (192 J/mol·K) and F_2 (202 J/mol·K).

The lesson here is that *degeneracy leads to increased entropy for a system.* To see why that is true, we'll work through the case of O_2 here. Remember that electronic states are so far apart in energy that in any real amount of O_2 only the ground state will be populated. This means that all of the n particles in the system will be in the lowest electronic state, and we have *generally*:

$$W_{elec} = \frac{N!}{N!} = 1$$

$$S_{elec} = k \ln W_{elec} = 0$$

Thus, usually there is no contribution to the molar entropy of a substance due to electronic energy. However, when there is a ground-state degeneracy, the N molecules present get split evenly among the multiple possibilities. For example, a mole of O_2 is really a mixture of 1/3 mol $O_2^{\uparrow\uparrow}$, 1/3 mol $O_2^{\downarrow\downarrow}$, and 1/3 mol $O_2^{\uparrow\downarrow}$. The effect of these additional possibilities is to increase W_{elec}, because now we have the system:

$$\overset{\displaystyle \overset{N/3}{\uparrow\uparrow}}{}\qquad \overset{\displaystyle \overset{N/3}{\downarrow\downarrow}}{}\qquad \overset{\displaystyle \overset{N/3}{\uparrow\downarrow}}{}$$

$$W_{elec} = \frac{N!}{(N/3)!\,(N/3)!\,(N/3)!}$$

and W_{elec} is *not* 1. In fact, we can show that $W_{elec} = 3^N$. The trick is to first calculate $\ln W_{elec}$ using an approximation for $\ln(N!)$ for large N called *Stirling's approximation*:[4]

$$\ln(N!) \approx N \ln(N) - N \quad \text{(Stirling's approximation)} \tag{7.8}$$

Here's a chance to practice your skills with natural logs. Specifically, we use the basic logarithmic relationships:

$$\ln(xy) = \ln(x) + \ln(y) \quad \ln(x/y) = \ln(x) - \ln(y) \quad \ln(x^a) = a \ln(x)$$

This calculation will show that for N particles in a three-fold degeneracy, $\ln W_{elec} = N \ln 3$, and thus $W_{elec} = 3^N$. The first job is to separate the different factorials into individual terms:

$$\ln W_{elec} = \ln \left(\frac{N!}{(N/3)!\,(N/3)!\,(N/3)!} \right)$$

$$= \ln N! - \ln(N/3)! - \ln(N/3)! - \ln(N/3)!$$

$$= \ln N! - 3 \ln(N/3)!$$

Next, we use Stirling's approximation to get rid of the factorials and regroup:

$$\ln W_{elec} = \ln N! - 3 \ln(N/3)!$$

$$= (N \ln N - N) - 3 \left(\frac{N}{3} \ln \frac{N}{3} - \frac{N}{3} \right)$$

$$= N \ln N - N - N \ln \frac{N}{3} + N$$

$$= N \ln N - N - N \ln N + N \ln 3 + N$$

$$= N \ln 3$$

Finally, bringing N into the logarithm as a power, we have the desired result:

$$\ln W_{elec} = N \ln 3 = \ln 3^N$$

$$W_{elec} = 3^N$$

How about that! Using $\ln W_{elec} = N \ln 3$ and letting N be a mole of particles, N_{av}, we simply multiply by k to get S_{elec} for a mole of O_2:

$$S_{elec} = k \ln W_{elec} = k(N_{av} \ln 3) = 9.1\,\text{J/K} \quad \text{(for one mole)}$$

Noting that $kN_{av} = R$ (Section 2.8), we can recast S_{elec} in a "per mole" way in terms of R:

$$S_{elec} = R \ln 3 = 9.1\,\text{J/mol·K}$$

Another way to get the same result with less trouble is to consider that each molecule is *independent*, and the number of ways a set of N molecules can be found when each

---

[4] Stirling discovered several approximations for $\ln(N!)$. His best is amazingly accurate—to one part in a million for all numbers greater than 5: $\ln(N!) \approx N \ln N - N + \ln(1 + 1/12N) + \frac{1}{2} \ln(2\pi N)$. Try to prove that one!

molecule has three possible ways of "being" is just:

$$W = 3 \times 3 \times 3 \times \cdots = 3^N$$

This method works just as well with coins as for molecules. Thus, the number of ways that 10 coins can be flipped (all heads, one tail, two tails, etc.) is 2^{10}:

$$W = 2 \times 2 \times 2 \times 2 \times 2 \times 2 \times 2 \times 2 \times 2 \times 2 = 2^{10} = 1024$$

It should be no surprise that for a two-fold degeneracy we have $W = 2^N$, and, in general, for a *p*-fold degeneracy involving N particles, we have $W = p^N$ and $S = k(N \ln p)$.

7.10 Determining the Standard Change in Entropy for a Chemical Reaction

The type of calculation Clausius performed for the change in entropy of the *surroundings*, where no chemical reactions occur, is only half the problem of calculating the entropy change of the universe, which must be positive for the process to occur:

$$\Delta S_{\text{universe}} = \Delta S_{\text{sur}} + \Delta S = \frac{q_{\text{sur}}}{T} + \Delta S > 0$$

If we are ever to make use of this most important law, we need to be able to measure or at least calculate the expected entropy change for our *system*, where the chemical reactions are occurring. Indeed, using Table 7.1 or the more extensive set of values for standard molar entropy given in Appendix D we can do just that, at least for the standard condition of 1 bar pressure and at the specified temperature of 298 K.[5]

Basically, since entropy is a state function, in the context of a chemical reaction, we can write the *standard reaction entropy* as

$$\text{reactants} \longrightarrow \text{products} \qquad \Delta_r S^\circ = S^\circ_{\text{products}} - S^\circ_{\text{reactants}}$$

Of course, we have to be careful to take into account the number of moles of substance in a balanced equation for the reaction. So, for example, for the reaction of nitrogen dioxide to form dinitrogen tetroxide, written

$$2\,NO_2(g) \longrightarrow N_2O_4(g) \qquad \Delta_r S^\circ = ?$$

we calculate

$$\Delta_r S^\circ = (1)S^\circ_{N_2O_4} - (2)S^\circ_{NO_2}$$

$$= (1)\frac{304\,\text{J}}{\text{mol} \cdot \text{K}} - (2)\frac{241\,\text{J}}{\text{mol} \cdot \text{K}} = -178\,\text{J/mol} \cdot \text{K}$$

(Technically, the numbers in parentheses here mean that many moles of reactant or product "per mole of reaction.") If we were to carry out this reaction starting with two moles of NO_2 at 1 bar pressure and ending with one mole of N_2O_4 also at 1 bar pressure (so, apparently, there would have to be a compression of the volume of the container), we expect a decrease in entropy on the order of 178 J/K.

In this particular case, we know that a bond is being made. Heat will be released (Section 4.8). The entropy of the surroundings will *increase*, since the heat coming out

---

[5] We will always consider these values to be close enough for any temperature around 298 K, and in Chapter 8 we will see how to adjust for pressure. For now we will just assume that any substance involved is at a partial pressure of 1 bar.

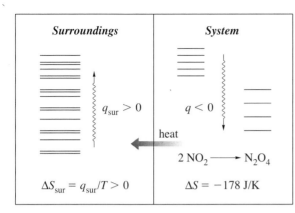

Figure 7.7
We know that for this reaction we are going to less closely spaced levels (or at the very least, fewer particles, either of which will decrease the entropy of the system). In addition, due to bond formation, heat will be released into the surroundings, and even though the entropy of the system decreases, the entropy of the surroundings must increase. Depending upon the temperature, the reaction may or may not happen. However, we can be quite certain that at a low enough temperature, we can get this reaction to be favorable, because as T decreases, q_{sur}/T will get larger and larger and become more and more important.

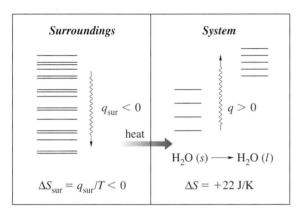

Figure 7.8
We know that for the melting of ice we are going to more closely spaced levels. The entropy of the system will increase. In addition, due to bond breaking, heat will be absorbed from the surroundings. Thus, the entropy of the surroundings must decrease. Below 0°C melting of ice will not happen, because the entropy change in the surroundings due to the reaction, q_{sur}/T, will be too negative.

of the system is going to have to go into the surroundings (Figure 7.7). But how much will the entropy of the surroundings increase? This depends upon the temperature. At low temperature, q_{sur}/T will be very positive (which is good); at high temperature, q_{sur}/T will be closer to 0. We expect that this reaction will be favorable only at low temperature! Once again we conclude that for a reaction which releases energy to be favorable, the temperature should be low. As another example, let's look again at the melting of ice:

$$H_2O(s) \longrightarrow H_2O(l)$$

From Table 7.1 we find $S_{298}^{\circ}[H_2O(l)] = 70$ J/mol·K. Although not given in the tables because solid water (ice) is not present at 298 K, the entropy for ice at 0°C, $S_{273}^{\circ}[H_2O(s)]$, is known to be 48 J/mol·K. Assuming there isn't very much of a difference between 273 K and 298 K (this turns out to be a reasonably good assumption), we can calculate the change in entropy at 298 K to be

$$\Delta_r S^{\circ} = (1)S_{H_2O(l)}^{\circ} - (1)S_{H_2O(s)}^{\circ}$$

$$= (1)\frac{70\,J}{mol \cdot K} - (1)\frac{48\,J}{mol \cdot K} = 22\,J/mol \cdot K$$

So melting ice increases its entropy. That's easy to believe. After all, liquid water has much less "constraint" than solid water, and its molecules can rotate, while those in solid ice cannot. Once again, we rely on our intuition regarding the heat involved in this reaction. Surely heat is going to be required to break those hydrogen bonds in ice and turn a mole of solid water into liquid water. Thus, here we have a situation something like that depicted in Figure 7.8.

Table 7.2
Calculation of $\Delta S_{universe}$ for the melting of water at different temperatures.

$T/^\circ C$	T/kelvin	$\Delta S/(J/K)$	$\Delta S_{sur}/(J/K)$	$\Delta S_{universe}/(J/K)$
-20	253	22	-24	-2
-10	263	22	-23	-1
0	273	22	-22	0
10	283	22	-21	1
30	303	22	-20	2

To a good approximation, the amount of heat required to melt a mole of water is 6010 J.[6] From this information, we can calculate the change in entropy of the universe at various temperatures around the melting point of water using $\Delta S_{sur} = q_{sur}/T$ (Table 7.2). A temperature of at least 0°C is required to melt ice. Above 0°C (the last two lines of the table), any amount of solid ice will melt, because as it melts, its entropy increases more than the entropy of the surroundings decreases, and $\Delta S_{universe} > 0$.

Below 0°C (first two lines), the decrease in entropy of the surroundings, ΔS_{sur}, is just too negative, and the Second Law is violated. In fact, below 0°C, the *reverse* reaction

$$H_2O(l) \longrightarrow H_2O(s)$$

will be the one with the positive $\Delta S_{universe}$, and any liquid present will freeze.

Most interesting, at exactly 0°C the loss in entropy of the surroundings just matches the gain in entropy of the system. This means that at this temperature any amount of liquid and solid water can be present. Putting heat into the system at 0°C does *not* increase the temperature. It simply converts solid ice into liquid!

7.11 Another Way to Look at ΔS

The change in entropy of the system, ΔS, measures the factor by which the system W is going to increase when a reaction occurs. But another way to look at what ΔS is all about is to look at the units of entropy, which are J/K. What does it mean that these are the units of "energy over temperature"? The answer is very informative: A value of "22 J/K" for ΔS, for example, means that *in order for this reaction to be spontaneous, a maximum of 22 J of heat may be drawn from the surroundings for every degree kelvin of temperature.*

So, for example, for the melting of 18 grams of ice ($\Delta S = 22$ J/K), at the very cold temperature of 1 K, only 22 J of heat may be withdrawn from the surroundings before the entropy of the surroundings becomes too negative. Well, that means the reaction will not happen at 1 K, because 6010 J of heat will be required to melt that amount of ice.

At 27°C (300 K), on the other hand, (22 J/K)(300 K) = 6600 J of heat can be removed from the surroundings and the reaction will still occur. Since the melting of a mole of ice requires only 6010 J of heat, melting of ice will occur at 27°C.

Thus, the units "J/K" are actually meaningful, and help us see that there is a connection between energy changes and entropy changes in a chemical reaction. The connection to energy is that energy changes in the system involve exchanges of heat with the surroundings, and that invariably changes the entropy of those surroundings.

---

[6] We are ignoring heat capacity, which is much higher for liquid water than for ice. This means that at higher temperatures, more heat is required to melt ice than at lower temperatures. There is a similar effect on ΔS. However, *together* these effects almost cancel out. It can be shown with some difficulty that the error in $\Delta S_{universe}$ due to ignoring heat capacity is less than 10%, which is well below the precision of these numbers.

7.12 Summary

We now know that energy in molecules is quantized and that there are four independent ways that energy can be organized: translational, rotational, vibrational, and electronic. We also know that energy will distribute itself automatically into the most probable distribution, and that the Boltzmann law governs what that is going to look like in terms of how many particles are doing what. The Second Law of Thermodynamics, coined by Clausius, states that for any *real* change, the entropy of the universe, which can be thought of as consisting of the system and the surroundings, must increase:

$$\Delta S_{universe} = \Delta S + \Delta S_{sur} > 0$$

Importantly, these two component changes are calculated two different ways. For the surroundings, where no chemical reactions are presumed to occur, we find that

$$\Delta S_{sur} = \frac{q_{sur}}{T}$$

And for the system, where chemical reactions are occurring, given a specific writing of the balanced chemical equation for the reaction, we calculate the *standard reaction entropy*:

$$\text{reactants} \longrightarrow \text{products} \qquad \Delta_r S^\circ = S^\circ_{products} - S^\circ_{reactants}$$

So far we know how to calculate only the *standard* change in entropy for a reaction. This requires the rather unusable requirement that all substances, particularly gases, be maintained at 1 bar partial pressure. We'll resolve this issue next, in Chapter 8.

Chapter 9 will then focus on the estimation of q_{sur} based on another new state function called *enthalpy*, H. It will turn out that ΔH for a reaction amounts to the heat required along the path of constant overall pressure, which is a pretty common way of doing things.

In the language of Boltzmann, we have

$$S = k \ln W$$

where W is the thermodynamic probability for the most probable distribution, and therefore

$$\Delta S = S_2 - S_1 = k \ln W_2 - k \ln W_1 = k \ln \frac{W_2}{W_1}$$

Since $\ln(1) = 0$, the change in entropy calculated this way is greater than 0 whenever W_2/W_1 is greater than 1, or $W_2 > W_1$. So saying that "entropy must increase" ($\Delta S > 0$) is simply another way of saying that "a change will always occur in the direction of the more probable distribution" ($W_2 > W_1$).

Whether or not a reaction "pays off" depends solely upon whether the overall entropy of the universe increases. In the end, all of chemistry is based on probability, and in the long run, the universal casino always wins.

Problems

The symbol Ⓐ indicates that the answer to the problem can be found in the "Answers to Selected Exercises" section at the back of the book.

The Definition of Entropy

7.1 *In*tensive properties, such as temperature, are those that are independent of the amount of substance present. *Ex*tensive properties, such as entropy, are those that do depend on the amount of substance present. List three intensive and three extensive properties.

7.2 Ⓐ Calculate U, W, and S for the three states shown below. One quantum of energy in this case equals 1.0 aJ (*atto* means 10^{-18}).

Quanta	State A	State B	State C
3	_____	_____	•
2	_____	_____	•
1	_____	•••	•••
0	••••••••	•••••	•••

7.3 For the three systems described below, (a) sketch each system, (b) determine U, W, and S. Recall that 1 aJ = 1×10^{-18} J. System A has energy levels spaced 5 aJ apart with the ground state at 0 aJ. In this system, 5 particles are in the lowest level, 4 particles are in the first excited state, 3 particles are in the second excited state, 2 particles are in the third excited state, and 1 particle is in the fourth excited state. System B has energy levels spaced 5 aJ apart with the ground state at 0 aJ. In this system, 10 particles are in the lowest level, 4 particles are in the first excited state, and 1 particle is in the second excited state. System C has energy levels spaced 5 aJ apart with the ground state at 0 aJ. In this system, 13 particles are in the lowest level and 2 particles are in the first excited state.

7.4 For the three systems described below, (a) sketch each system, (b) determine U, W, and S. Recall that 1 aJ = 1×10^{-18} J. System A has energy levels spaced 2 aJ apart with the ground state at 0 aJ. In this system, 5 particles are in the lowest level, 4 particles are in the first excited state, 3 particles are in the second excited state, 2 particles are in the third excited state and one particle is in the fourth excited state. System B has energy levels spaced 2 aJ apart with the ground state at 0 aJ. In this system, 10 particles are in the lowest level, 4 particles are in the first excited state and 1 particle is in the second excited state. System C has energy levels spaced 2 aJ apart with the ground state at 0 aJ. In this system, 13 particles are in the lowest level and 2 particles are in the first excited state.

7.5 For the three systems described below, (a) sketch each system, (b) determine U, W, and S. Recall that 1 aJ = 1×10^{-18} J. System A has energy levels spaced 2 aJ apart with the ground state at 0 aJ. In this system, 11 particles are in the lowest level and 4 particles are in the fifth excited state. System B has energy levels spaced 2 aJ apart with the ground state at 0 aJ. In this

system, 12 particles are in the lowest level and 3 particles are in the second excited state. System C has energy levels spaced 2 aJ apart with the ground state at 0 aJ. In this system, 14 particles are in the lowest level and 1 particle is in the second excited state.

Standard Entropies

7.6 Ⓐ In each case below, identify the substance that should have the higher entropy at 298 K. Explain your reasoning in terms of differences in the effect of increasing temperature on the translational, rotational, and vibrational energy level systems for the two indicated substances.
  (a) $SiBr_4(l)$ vs. $SiBr_4(g)$
  (b) $GeH_4(g)$ vs. $Kr(g)$
  (c) $H_2S(g)$ vs. $H_2Se(g)$

7.7 In each case below, identify the substance that should have the higher entropy at 298 K. Explain your reasoning in terms of differences in the effect of increasing temperature on the translational, rotational, and vibrational energy level systems for the two indicated substances.
  (a) $CH_4(g)$ vs. $SiH_4(g)$
  (b) $Br_2(l)$ vs. $Br_2(g)$
  (c) $SiH_4(g)$ vs. $Ar(g)$

7.8 Ⓐ In each case below, identify the substance that should have the higher entropy at 298 K. Explain your reasoning in terms of differences in the effect of increasing temperature on the translational, rotational, and vibrational energy level systems for the two indicated substances.
  (a) $BaCl_2(s)$ vs. $CaCl_2(s)$
  (b) $CCl_4(g)$ vs. $CCl_4(l)$
  (c) $CO(g)$ vs. $CO_2(g)$

7.9 In each case below, identify the substance that should have the higher entropy at 298 K. Explain your reasoning in terms of differences in the effect of increasing temperature on the translational, rotational, and vibrational energy level systems for the two indicated substances.
  (a) $HF(g)$ vs. $HBr(g)$
  (b) $I_2(g)$ vs. $I_2(s)$
  (c) $FeCl_2(s)$ vs. $FeCl_3(s)$

7.10 Why does S_{298}° for $H_2O(s)$ not appear in Table 7.1?

7.11 Ⓐ Rationalize the general trend in standard molar entropy when comparing solid, liquid, and gas for the same substance. Give three examples from Table 7.1.

7.12 What is the general trend in entropy when comparing different gases? Explain one exception to this trend.

7.13 Ⓐ Compare S_{298}° for the three forms of solid phosphorus that appear in Table 7.1. What conclusions can you draw about relative spacings between energy levels and the arrangement of the atoms in each of these phases?

7.14 Compare S_{298}° for the two forms of solid mercury sulfide that appear in Appendix D. What conclusions can you draw about relative spacings between energy levels and the arrangement of the atoms in each of these phases?

7.15 Compare S_{298}° for the two forms of solid tin that appear in Table 7.1. What conclusions can you draw about relative spacings between energy levels and the arrangement of the atoms in each of these phases?

7.16 Ⓐ Compare S_{298}° for the two forms of lead oxide that appear in Appendix D. What conclusions can you draw about relative spacings between energy levels and the arrangement of the atoms in each of these phases?

7.17 Find S_{298}° for gaseous $SiCl_4$ in the NIST Chemistry WebBook[7] and compare it with S_{298}° for gaseous carbon tetrachloride (CCl_4). Explain this trend in S_{298}° as you move down a column of the periodic table based on your knowledge of translational energy.

7.18 Ⓐ Find S_{298}° for gaseous CF_4 in the NIST Chemistry WebBook[7] and compare it with S_{298}° for gaseous silicon tetrafluoride (SiF_4). Explain this trend in S_{298}° as you move down a column of the periodic table based on your knowledge of translational energy.

7.19 Find S_{298}° for gaseous CH_4 and compare it with S_{298}° for gaseous silane (SiH_4). Explain this trend in S_{298}° as you move down a column of the periodic table based on your knowledge of translational energy.

7.20 Using Stirling's approximation show that for a two-fold degeneracy, $S = R \ln 2$.

Reaction Entropies

7.21 Ⓐ For each of the two changes in state, A to B and B to C, in Problem 7.2: (a) Calculate q and ΔS. (b) Rationalize the sign of ΔS based on what happens to W. (c) Rationalize the sign of ΔS based on the sign of q.

7.22 For each of the two changes in state, A to B and B to C, in Problem 7.3: (a) Calculate q and ΔS. (b) Rationalize the sign of ΔS based on what happens to W. (c) Rationalize the sign of ΔS based on the sign of q.

7.23 Ⓐ For each of the two changes in state, A to B and B to C, in Problem 7.4: (a) Calculate q and ΔS. (b) Rationalize the sign of ΔS based on what happens to W. (c) Rationalize the sign of ΔS based on the sign of q.

7.24 For each of the two changes in state, A to B and B to C, in Problem 7.5: (a) Calculate q and ΔS. (b) Rationalize the sign of ΔS based on what happens to W. (c) Rationalize the sign of ΔS based on the sign of q.

7.25 Ⓐ Based on your knowledge of electronic, vibrational, rotational, and translational energy, predict the value of $\Delta_r S^\circ$ (*very positive*, *very negative*, or *small*) for the following phase changes:

    (a) $I_2(s) \longrightarrow I_2(g)$

    (b) $Br_2(g) \longrightarrow Br_2(l)$

    (c) $Hg(l) \longrightarrow Hg(g)$

7.26 Based on your knowledge of electronic, vibrational, rotational, and translational energy, predict the value of $\Delta_r S^\circ$ (*very positive*, *very negative*, or *small*) for the following dissociation reactions, as written:

    (a) $H_2(g) + D_2(g) \longrightarrow 2\ HD(g)$

    (b) $2\ H_2O(g) \longrightarrow 2\ H_2(g) + O_2(g)$

    (c) $2\ HCl(g) \longrightarrow H_2(g) + Cl_2(g)$

7.27 Ⓐ For the following three dissociation reactions, calculate $\Delta_r S_{298}^\circ$ using Table 7.1 and rationalize its value (*very positive*, *very negative*, or *small*):

    (a) $H_2(g) + D_2(g) \longrightarrow 2\ HD(g)$

    (b) $2\ H_2O(g) \longrightarrow 2\ H_2(g) + O_2(g)$

    (c) $2\ HCl(g) \longrightarrow H_2(g) + Cl_2(g)$

7.28 For the following four phase changes, as written, calculate $\Delta_r S_{298}^\circ$ using Table 7.1 and rationalize its value (*very positive*, *very negative*, or *small*) based on your knowledge of electronic, vibrational, rotational, and translational energy:

    (a) $I_2(s) \longrightarrow I_2(g)$

    (b) $Br_2(g) \longrightarrow Br_2(l)$

    (c) $Hg(l) \longrightarrow Hg(g)$

    (d) $P(red) \longrightarrow P(black)$

Brain Teasers

7.29 The actual heat required to melt water and the entropy change of the system upon melting at different temperatures, taking into consideration heat capacity (C), are

$$q = q(0) + \Delta C \delta T$$

$$\Delta S = \Delta S(0) + \Delta C \ln \frac{T}{273}$$

where $q(0)$ is 6010 J/mol, $\Delta S(0)$ is 22 J/mol·K, ΔC is the difference in heat capacity between solid and liquid water (37.7 J/mol·K), and δT is $T - 273$ K. Show that the actual change in entropy of the universe upon melting is

$$\Delta S_{universe} = \frac{\delta T}{T} \Delta S(0) + \Delta C \left(\ln \frac{T}{273} - \frac{\delta T}{T} \right)$$

$$\approx \frac{\delta T}{T} \Delta S(0) + \frac{\Delta C}{2} \left(\frac{\delta T}{T} \right)^2$$

Also show that with or without including heat capacity, $\Delta S_{universe}$ is positive above 273 K and negative below 273 K and that the error for ignoring heat capacity is less than 10% in the temperature range $0 \pm 20°C$.

---

[7] *http://webbook.nist.gov/chemistry*. To look up a compound, search for its molecular formula, then request "thermodynamic data—gas phase".

CHAPTER 8

The Effect of Pressure and Concentration on Entropy

[A law] is more impressive the greater the simplicity of its premises, the more different are the kinds of things it relates, and the more extended its range of applicability. Therefore, the deep impression which classical thermodynamics made on me. It is the only physical theory of universal content, which I am convinced, that within the framework of applicability of its basic concepts will never be overthrown.

—*Albert Einstein, quoted in M. J. Klein, "Thermodynamics in Einstein's Universe," in* Science, *157 (1967), p. 509.*

8.1 Introduction

In Chapter 7 it was argued that the following two statements are equivalent:

1. All processes progress toward the most probable distribution of energy.
2. For a process to be observed, the entropy of the universe must increase.

Furthermore, and very importantly, we dissected the universe into two independent subsets: system and surroundings. The system is where the chemistry occurs; the surroundings is everything else. In mathematical terms, we write the Second Law of Thermodynamics:

$$\Delta S_{universe} = \Delta S + \Delta S_{sur} > 0$$

The change in entropy of the surroundings, ΔS_{sur}, can be determined based on the fact that it is just being heated or cooled by the amount q_{sur}:

$$\Delta S_{sur} = \frac{q_{sur}}{T}$$

The change in entropy of the system, ΔS, on the other hand, is a bit trickier. We saw how from a table of standard molar entropies we could at least calculate the *standard reaction entropy*, $\Delta_r S°$, as

$$\text{reactants} \longrightarrow \text{products} \qquad \Delta_r S° = S°_{products} - S°_{reactants}$$

This requires first writing a balanced chemical equation for the reaction, looking up the standard molar entropies on a table, and taking into consideration how many moles of reactants and products are in the chemical equation "per mole of reaction" (the *stoichiometric amounts*).

However, most reactions are not carried out under the standard condition of 1 bar pressure, and we must somehow determine how to account for nonstandard pressures. In this chapter we find that the correction to standard molar entropy for a gas "X" is simply

$$S_x = S_x^\circ - R \ln(P_x/\text{bar}) \tag{8.1}$$

where P_x is the partial pressure of gas X. The "/bar" here means that the pressure must be in bars, and, prior to taking the natural logarithm, we are removing the units. (No units of any kind are ever allowed in logarithmic or exponential terms.)

So we will see that the entropy of a gas decreases with increasing pressure. This should sound reasonable. As we compress a gas, we push it into a smaller and smaller volume, constraining it more and more. Constraining a system always leads to some sort of increase in energy level spacings, and that always leads to a decrease in thermodynamic probability W, and likewise entropy S, which equals $k \ln W$.

In addition, we will see that for substances in solution (for example, dissolved in water), the standard state is defined as 1 M concentration, and the correction necessary amounts to

$$S_x = S_x^\circ - R \ln[\text{X}]/M \tag{8.2}$$

Finally, we will find that when we take into consideration all of the corrections to standard pressure and concentration in a chemical reaction, we can write that overall

$$\Delta_r S = \Delta_r S^\circ - R \ln Q \tag{8.3}$$

where the little subscript "r" indicates that we are indicating a value that is "per mole of reaction" and where Q is the "reaction quotient" (Section 1.12) for a particular writing of the balanced chemical equation representing the reaction.

Although several other ideas are presented in this chapter, these three equations form the basis of all we will need regarding entropy for the rest of our discussion of thermodynamics. In order to understand where these equations come from, we focus not on energy but on probability. Our analysis begins again with thermodynamic probability, W, which, remember, is simply a count of all of the "number of ways" in which a system can be found. The argument we use was developed by Albert Einstein, who used it in his landmark paper of 1905 to calculate the entropy of radiation.

8.2 *Impossible?* or Just *Improbable?*

Although entropy may seem somewhat of a nebulous concept, you can see entropy at work all around you, if you know where to look for it. Here's a little experiment you can do anywhere: Fill a glass with water, add a drop of red food coloring, and stir gently. What do you expect to see? Will the drop of red just drop to the bottom like a stone? Will it float on the top like a sesame seed? No, of course not. The red color of the dye in the food coloring will be seen to spread uniformly throughout the solution. Why did it do this?

Now imagine doing it backward. You are sitting, looking at a glass of cranberry juice, when, suddenly, all the red coalesces into one tiny spot and the rest of the liquid is colorless. What would you do? Is it a dream? Surely, this *though possible*, is an extremely improbable event.

Imagine you've graduated and have a fine job and a house in the suburbs with a swimming pool in the back yard. You're enjoying a nice summer's day at the edge of your pool. Suddenly all the water jumps out and flies over your fence into your neighbor's yard. Impossible? Is it? Or is it just *improbable*? Why do you think so?

Just as with the swimming pool water, we would never expect to shuffle a deck and end up with 26 red cards first and 26 black cards next. The probability must be incredibly low and must be related to entropy. How can we measure how much more disordered a well shuffled deck is than one with 26 red cards on top and 26 black cards on the bottom? The number of ways, W_2, of the deck of cards being ordered like this, although large, is much, much less than the number of ways the deck could be arranged in all (W_1).

In going from the state of "randomly distributed cards" with W_1 ways to the state of "26 red cards on top and 26 black cards on the bottom" with W_2 ways, we are going from the most probable state to an extremely improbable state, and $W_2 \ll W_1$. If we use Boltzmann's definition of entropy as $k \ln W$, then the change in entropy,

$$\Delta S = k \ln \frac{W_2}{W_1}$$

will be negative, since $W_2/W_1 \ll 1$. *Spontaneous formation of a deck of 26 red cards on top and 26 black cards on the bottom by random shuffling in effect violates the Second Law of Thermodynamics!*

Similarly, in Chapter 1 we focused on the isotope exchange reaction between H_2 and D_2:

$$H_2 + D_2 \rightleftharpoons 2\,HD$$

Although there is a tiny energy component to this equilibrium, this exchange seems to have essentially nothing to do with energy (except in the sense that we need to have *some* energy to get the reaction going). All we are doing here is shuffling isotopes. There are many other situations in chemistry such as these where the difference in energy does not seem to be the issue. We must once again find a way to apply the probability involved in shuffling cards to chemical systems.

8.3 Ideal Gases and Ideal Solutions

To see how changes in entropy relate to chemical systems, we'll work exclusively with ideal gases and ideal solutions. The word *ideal* here simply means that all of the energy of the system can be thought of as based in the molecules of gas or solute themselves, not in the relationship *between* those molecules. Ideal gases behave according to the *ideal-gas equation*:

$$PV = nRT$$

Note that the units on both sides of this equation are the units of energy. In an ideal gas, spreading the molecules apart or pushing them together does not affect the energy of the system as long as no walls move and no work is done. Thus, for an ideal gas, expansion from "P_1V_1" to "P_2V_2" involves no change of temperature as long as no work is done on the surroundings. The way this is accomplished is to expand the gas *into a vacuum* (Figure 8.1). Since no walls move in this sort of expansion, no work is done. Both the temperature and the internal energy of the system remain constant.

Real gases can be slightly more complicated. In any real gas, such as H_2, there are tiny forces of attraction between the molecules, even though they are relatively far apart, on average.[1] These interactions become more important for gases that are under great pressure or when the total energy of the system is very low (at low temperature). In all the cases we will be interested in, the ideal-gas equation will be a very good approximation.

Similarly, in an ideal solution, making the solution more dilute by adding solvent (water) has no effect on the energy of the *solute* itself. This approximation is somewhat less valid than the assumption that a gas is ideal. Even 1 *M* solutions can be significantly nonideal, especially if ions (which strongly attract and repel each other) are involved. Real solutions have interesting properties such as the fact that due to those interactions, adding

---

[1] If you were to imagine yourself a CO_2 molecule, with your head the C and your two hands at arm's length the O atoms, then at room temperature and one bar pressure, your nearest neighbor would be about 25 meters away, on average.

Figure 8.1
On the left (Flask A) is a flask containing the ideal gas xenon. On the right is a total vacuum (Flask B), totally void of molecules of any type. In between there is a stopcock that is closed but, when opened, allows passage of gas both ways. The volume of Flask A is V_A, and the volume of Flask B is V_B, for a total combined volume of $V_A + V_B$. The temperature is constant at temperature T.

A   B

10.00 mL of one solution to 10.00 mL of another does *not* result in 20.00 mL of solution in the end. Nonetheless, for our purposes we will just assume that when two solutions are mixed or water is added to a solution to dilute it, no heat is required, no heat is released, and the final volume is the sum of the two original volumes.

8.4 The Volume Effect on Entropy: $\Delta S = nR \ln(V_2/V_1)$

Consider the setup in Figure 8.1. We have a flask on the left ("Flask A") with some xenon gas in it and a flask on the right with absolutely nothing in it ("Flask B" is a total vacuum). The flask with the xenon in it has a volume V_A, and the flask without has a volume V_B, for a total system volume of $V_A + V_B$. There's a stopcock between the flasks that is closed. Let's calculate what the probability is of the xenon *remaining in Flask A* when we open the stopcock.

We can easily argue that for *each particle* the probabilities of being in Flask A or Flask B once the stopcock is opened and equilibrium has been reached are

$$\text{Prob}_A = \frac{V_A}{V_A + V_B} \qquad \text{Prob}_B = \frac{V_B}{V_A + V_B}$$

Clearly these are the only probabilities. $\text{Prob}_A + \text{Prob}_B = 1$, and, besides, where else could the particles be?

Now, there is not just one particle of xenon in Flask A. The probabilities calculated above are for *each* particle. We want to find the probability now of *all* the particles being in Flask A (which has, say N particles in it). That is, we want to know the probability of the first particle being in Flask A *and* the probability of the second particle being in Flask A *and* the probability of the third particle being in Flask A *and* . . . . Clearly we have to multiply these probabilities to get the total probability that *all* the particles will be found on the left. We then have

$$\text{Prob}_{\text{all in A}} = \left(\text{Prob}_A\right)_1 \times \left(\text{Prob}_A\right)_2 \times \left(\text{Prob}_A\right)_3 \times \cdots \times \left(\text{Prob}_A\right)_N$$

$$= \left(\text{Prob}_A\right)^N$$

$$= \left(\frac{V_A}{V_A + V_B}\right)^N$$

For example, if there were a mole of xenon particles in Flask A, and the volumes of Flasks A and B were both 1 L, we would have a probability of all the particles being found in Flask A *at the same time* as being:

$$\text{Prob}_{\text{all in A}} = \left(\frac{V_A}{V_A + V_B}\right)^{N_{\text{av}}} = \left(\frac{1}{2}\right)^{6.022 \times 10^{23}}$$

This is truly an unimaginably small number. Now consider the following two states:

State 1: n moles of xenon atoms all in Flask A, volume $V_1 = V_A$
State 2: n moles of xenon atoms distributed evenly throughout volume $V_2 = V_A + V_B$

State 1 is the state just after we open the stopcock in Figure 8.1 but before any diffusion occurs. State 2 is what we expect to find in the end. We want to determine the entropy change in going from State 1 to State 2. For that we will need the relative probabilities of these two states, $\text{Prob}_2 / \text{Prob}_1$, because that ratio is equivalent to the ratio of thermodynamic probabilities we use in calculating entropy changes:

$$\frac{W_2}{W_1} = \frac{\text{Prob}_2}{\text{Prob}_1}$$

We know that the probability of State 1 is the probability we have just determined, this time for nN_{av} particles of xenon. What about State 2? As we saw in Chapter 1, once we go to systems having a large number of particles (and a mole is *certainly* a large number) the fluctuations become far too small to observe. In this case, that means that State 2, the expected observation, has, effectively, a probability of one. It isn't really one, but the probability of finding this state along with all the little fluctuations we would call the same observation, is close enough to one that we can call it that. We write

$$\frac{W_2}{W_1} = \frac{\text{Prob}_2}{\text{Prob}_1} = \frac{1}{\text{Prob}_{\text{all in A}}} = \left(\frac{V_A + V_B}{V_A}\right)^{nN_{av}} = \left(\frac{V_2}{V_1}\right)^{nN_{av}}$$

And the change in entropy in going from State 1 to State 2 is

$$\Delta S = k \ln \frac{W_2}{W_1} = k \ln \left[\left(\frac{V_2}{V_1}\right)^{nN_{av}}\right] = k n N_{av} \ln \frac{V_2}{V_1} = n R \ln \frac{V_2}{V_1}$$

Here we use again the fact that the ideal-gas constant, $k N_{av}$, equals $R = 8.31451 \, \text{J/mol} \cdot \text{K}$. Thus, for example, if Flask A were 2.0 L and Flask B were 3.0 L, and initially there were 0.50 mol of xenon in Flask A, then the total volume would be 5.0 L, and the change in entropy upon opening the stopcock and allowing diffusion would be

$$\Delta S = n R \ln \frac{V_2}{V_1} = 0.50 \, \text{mol} \times \frac{8.31451 \, \text{J}}{\text{mol} \cdot \text{K}} \ln \left(\frac{5.0 \, \text{L}}{2.0 \, \text{L}}\right) = 3.8 \, \text{J/K}$$

Notice that V_2 is the final volume, which includes *both* flasks. Note also that ΔS here is extensive—that is, it is proportional to the number of moles of particles present. Diffusion from a smaller volume into a larger volume is always predicted to be favorable ($\Delta S > 0$), because it is in the direction leading to the most probable distribution.

In short, we have for any such process, the change in entropy for n moles of particles diffusing from one volume, V_1, into a new total volume, V_2, is

$$\Delta S = n R \ln \frac{V_2}{V_1} \tag{8.4}$$

8.5 The Entropy of Mixing Is Just the Entropy of Expansion

It is interesting to consider the case where there is some *other* gas, such as argon, in Flask B to start with. When the stopcock is opened, the two gases mix thoroughly. The total entropy change would just be the sum of the entropy change for each gas independently, one going from $V_1 = V_A$ and the other going from $V_1 = V_B$, both into a total volume $V_2 = V_A + V_B$.

If there were n_A moles of gas in Flask A to begin with and n_B moles of gas in Flask B to begin with, we would have

$$\Delta S = \Delta S_A + \Delta S_B = n_A R \ln \frac{V_A + V_B}{V_A} + n_B R \ln \frac{V_A + V_B}{V_B} \qquad (8.5)$$

That is, for more than one gas, you simply calculate the change in entropy for each gas as though it were expanding into a vacuum and then add all the entropy changes.

For the **very special case** when the pressure is the same in both flasks, the entropy change is sometimes called the *entropy of mixing*. There is nothing magical about mixing. The entropy of mixing is simply the sum of the entropy changes for each component that is being mixed. However, in this case **where the pressure is the same in both flasks**, the entropy change on opening the stopcock can be recast in terms of *mole fractions*, x:

$$x_A = \frac{n_A}{n} \qquad (8.6)$$

where n is the total number of particles, $n_A + n_B$. For example, if there were 1 mol of A and 2 mol of B, then the mole fraction of A, x_A, would be 1/3, and the mole fraction of B, x_B, would be 2/3.

Due to the fact that the pressure is the same in both flasks, we can apply the ideal-gas equation to show that volume ratios are the same as mole ratios:

$$\frac{PV_A}{PV_B} = \frac{n_A RT}{n_B RT} \quad \text{so} \quad \frac{V_A}{V_B} = \frac{n_A}{n_B}$$

(The trick here is that we are dividing the left side and the right side of "$PV_A = n_A RT$" by the same quantity, since $PV_B = n_B RT$.) This allows us to convert the two volume ratios found in Equation 8.5 to $1/x_A$ and $1/x_B$:

$$\frac{V_A + V_B}{V_A} = 1 + \frac{V_B}{V_A} = 1 + \frac{n_B}{n_A} = \frac{n_A + n_B}{n_A} = \frac{n}{n_A} = \frac{1}{x_A}$$

$$\frac{V_A + V_B}{V_B} = \frac{V_A}{V_B} + 1 = \frac{n_A}{n_B} + 1 = \frac{n_A + n_B}{n_B} = \frac{n}{n_B} = \frac{1}{x_B}$$

Finally, substituting back into Equation 8.5 and letting n be the total number of molecules being mixed, we have

$$\Delta S_{\text{mix}} = n_A R \ln \frac{1}{x_A} + n_B R \ln \frac{1}{x_B}$$

$$= x_A n R \ln \frac{1}{x_A} + x_B n R \ln \frac{1}{x_B}$$

which becomes

$$\Delta S_{\text{mix}} = n R \left[x_A \ln \frac{1}{x_A} + x_B \ln \frac{1}{x_B} \right] \qquad (8.7)$$

This equation is the general expression for the entropy change when two gases **initially at the same partial pressure** are mixed. For example, if 1 mol of argon at 1 bar pressure is mixed with 2 mol of neon also at 1 bar pressure, then the change in entropy is

$$\Delta S = 3 \text{ mol} \times \frac{8.31451 \text{ J}}{\text{mol} \cdot \text{K}} \left[\frac{1}{3} \ln \left(\frac{1}{1/3} \right) + \frac{2}{3} \ln \left(\frac{1}{2/3} \right) \right] = 15.9 \text{ J/K}$$

It shouldn't be too hard to see that if more than two gases are mixed, we just have to add more terms of the form $x \ln (1/x)$. Caution, though! Equation 8.5 is far more general. Equation 8.7 requires that the two gases were at the same pressure BEFORE being mixed.

8.6 The Pressure Effect for Ideal Gases: $\Delta S = -nR \ln(P_2/P_1)$

In the case of an ideal gas, we can turn the volume ratio in Equation 8.4 into a pressure ratio by applying the ideal-gas equation again:

$$\frac{P_2 V_2}{P_1 V_1} = \frac{nRT}{nRT} = 1 \quad \Rightarrow \quad \frac{V_2}{V_1} = \frac{P_1}{P_2}$$

And, since $\ln(x/y) = -\ln(y/x)$, we have

$$\Delta S = nR \ln \frac{V_2}{V_1} = nR \ln \frac{P_1}{P_2} = -nR \ln \frac{P_2}{P_1} \qquad (8.8)$$

This last form is preferred, because it retains the "product − reactant" aspect of our Δ notation.

Notice that for a *compression* of an ideal gas at constant temperature, the pressure will go up, so P_2/P_1 will be greater than 1, and $\Delta S < 0$. The entropy of a gas decreases if it is compressed while keeping the temperature constant (by dumping heat into the surroundings).

8.7 Concentration Effect for Solutions: $\Delta S = -nR \ln([X]_2/[X]_1)$

A solution of a substance in a solvent is very similar to what we have been discussing so far. Instead of the ideal-gas equation, we use the fact that concentration times volume gives number of moles, $[X]V_x = n_x$:

$$\frac{[X]_2 V_2}{[X]_1 V_1} = \frac{n}{n} = 1 \quad \Rightarrow \quad \frac{V_2}{V_1} = \frac{[X]_1}{[X]_2}$$

Then we have

$$\Delta S = nR \ln \frac{V_2}{V_1} = nR \ln \frac{[X]_1}{[X]_2} = -nR \ln \frac{[X]_2}{[X]_1} \qquad (8.9)$$

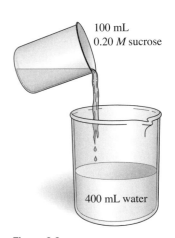

100 mL
0.20 *M* sucrose

400 mL water

Figure 8.2
Diluting an aqueous sugar solution with water is much the same as allowing a gas to expand in terms of the increase in entropy.

in exact analogy to ideal gases. In the common case of *dilution*, we always have $[X]_2 < [X]_1$, so the change in entropy for dilution is always positive. For example, say you were carrying out the dilution shown in Figure 8.2: What is the entropy change when the two solutions are mixed?

We can do this calculation either using the volume ratio or using concentration. In either case we need to know how much sucrose is present. From the information, we know that the smaller beaker contains $(0.100 \text{ L})(0.20 \text{ mol/L}) = 0.020$ mol sucrose. Using volumes, we get

$$\Delta S = nR \ln \frac{V_2}{V_1} = 0.020 \text{ mol} \times \frac{8.31451 \text{ J}}{\text{mol·K}} \ln \frac{500 \text{ mL}}{100 \text{ mL}} = 0.27 \text{ J/K}$$

Alternatively, we could get the same result using concentration instead of volume. We first calculate that the final molarity is 0.020 mol/0.500 L = 0.040 M. Note the minus sign in this equation:

$$\Delta S = -nR \ln \frac{[\text{sucrose}]_2}{[\text{sucrose}]_1} = -0.020 \text{ mol} \times \frac{8.31451 \text{ J}}{\text{mol·K}} \ln \frac{0.040 \text{ M}}{0.20 \text{ M}} = 0.27 \text{ J/K}$$

8.8 Adjustment to the Standard State: $S_x = S_x^\circ - R \ln P_x$ and $S_x = S_x^\circ - R \ln[X]$

Remember that the standard state for molar entropy is 1 bar pressure. Now consider the change in entropy in going *from* the standard state (1 bar) to the "actual" state (P_x or [X]). In the case of gases, using Equation 8.8, we write

$$S_x - S_x^\circ = \Delta S = -R \ln \frac{P_x}{(1 \text{ bar})} = -R \ln(P_x/\text{bar})$$

where "/bar" remains to remind us that we are comparing this pressure to the standard state, which is at 1 bar. Rearranging, we get Equation 8.1:

$$S_x = S_x^\circ - R \ln(P_x/\text{bar})$$

Similarly, we can derive Equation 8.2 by extending this idea of the standard state and call 1 M concentration the standard state for solutions:[2] Note that it is *critical* that we express our pressure in bars and our concentrations in mol/L, as those are the units that were used for the standard state! Always convert pressures to bar and concentrations to mol/L prior to doing these calculations!

$$S_x - S_x^\circ = \Delta S = -R \ln \frac{[X]}{(1 \text{ M})} = -R \ln([X]/M)$$

$$S_x = S_x^\circ - R \ln([X]/M)$$

Also note that these values are "per mole" just as are the standard table values. For example, to determine the entropy of 2.0 mol of gaseous NO_2 (S_{298}° = 240 J/mol-K) at a pressure of 1.0 torr (0.00133 bar), we would calculate the correct *molar* entropy to be

$$S_{NO_2} = S_{NO_2}^\circ - R \ln(P_{NO_2}/\text{bar})$$

$$= 240 \text{ J/mol·K} - \frac{8.31451 \text{ J}}{\text{mol·K}} \ln 0.00133 = 300 \text{ J/mol·K}$$

And since we have 2.0 mol, the total entropy of this system would be 600 J/K.

8.9 The Reaction Quotient: $\Delta_r S = \Delta_r S^\circ - R \ln Q$

When a system is at equilibrium (Section 1.13), we say that "$Q = K$." The right-hand side of this equivalence, the equilibrium constant K, is a pure unitless number, a *constant*. The left-hand side, Q, is called the *reaction quotient*, which involves concentrations or partial pressures of product and reactants taken to stoichiometric powers (that is, the coefficients of a balanced chemical equation). For example, a perfectly fine balanced equation for the dissociation of N_2O_4 and its associated reaction quotient is

$$N_2O_4(g) \longrightarrow 2\,NO_2(g) \qquad Q = \frac{P_{NO_2}^2}{P_{N_2O_4}}$$

We will now see that this "magic ratio" Q comes from none other than the correction to the standard reaction entropy for a chemical reaction. The idea is to imagine a system

---

[2] Because of the fact that ions in solution always come as +/− pairs, ion entropies are all relative and are given in reference to $H^+(aq)$, which has a defined standard entropy of 0 J/K. This results in S° for some ions being negative. For example $S^\circ(OH^-) = -10.7$ J/mol·K.

containing any number of moles of N_2O_4 and NO_2 at any partial pressures whatsoever. The question is then:

How much would the entropies of the system and the surroundings change if one mole of N_2O_4 reacted to give two moles of NO_2 under these specific conditions? Note that we aren't requiring that a whole mole of N_2O_4 has to actually react. This is just hypothetical, a thought experiment. We just want a "per this stoichiometric number of moles" value for ΔS that corresponds to our chemical equation and is appropriate for our specific nonstandard conditions. In fact, to say that a whole mole of N_2O_4 could react "under these conditions" is impossible. "These conditions" include specific partial pressures of N_2O_4 and NO_2, and when the reaction occurs, those partial pressures are bound to change!

Nonetheless, we already know that this is a very important question, because if we know ΔS and ΔS_{sur}, then we will know how much change to expect overall, $\Delta S_{universe}$. In fact, it isn't the actual magnitude of $\Delta S_{universe}$ that interests us. All we really want to know is its sign. If $\Delta S_{universe} > 0$ for the stoichiometric amounts, then we can expect the reaction to go forward spontaneously for a mole or a micromole or a picomole or a molecule. The stoichiometric amounts simply give us a handy reference. The surroundings part of this question is addressed in Chapter 9; the system part, which will show the importance of Q, we answer now.

The approach we take is a general one. We presume some pressures for each substance in the reaction. These pressures could be anything. Determining the hypothetical change in entropy for the stoichiometric number of moles of reactant going to the stoichiometric number of moles of product (we are gaining 2 moles of NO_2 at the expense of 1 mole of N_2O_4) under these nonstandard conditions we have

$$\Delta S = S_{products} - S_{reactants}$$

$$= (2\ mol)S_{NO_2} - (1\ mol)S_{N_2O_4}$$

$$= (2\ mol)[S^\circ_{NO_2} - R\ln(P_{NO_2}/bar)] - (1\ mol)[S^\circ_{N_2O_4} - R\ln(P_{N_2O_4}/bar)]$$

$$= [(2\ mol)S^\circ_{NO_2} - (1\ mol)S^\circ_{N_2O_4}]$$

$$\qquad - [(2\ mol)R\ln(P_{NO_2}/bar) - (1\ mol)R\ln(P_{N_2O_4}/bar)]$$

Notice that the "2 mol" and "1 mol" are present in the equation because we are dealing with a stoichiometric amount of reactant and product based on a specific writing of the chemical equation for this reaction. Both the standard molar entropy, S°, and the ideal-gas constant, R, are *per mole* quantities. Multiplying in each case by "n mol" cancels out the units of 1/mol present in both S° and R.

Now, the first of these two bracketed quantities is simply ΔS° for the equation:

$$\Delta S^\circ = [(2\ mol)S^\circ_{NO_2} - (1\ mol)S^\circ_{N_2O_4}]$$

which has units of J/K. The second bracketed quantity, involving the natural logarithms, can be rearranged to give[3]

$$(2\ mol)R\ln(P_{NO_2}/bar) - (1\ mol)R\ln(P_{N_2O_4}/bar)$$

$$= (1\ mol)R[2\ln(P_{NO_2}/bar) - \ln(P_{N_2O_4}/bar)]$$

$$= (1\ mol)R[\ln(P_{NO_2}/bar)^2 - \ln(P_{N_2O_4}/bar)]$$

$$= (1\ mol)R\ln\frac{(P_{NO_2}/bar)^2}{P_{N_2O_4}/bar}$$

---

[3] The tricks used here are that $a\ln x = \ln x^a$ and that $\ln x - \ln y = \ln(x/y)$.

Thus, overall, we have

$$\Delta S = \Delta S^\circ - (1 \text{ mol})R \ln \frac{(P_{NO_2}/\text{bar})^2}{P_{N_2O_4}/\text{bar}}$$

or

$$\Delta S/(1 \text{ mol}) = \Delta S^\circ/(1 \text{ mol}) - R \ln \frac{(P_{NO_2}/\text{bar})^2}{P_{N_2O_4}/\text{bar}}$$

We now define the *reaction entropy*, $\Delta_r S$, to be $\Delta S/(1 \text{ mol})$, and we define the *standard reaction entropy*, $\Delta_r S^\circ$, to be $\Delta S^\circ/(1 \text{ mol})$. This just cleans up the equation a bit:

$$\Delta_r S = \Delta_r S^\circ - R \ln \frac{(P_{NO_2}/\text{bar})^2}{P_{N_2O_4}/\text{bar}}$$

The *reaction quotient*, Q, for the equation is now *defined* to be this ratio of pressures resulting from the grouping of the natural log terms. It is common to drop the "/bar" and demand that "all pressures be in bars." This gives the standard form for Q:

$$\Delta_r S = \Delta_r S^\circ - R \ln Q \qquad Q = \frac{P_{NO_2}^2}{P_{N_2O_4}} \tag{8.10}$$

Notice that the values of $\Delta_r S$, $\Delta_r S^\circ$, and Q all depend upon the way we have chosen to write the balanced chemical equation for the reaction. You should verify for yourself that the following alternative and perfectly correct equation for the very same reaction gives a perfectly fine *but different* expression for Q:

$$\tfrac{1}{2}N_2O_4(g) \longrightarrow NO_2(g) \qquad Q = \frac{P_{NO_2}}{P_{N_2O_4}^{1/2}}$$

It is also important for you to realize that although generally first introduced to beginning students associated with equilibrium, Q is a far more general quantity. It derives from this need to determine the change in entropy for a reaction under nonstandard conditions of pressure.

For example, if we knew that the partial pressure of N_2O_4 is 0.010 bar and the partial pressure of NO_2 is 3.0 bar, we could write *for this particular balanced equation* and *for these particular conditions*:

$$N_2O_4(g) \longrightarrow 2\,NO_2(g)$$

$$\Delta_r S^\circ = (2)\frac{240 \text{ J}}{\text{mol} \cdot \text{K}} - (1)\frac{304 \text{ J}}{\text{mol} \cdot \text{K}} = 176 \text{ J/mol} \cdot \text{K}$$

$$Q = \frac{P_{NO_2}^2}{P_{N_2O_4}} = \frac{(3.0)^2}{(0.010)} = 900$$

$$\Delta_r S = \frac{176 \text{ J}}{\text{mol} \cdot \text{K}} - \frac{8.31451 \text{ J}}{\text{mol} \cdot \text{K}} \ln 900 = 119 \text{ J/mol} \cdot \text{K}$$

Here the "per mole" in the answer refers to *per mole of reaction*—that is, per 1 mole of N_2O_4 reacting to give 2 moles of NO_2.

Unless instructed otherwise, we are perfectly free to write any correctly balanced equation for this reaction. We just have to be sure that when we write Q and $\Delta_r S^\circ$ we are careful to use the stoichiometric coefficients for the exact chemical equation we have written. For example, we might have written

$$\tfrac{1}{2} N_2O_4(g) \longrightarrow NO_2(g)$$

In this case we would have concluded that

$$\Delta_r S^\circ = (1)\frac{240 \text{ J}}{\text{mol} \cdot \text{K}} - (1/2)\frac{304 \text{ J}}{\text{mol} \cdot \text{K}} = 88 \text{ J/mol} \cdot \text{K}$$

$$Q = \frac{P_{NO_2}}{P_{N_2O_4}^{1/2}} = \frac{(3.0)}{(0.010)^{1/2}} = 30$$

$$\Delta_r S = \frac{88 \text{ J}}{\text{mol} \cdot \text{K}} - R \ln 30 = 60 \text{ J/mol} \cdot \text{K}$$

Both of these results are identical. The first says, "The change in entropy for the system will be 119 J/K for every mole of reaction (i.e., 119 J/K for every 2 moles of NO_2 produced)." The second says, "The change in entropy for the system will be 60 J/K for every mole of reaction (i.e., 60 J/K for every 1 mole of NO_2 produced)." These two statements are the same. *Which* equation we use is not important. The important thing is to use the same numbers appearing as coefficients in the balanced chemical equation as in our calculation of $\Delta_r S^\circ$ and Q.

Finally, note that the reaction quotient Q has no units, because these P values are all unitless "P/bar" or "pressures in bars" themselves. All quantities in exponential and logarithmic functions must always be unitless.

An exactly similar process would have worked for substances in solution. In that case, we would use concentrations in moles per liter instead of pressures:

$$HF(aq) \longrightarrow H^+(aq) + F^-(aq) \qquad Q = \frac{[H^+][F^-]}{[HF]}$$

And for mixed systems, involving both gases and solutes, we go ahead and mix pressures with concentrations in writing Q:

$$CO_2(g) + OH^-(aq) \longrightarrow HCO_3^-(aq) \qquad Q = \frac{[HCO_3^-]}{P_{CO_2}[OH^-]}$$

This may look strange, but it is correct. It is based on following through the calculation of entropy based on Equations 8.1 and 8.2.

Understanding how we arrived at Q is immensely important! It is the basis of all calculations you might have ever done relating to equilibrium, and it comes from considering the effect of nonstandard pressures on the entropy of ideal gases and nonstandard concentrations on the entropy of ideal solutions. To reiterate, the derivation went as follows:

1. The actual molar entropies of products and reactants were put in terms of their *standard* values minus a logarithmic adjustment for nonstandard pressures and concentrations.
2. Reactant entropies were subtracted from product entropies, using stoichiometric amounts of each.
3. Terms were regrouped. The standard reaction entropy was identified as $\Delta_r S^\circ$; the logarithmic terms were combined and called Q.

8.10 Solids and Liquids Do Not Appear in the Reaction Quotient

What about solids and liquids? Well, solids and liquids don't come into play here at all—their entropies are *always* their standard values! Let's see how it works for the evaporation of water:

$$H_2O(l) \longrightarrow H_2O(g) \qquad Q = P_{H_2O}$$

In this case the reactant is a liquid and the product is a gas. We must be careful. Only the gas has the pressure effect on its entropy. The liquid's molar entropy is always its standard entropy, or at least very, very close to it. We have

$$S_{H_2O(l)} = (1\,mol)\,S^{\circ}_{H_2O(l)}$$

$$S_{H_2O(g)} = (1\,mol)[S^{\circ}_{H_2O(g)} - R\,\ln(P_{H_2O(g)}/bar)]$$

Now we need to calculate ΔS:

$$\Delta S = S_{products} - S_{reactants}$$

$$= S_{H_2O(g)} - S_{H_2O(l)}$$

$$= (1\,mol)[S^{\circ}_{H_2O(g)} - R\,\ln(P_{H_2O(g)}/bar)] - (1\,mol)\,S^{\circ}_{H_2O(l)}$$

$$= [(1\,mol)\,S^{\circ}_{H_2O(g)} - (1\,mol)\,S^{\circ}_{H_2O(l)}] - (1\,mol)\,R\,\ln(P_{H_2O(g)}/bar)$$

$$\Delta S = \Delta S^{\circ} - (1\,mol)\,R\,\ln(P_{H_2O(g)}/bar)$$

or

$$\Delta_r S = \Delta_r S^{\circ} - R\,\ln(P_{H_2O(g)}/bar)$$

And again, we simply *define* the reaction quotient Q to be the logarithmic term and demand that the pressure of the gaseous water be expressed as "pressure in bars." It isn't so much that the concentration of the liquid is constant, as some books suggest. Rather, there never was a term for the liquid in the first place!

8.11 The Evaporation of Liquid Water

This last result is for the reaction

$$H_2O(l) \longrightarrow H_2O(g)$$

The correction to the standard reaction entropy,

$$S = \Delta_r S^{\circ} - R\,\ln P_{H_2O(g)}/bar$$

is an extremely useful and general result. In Chapter 12 it will be used over and over as part of an explanation of many diverse applications of thermodynamics. For example, it explains why (and when) a drop of water evaporates spontaneously, how pressure cookers work, and why fog and dew form on some nights but not on others.

   All of these phenomena are associated with heat, temperature, and water vapor pressure, and while Chapter 9 presents the basic elements we need in order to understand how to *calculate* the heat part of the equation, we have enough pieces of the puzzle already to at least start to understand what is going on. Based on this last finding and the fact that the change in entropy of the surroundings is q_{sur}/T, we have for the change in entropy of the *universe* when one mole of liquid water evaporates:

$$\Delta S_{universe} = \Delta S + \Delta S_{sur} = (1\,mol)[\Delta_r S^{\circ} - R\,\ln(P_{H_2O(g)}/bar)] + \frac{q_{sur}}{T}$$

Table 8.1 gives precise values known for ΔS° and q_{sur} for the evaporation of one mole of liquid water at four different temperatures. The values for standard molar entropy given in Appendix D and Table 7.1 are for the specific temperature of 25°C. At 0°C these values would be lower, because at lower temperature more particles would be in the ground state, especially for the liquid, which is almost solid at that temperature. At 100°C, the values would be higher, again especially for the liquid. The result is that ΔS° for the conversion

Table 8.1
$\Delta S_{\text{universe}}$ for the evaporation of a mole of liquid water

			Water Vapor Pressure/bar			
$T/°C$	$\Delta_r S°/(\text{J/mol·K})$	q_{sur}/kJ	0.01	0.1	1.0	2.0
0	122.6	−45.07	−4	−23	−42	−48
25	119.3	−44.13	10	−10	−29	−35
70	113.6	−42.34	28	9	−10	−16
100	110.8	−41.34	38	19	0	−6

of liquid to gas is especially large at 0°C and especially small at 100°C. A similar effect is seen for q_{sur}, which is related to the amount of energy required to break the hydrogen bonds holding the liquid together. At low temperature, these bonds are many, but as the temperature increases, and more and more particles are vibrating and rotating, hydrogen bonding becomes less effective. Thus, at higher temperatures less heat is required to be pulled from the surroundings to turn a mole of liquid water into gas. From these we can calculate the change in entropy of the universe when a mole of water is evaporated at various temperatures and vapor pressures. For example, for 25°C and 0.01 bar water vapor pressure, we have

$$\Delta S_{\text{universe}} = (1\,\text{mole})[\Delta_r S° - R \ln(P_{\text{H}_2\text{O}(g)}/\text{bar})] + \frac{q_{\text{sur}}}{T}$$

$$= \frac{119.3\,\text{J}}{\text{K}} - \frac{8.31451\,\text{J}}{\text{K}} \ln 0.01 + \frac{-44130\,\text{J}}{298\,\text{K}}$$

$$= 9.5\,\text{J/K}$$

Now, the reaction is going to be spontaneous only when the entropy change of the universe is positive. So when the vapor pressure of water is just 0.01 bar (which we call *low humidity*), evaporation of water will occur. But, at 25°C, if the vapor pressure of water rises to just 0.1 bar, evaporation will *not* occur, because under those conditions, $\Delta S_{\text{universe}} = -10$ J/K. (In fact, the reverse reaction, condensation, will occur instead, because the change in entropy for the universe when it occurs is +10 J/K.)

According to Table 8.1, at 0°C even if the humidity is very low, with $P_{\text{vap}} = 0.01$ bar, water will still not evaporate, and at 100°C (the "standard" boiling point of water), the vapor pressure of water can be as high as 1 bar ("standard" atmospheric pressure) and evaporation will still occur. And at 125°C (not shown in Table 8.1), the vapor pressure can get as high as 2 bar, and liquid water will *still* evaporate.

These differences in the behavior of water at different vapor pressures and temperatures are all due to entropy. The water vapor pressure affects how much the entropy of the *system* is going to change when the reaction occurs; the temperature affects how much the entropy of the *surroundings* is going to change. Together, these two factors determine how much the entropy of the *universe* would change were evaporation to occur.

8.12 A Microscopic Picture of Pressure Effects on Entropy

Interestingly (and importantly) the result that compression of a gas at constant temperature and energy decreases its entropy:

$$S_x = S_x° - R \ln(P_x/\text{bar})$$

is exactly what we would conclude (at least qualitatively) from just looking at the energy levels for a Boltzmann distribution (Figure 8.3). Compression of a gas usually results in it

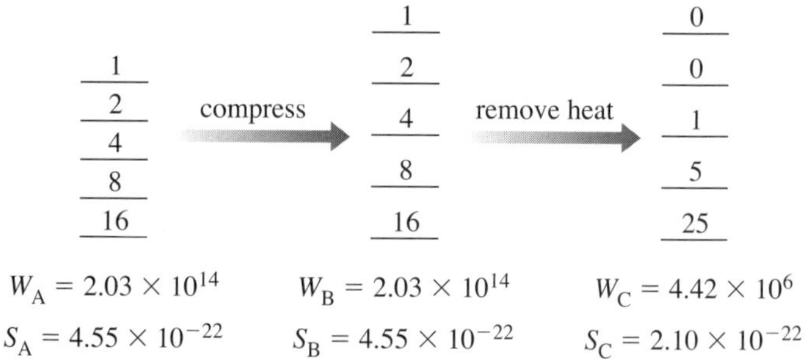

Figure 8.3
Since entropy is a state function, its change is independent of the actual path taken. Hess's law allows us to consider the change in entropy for a constant-energy compression as the sum of a step involving work and a step removing heat. The work step involves no change in entropy; the removal of heat reduces the entropy of the system. The unit of S is J/K.

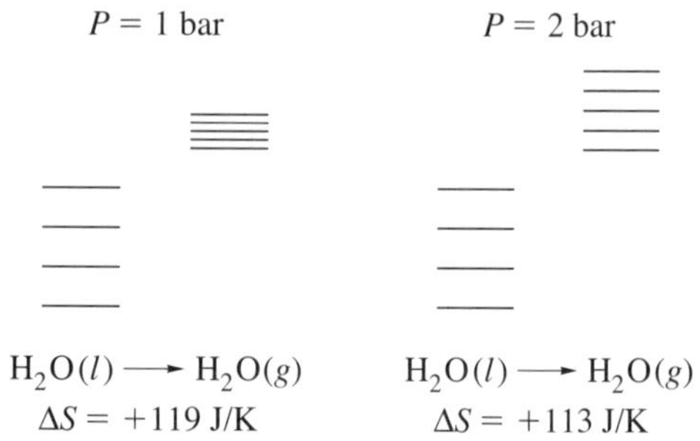

$$\begin{array}{ccc} \dfrac{1}{\begin{array}{c}2\\4\\8\\16\end{array}} & \xrightarrow{\text{compress}} & \dfrac{\begin{array}{c}1\\2\end{array}}{\begin{array}{c}4\\8\\16\end{array}} & \xrightarrow{\text{remove heat}} & \dfrac{\begin{array}{c}0\\0\\1\end{array}}{\begin{array}{c}5\\25\end{array}} \end{array}$$

$$W_A = 2.03 \times 10^{14} \qquad W_B = 2.03 \times 10^{14} \qquad W_C = 4.42 \times 10^{6}$$
$$S_A = 4.55 \times 10^{-22} \qquad S_B = 4.55 \times 10^{-22} \qquad S_C = 2.10 \times 10^{-22}$$

Figure 8.4
The evaporation of liquid water is depicted both at lower vapor pressure (on the left) and at higher vapor pressure (on the right). In both cases the system goes from relatively widely spaced energy levels to relatively closely spaced levels, resulting in an increase in entropy. However, at higher pressure the translational levels of the gas are more widely spaced, resulting in less of an increase in entropy when the reaction occurs.

$$P = 1 \text{ bar} \qquad\qquad P = 2 \text{ bar}$$

$$H_2O(l) \longrightarrow H_2O(g) \qquad H_2O(l) \longrightarrow H_2O(g)$$
$$\Delta S = +119 \text{ J/K} \qquad\qquad \Delta S = +113 \text{ J/K}$$

heating up (Section 4.3), because compressing the gas pushes the translational energy levels apart and increases its internal energy and temperature. But if we do that compression in such a way that we draw off the generated heat and keep that temperature constant, then *at the same time* the particles must drop down to lower levels and expel heat energy. This is what we are doing here. Reducing the number of excited particles always reduces the entropy. Now we have an equation that tells us exactly by how much to expect the entropy to drop when the gas is compressed at constant temperature.

The effect of pressure on entropy changes in reactions can also be seen in microscopic terms involving energy level spacings. For example, consider again the evaporation of water (Figure 8.4).

$$H_2O(l) \longrightarrow H_2O(g) \qquad Q = P_{H_2O(g)}$$

As the liquid evaporates, particles move from levels that are relatively widely spaced in the liquid to levels that are relatively closely spaced in the gas. This amounts to a decrease in the energy level spacings for the system and must lead to an increase in both W and S. At higher vapor pressure for the gas, the volume a mole of gas requires is small, which means the spacings of the translational energy levels in the gas will be especially far apart. (The gas is more constrained by the small volume that is associated with the higher pressure.) The result is that at high vapor pressure, although both W and S increase upon evaporation, they do not increase as much as at lower pressure.

Thus, there is a qualitative correlation between what was derived in this chapter using probability:

$$S_x = S_x^\circ - R \ln(P_x/\text{bar})$$

$$\text{reactants} \longrightarrow \text{products} \qquad \Delta_r S = \Delta_r S^\circ - R \ln Q$$

and what we have learned prior to this regarding energy level spacings.

8.13 Summary

The molar entropy of a gas depends upon its pressure. At a given temperature, as the pressure increases, the molar entropy of the gas decreases due to the fact that the volume containing a mole of the gas gets smaller. In effect, the gas is more tightly constrained. In terms of Boltzmann distributions, what we are seeing is the effect of the box size, d, in the denominator of the translational energy equation becoming smaller as the gas is compressed. The effect is to increase the spacings of the translational energy levels, thus *decreasing* the number of excited particles and *decreasing* the entropy. The adjustment to the standard molar entropy is simply to subtract $R \ln P_x$, where P_x is the nonstandard pressure in bar (Equation 8.1):

$$S_x = S_x^{\circ} - R \ln(P_x/\text{bar})$$

Similarly for a solute in solution, as its concentration increases, the molar entropy decreases because the volume containing exactly one mole decreases. We have not dealt rigorously with Boltzmann distributions involving solutes, but simple probability arguments presented here argue for the same sort of effect as seen for gases. The adjustment to the standard molar entropy is to subtract $R \ln [X]$, where $[X]$ is the nonstandard concentration in mol/L, Equation 8.2:

$$S_x = S_x^{\circ} - R \ln[X]/M$$

Extending these ideas to chemical reactions, we arrive at one of the most important results in this discussion: the origin of the reaction quotient, Q, which fits into the puzzle as a correction to the standard reaction entropy based on a specific chemical equation, Equation 8.3:

$$\Delta_r S = \Delta_r S^{\circ} - R \ln Q$$

"Rules" for writing Q all arise from Equations 8.1 and 8.2 and include

1. Always start with a balanced chemical equation.
2. Gases are included as partial pressures in bar.
3. Solutes are included as concentrations in mol/L.
4. Liquids and solids do not appear because their entropy is considered standard in this approximation.
5. Products are in the numerator; reactants are in the denominator.
6. All substances appear to their stoichiometric power.

This finding gives us all we need to know to determine (or at least estimate) the change in entropy for the system. In the next chapter we focus on the surroundings and see how an additional state function called *enthalpy*, H, gets us the surroundings part of the entropy-change puzzle. Finally, in Chapter 11, we will see that at equilibrium, Q is a specific temperature-dependent number we call the *equilibrium constant*, K.

Problems

The symbol A *indicates that the answer to the problem can be found in the "Answers to Selected Exercises" section at the back of the book.*

In all of these problems, gases and solutions are assumed to be ideal. Use the ideal gas law, $PV = nRT$, as needed.

Changes in Entropy

8.1 A Two flasks are joined, separated by a stopcock, as in Figure 8.1. The volume of Flask A is 250 mL, and the volume of Flask B is 750 mL. Initially the stopcock is closed and 0.25 mol of N_2 gas is confined to Flask A. (a) What is the probability that any given N_2 molecule will be found in Flask A after the stopcock is opened and the system has reached its most probable state? (b) Determine the change in entropy which occurs in going to this most probable distribution. (c) What would be the change in entropy if the N_2 molecules then somehow suddenly all went to Flask B at the same time?

8.2   In the same apparatus as described in Problem 8.1 (250 mL for Flask A and 750 mL for Flask B), 0.50 mol of argon gas is initially confined to Flask A, and 1.50 mol of krypton gas is confined to Flask B. Determine the change in entropy of each gas when the stopcock is opened (a) based on volumes, (b) based on concentrations, and (c) based on pressures if the temperature is 300 K.

8.3 A Two flasks are joined, separated by a stopcock, as in Figure 8.1. The volume of Flask A is 500 mL, and the volume of Flask B is 1.25 L. Initially the stopcock is closed and 0.25 mol of N_2 gas is confined to Flask A. (a) What is the probability that any given N_2 molecule will be found in Flask A after the stopcock is opened and the system has reached its most probable state? (b) Determine the change in entropy that occurs in going to this most probable distribution. (c) What would be the change in entropy if the N_2 molecules then somehow suddenly all went to Flask B at the same time?

8.4 A In the same apparatus as described in the previous problem (500 mL for Flask A and 1.25 L for Flask B), 0.50 mol of argon gas is initially confined to Flask A, and 1.50 mol of krypton gas is confined to Flask B. Determine the change in entropy of each gas when the stopcock is opened (a) based on volumes, (b) based on concentrations, (c) based on pressures if the temperature is 300 K, and (d) based on mole fraction.

8.5   Two flasks are joined, separated by a stopcock, as in Figure 8.1. The volume of Flask A is 250 mL, and the volume of Flask B is 750 mL. Initially the stopcock is closed and 0.75 mol of N_2 gas is confined to Flask A. (a) What is the probability that any given N_2 molecule will be found in Flask A after the stopcock is opened and the system has reached its most probable state? (b) Determine the change in entropy that occurs in going to this most probable distribution. (c) What would be the change in entropy if the N_2 molecules then somehow suddenly all went to Flask B at the same time?

8.6   In the same apparatus as described in the previous problem (250 mL for Flask A and 750 mL for Flask B), 1.50 mol of argon gas is initially confined to Flask A, and 1.50 mol of krypton gas is confined to Flask B. Determine the change in entropy of each gas when the stopcock is opened (a) based on volumes, (b) based on concentrations, (c) based on pressures if the temperature is 300 K, and (d) based on mole fraction.

8.7   Show that the pressures in both flasks are the same and that the same overall entropy change in Problem 8.2 is obtained using ideas of mole fraction.

8.8   Derive a formula for ΔS_{mix} in terms of mole fractions of gases if three different ideal gases all at the same pressure are allowed to mix. Imagine an apparatus similar to the one in Figure 8.1, except there are three gas bulbs instead of two.

8.9 A Calculate the change in entropy in the following situations: (a) 250 mL of 12 *M* acetic acid is mixed with 750 mL of pure water. (b) 75.0 mL of 0.100 *M* NaCl is mixed with 25.0 mL of 0.050 *M* $MgBr_2$. [Hint: All species of a strong electrolyte must be accounted for independently!]

8.10   Calculate the change in entropy for the following situations: (a) 500 mL of 9.0 *M* ammonia is mixed with 500 mL of pure water. (b) 50.0 mL of 0.25 *M* nitric acid is mixed with 50 mL of 0.50 *M* sodium nitrate.

8.11   Calculate the change in entropy for the following situations: (a) 150 mL of 2.0 *M* sucrose is mixed with 500 mL of pure water. (b) 75.0 mL of 0.15 *M* nitric acid is mixed with 50.0 mL of 0.50 *M* sodium nitrate.

8.12   Calculate the change in entropy for the following situations: (a) 500 mL of 9.0 *M* ammonia is mixed with 500 mL of pure water. (b) 50.0 mL of 0.25 *M* Na_2CO_3 is mixed with 50.0 mL of 0.75 *M* potassium nitrate.

8.13 A Calculate the entropy for the following substances at the pressure indicated: (a) a mole of gaseous water at a partial pressure of 760 torr. (b) 1.8 g of gaseous water at a partial pressure of 730 torr. (c) 2.00 mol of gaseous $SiCl_4$ at a partial pressure of 29.92 inHg.

8.14   Calculate the entropy for the following substances at the pressure indicated: (a) 1.00 moles of gaseous water at a partial pressure of 30.25 inHg. (b) 0.841 g of gaseous bromine at a partial pressure of 750 torr. (c) 2.00 mol of gaseous CH_4 at a partial pressure of 0.987 atm.

8.15 A Calculate the total entropy of the gas in a 1.0-L flask containing 0.25 mol of neon and 0.75 mol of krypton. [*Hint:* Where on the Web can we go to get standard molar entropy values?]

8.16   Calculate the total entropy of the gas in a 1.0-L flask containing 0.50 mol of neon and 1.25 mol of krypton.

8.17 A Calculate the total entropy of the gas in a 2.0-L flask containing 0.25 mol of neon and 0.75 mol of krypton.

8.18   Calculate the total entropy of the gas in a 2.0-L flask containing 0.50 mol of neon and 1.25 mol of krypton.

8.19   What is the total entropy of the gas in a 2.0-L flask containing 0.50 mol of argon and 1.5 mol of neon at 25°C?

Reaction Entropy and the Reaction Quotient

8.20 ▣ Determine $\Delta_r S$ for the following equations, assuming no significant change in concentrations or pressures when the reactions occur. Conditions are (a) $Q = 4.34 \times 10^{-3}$; (b) $P(PCl_3) = 0.378$ bar, $P(Cl_2) = 0.678$ bar, and $P(PCl_5) = 1.30$ bar.

   (a) $3\,H_2(g) + N_2(g) \longrightarrow 2\,NH_3(g)$
   (b) $PCl_3(g) + Cl_2(g) \longrightarrow PCl_5(g)$

8.21 What would be the entropy change of the system if 2.5 g of H_2 were reacted under the conditions of Problem 8.20(a), provided the partial pressures of H_2, N_2, and NH_3 are kept approximately constant?

8.22 Explain why it is that the reaction quotient Q is a unitless quantity.

8.23 ▣ Using algebra along the lines of the discussion in Section 8.9, derive the expression for the reaction quotient, Q, in terms of pressures and concentrations, as appropriate, for each of the following equations:

   (a) $3\,H_2(g) + N_2(g) \longrightarrow 2\,NH_3(g)$
   (b) $PCl_3(g) + Cl_2(g) \longrightarrow PCl_5(g)$
   (c) $H^+(aq) + OH^-(aq) \longrightarrow H_2O(l)$

8.24 Using algebra along the lines of the discussion in Section 8.9, derive the expression for the reaction quotient, Q, in terms of pressures and concentrations, as appropriate, for each of the following equations:

   (a) $CaCO_3(s) \longrightarrow CaO(s) + CO_2(g)$
   (b) $SO_2(g) + \frac{1}{2}\,O_2(g) \longrightarrow SO_3(g)$
   (c) $CO_2(g) + OH^-(aq) \longrightarrow HCO_3^-(aq)$

8.25 Using algebra along the lines of the discussion in Section 8.9, derive the expression for the reaction quotient, Q, in terms of pressures and concentrations, as appropriate, for each of the following equations:

   (a) $P_4(g) + 5\,O_2(g) \longrightarrow P_4H_{10}(s)$
   (b) $2\,H_2S(g) \longrightarrow 2\,H_2(g) + S_2(g)$
   (c) $CaCO_3(s) \longrightarrow Ca^{2+}(aq) + CO_3^{2-}(aq)$

Brain Teasers

8.26 The sorts of expansions dealt with in this chapter take place in fixed flasks totally isolated from their surroundings so that $q = 0$, $w = 0$, $\Delta U = 0$, and $\Delta T = 0$. However, these results apply equally well for any system any time $\Delta U = 0$ and $\Delta T = 0$ *overall*, regardless of how many steps are involved or whether heat or work are exchanged with the surroundings along the way. This is one more manifestation of the fact that entropy is a state function. Calculate the change in entropy **for the system** for each of the following three changes. In each case, rationalize the sign of ΔS in terms of what must happen to particles in energy level spacings and populations, and the effect that must have on thermodynamic probability, W:

   (a) The stopcock in Figure 8.1 is opened. (That is, explain why the entropy must increase based on what must be happening to the energy levels and their populations when both $\Delta T = 0$ and $\Delta U = 0$.)
   (b) A 2.5-mol sample of O_2 is compressed from 1.0 bar to 35 bar slowly, keeping the temperature constant and allowing heat to be lost to the surroundings.
   (c) A 2.5-mol sample of O_2 is compressed from 1.0 bar to 35 bar rapidly, so the temperature rises, but there is no time for heat to be transferred to the surroundings.

8.27 How many ways are there of distributing the cards in a standard deck of 52 cards? How many ways are there of finding that deck with all 26 red cards on top after shuffling? (Careful, don't forget those black cards!) Using these results, calculate the change in entropy upon going from the state of "26 red cards on top" to the state of "randomly distributed cards" based on the idea that:

$$\frac{W_2}{W_1} = \frac{\text{Prob}_2}{\text{Prob}_1} = \frac{1}{\text{Prob}_{\text{all 26 red cards on top}}}$$

Also express this result in terms of factorials and show using Stirling's approximation (Section 7.9) that in this case:

$$\ln \frac{W_2}{W_1} \approx 52 \ln 2$$

Recalculate the entropy change using this approximation and show that this is exactly the same result as would be obtained using the ideas in Section 8.5 dealing with the entropy of mixing. Why does this work?

CHAPTER 9

Enthalpy (*H*) and the Surroundings

The general struggle for existence of all living beings is not the struggle for the fundamental substances, for these fundamental substances indispensable for all living creatures exist abundantly in the air, the water, and the soil. This struggle is not a struggle for the energy which in the form of heat, unfortunately not utilizable, is present in a great quantity in every object, but it is a struggle for entropy, which is available when energy passes from the hot sun to the cold earth.

—*Ludwig Boltzmann, 1900*

9.1 Heat Is Not a State Function

In order to make anything quantitative of the Second Law of Thermodynamics:

$$\Delta S_{universe} = \Delta S + \Delta S_{sur} > 0$$

we need to quantify the change in entropy in the surroundings that occurs as a result of the heat coming in or going out of a reacting chemical system. Since energy is conserved, we can say $q_{sur} = -q$. We also know that heat flow results in a change in entropy of the surroundings (Section 7.5):

$$\Delta S_{sur} = \frac{q_{sur}}{T} = \frac{-q}{T}$$

Substituting this expression for ΔS_{sur} into the Second Law, we get the key relationship we need in this chapter:

$$\Delta S_{universe} = \Delta S - \frac{q}{T} > 0$$

However, there is a slight problem here. Heat is not a state function. Instead, the heat given off or absorbed by a system in going from one state to another is dependent upon the path taken (Section 4.5). How can the change in entropy of the universe (a state function itself) be dependent upon the path?

The answer to this quandary lies in the fact that our very first derivation of $\Delta S_{sur} = q_{sur}/T$ did actually require a certain path. We were assuming that the surroundings were in a

Boltzmann distribution at all times. This requirement "sets the path" and effectively makes q a state function under these conditions.[1] The name of this path assumed by Boltzmann is the *reversible* path.

9.2 The Definition of Enthalpy: $H = U + PV$

It turns out that there are two common reversible paths along which many reactions are carried out: the *constant pressure* path and the *constant volume* path. If we were to assume a constant volume, where no work is done, we could simply write that $q = \Delta U$ and we would be done. Heat, q, would be a state function because U is a state function already. But to require a constant volume is too limiting. Conditions involving constant volume (as in a bomb calorimeter) are not particularly common.

By defining a new state function, *enthalpy*, we can work with a more common sort of path, the path of constant pressure. Enthalpy is given the letter H, to remind us that it's very closely related to heat. This is a little tricky, but see if you can follow along. First, we simply define enthalpy as

$$H = U + PV \tag{9.1}$$

Remember, this means nothing in and of itself. But U, P, and V are all state functions, so H must be, too. What about changes in enthalpy? We have

$$\Delta H = \Delta U + \Delta(PV)$$

Consider the condition of constant pressure, where $\Delta(PV) = P\Delta V$. Then we have

$$\Delta H = \Delta U + P\Delta V$$

Work, w, is defined as $-P\Delta V$ (Section 4.7), so $P\Delta V = -w$. In addition, the change in internal energy, ΔU, equals $q + w$ under all circumstances (Section 4.5). So we have, assuming constant pressure throughout the change in state

$$\Delta H = \Delta U + P\Delta V$$
$$= (q + w) - w$$

and so

$$\Delta H = q \tag{9.2}$$

And there we have it! Under the conditions of constant pressure, q is a state function, a very specific number, which only depends upon the starting and ending states, not anything else about the path. The path, in effect, is defined enough by the statement that "pressure is always constant" to ensure reversibility.

This is the key finding of this chapter, for with it we can redefine the change in entropy of the surroundings in terms of changes in the system:

$$\Delta S_{\text{sur}} = \frac{q_{\text{sur}}}{T} = \frac{-q}{T} = \frac{-\Delta H}{T} \tag{9.3}$$

The Second Law restated in terms only of state functions of the *system*, specifically when the pressure and temperature are held constant, then, is

---

[1] It is certainly possible to *not* have a Boltzmann distribution. If heat is flowing fast enough, there may not have been time for the most probable distribution or one even very close to it to have been found through random collisions.

$$\Delta S_{\text{universe}} = \Delta S + \Delta S_{\text{sur}} = \Delta S - \frac{\Delta H}{T} > 0 \qquad (9.4)$$

It is very important to understand that this second term, $-\Delta H/T$, is just the entropy change of the surroundings.

9.3 Standard Enthalpies of Formation, $\Delta_f H°$

We already know how to estimate the change in entropy of a chemical reaction. That involves writing a balanced chemical equation and using tables of standard molar entropies. You might think there would be tables of standard molar enthalpies as well, but you would be only partially right. One of the big differences between energy and entropy is that energy is relative, while entropy is absolute. It makes sense to talk about $S°$ for a substance, because we have the reference point of absolute zero, where all particles are in the ground state and $S = 0$.

The same cannot be said about energy in any of its forms. Consider motion. Is a rock moving? It depends upon your frame of reference. Relative to the Earth, maybe not; relative to the Sun or the Moon, yes. If you tell me a molecule has an energy, I'm going to ask you, "In reference to what? Isolated atoms? Some other molecule?" You have to pick a standard reference point for energy and stick with it.

The current international standard reference for enthalpy is for the *elements* in their standard state at a temperature of 298 K and a pressure of 1 bar. For example, the standard state of oxygen is $O_2(g)$, and the standard state of mercury is $Hg(l)$. We will see that the "standard state" is very handy, because many of the substances we work with actually are elements in their standard states. *Standard enthalpies of formation*, as they are called, are referred to as "$\Delta_f H°$," where the subscript "f" stands for "formation from the elements."

The standard enthalpy of formation of liquid water is -285.83 kJ/mol. This means we can write

$$H_2(g) + \tfrac{1}{2} O_2(g) \longrightarrow H_2O(l) \qquad \Delta_r H° = -285.83 \text{ kJ/mol}$$

Similarly, the standard enthalpy of formation of liquid hydrogen peroxide, $H_2O_2(l)$, is -187.78 kJ/mol. We can write

$$H_2(g) + O_2(g) \longrightarrow H_2O_2(l) \qquad \Delta_r H° = -187.78 \text{ kJ/mol}$$

Interestingly, using Hess's law, we can combine standard enthalpies of formation to get the change in enthalpy for a reaction of interest. For example, for the decomposition of hydrogen peroxide we have

$$
\begin{aligned}
H_2(g) + \tfrac{1}{2} O_2(g) &\longrightarrow H_2O(l) & \Delta_r H° &= -285.83 \text{ kJ/mol} \\
H_2O_2(l) &\longrightarrow H_2(g) + O_2(g) & \Delta_r H° &= +187.78 \text{ kJ/mol} \\
\hline
H_2O_2(l) &\longrightarrow H_2O(l) + \tfrac{1}{2} O_2(g) & \Delta_r H° &= -98.05 \text{ kJ/mol}
\end{aligned}
$$

Several representative heats of formation are given in Table 9.1, and many more can be found in Appendix D. Although it is only proper to call these enthalpies of formation, many chemists refer to them as "standard heats of formation," because enthalpy and heat are so closely related. Several aspects of Table 9.1 deserve mentioning:

For each element there is one standard molar enthalpy of formation that is 0. This is because there must be a standard reference for each element so that other enthalpies of formation can be measured "from the elements." Note that many elements are polyatomic in their standard state. For example, the standard state of oxygen is $O_2(g)$, not $O(g)$ or $O_3(g)$.

Table 9.1
Standard molar enthalpies of formation (kJ/mol) at 298 K ($\Delta_f H^\circ_{298}$).

Solids		Liquids		Gases			
C(diamond)	1.895	$Br_2(l)$	0	$Br_2(g)$	30.91	HCl(g)	−92.31
C(graphite)	0	$CH_3OH(l)$	−238.66	$CH_4(g)$	−74.81	Hg(g)	60.83
Fe(s)	0	Hg(l)	0	$C_2H_6(g)$	−85.68	$H_2O(g)$	−241.82
$I_2(s)$	0	$HNO_3(l)$	−174.10	CO(g)	−110.53	$I_2(g)$	62.44
LiCl(s)	−408.71	$H_2O(l)$	−285.83	$CO_2(g)$	−393.51	$N_2(g)$	0
NaCl(s)	−410.9	$H_2O_2(l)$	−187.78	$Cl_2(g)$	0	$NH_3(g)$	−46.11
P(red)	−17.6	$H_2SO_4(l)$	−813.99	$F_2(g)$	0	NO(g)	90.25
P(white)	0			H(g)	217.97	$NO_2(g)$	33.18
$SiO_2(q)$	−910.94			$H_2(g)$	0	$N_2O_4(g)$	9.16
Sn(gray)	−2.09			HF(g)	−271.1	O(g)	249.17
Sn(white)	0			HBr(g)	199	$O_2(g)$	0
						$O_3(g)$	142.7

Some elements have more than one possible state, but only one is *standard*. Examples include bromine, carbon, iodine, mercury, and phosphorus. Only the state of the element that predominates at 298 K is assigned 0 as its standard enthalpy of formation. Other states which may coexist at 298 K but are not as prevalent generally have standard enthalpies of formation which are greater than zero. (Tin and phosphorus are exceptions.)

Most compounds have standard enthalpies of formation that are negative. This is because in general, bonding is stronger in compounds than in elements, and heat is released in forming compounds from their elements. A notable exception is the element N_2, which itself has such a strong bond that many of its compounds have positive standard molar enthalpies of formation.

Note also that the numbers in these tables, like the ones given on tables of standard molar entropy, are "official" only for 298 K. Enthalpy, being closely related to internal energy and heat, will certainly increase with temperature. However, we are never going to take that into account.[2] It can be shown that around 298 K, at least, these numbers will be quite sufficient, and pressure changes have little effect on ΔH. Thus, in all work in this book, we will always simply say that for our purposes $q = \Delta H = \Delta H^\circ$ when the pressure is constant.

9.4 Using Hess's Law and $\Delta_f H^\circ$ to Get $\Delta_r H^\circ$ for a Reaction

According to Hess's law (Section 5.2), an overall reaction for which we don't know a state function (such as $\Delta_r H^\circ$) can be broken down into individual steps for which we do. We can then add the various $\Delta_r H^\circ$ for the reactions to get the overall change. Consider again the burning of natural gas (methane):

---

[2] If we wanted to, we could make a correction to standard enthalpy of formation using heat capacity:

$$H^\circ = H^\circ_{298} + \int_{298}^{T} C\,dT$$

$$CH_4(g) + 2\,O_2(g) \longrightarrow CO_2(g) + 2\,H_2O(g)$$

If we write out the equations for both the formation of products $CO_2(g)$ and "2 $H_2O(g)$" from their elements along with the *reverse* for reactants $CH_4(g)$ and "2 $O_2(g)$" we get the following four equations, with their associated standard enthalpy changes:

1.	$C(graphite) + O_2(g) \longrightarrow \mathbf{CO_2(g)}$	$\Delta_r H° = -393.5$ kJ/mol
2.	$2\,H_2(g) + O_2(g) \longrightarrow \mathbf{2H_2O(g)}$	$\Delta_r H° = -484$ kJ/mol
3.	$\mathbf{CH_4(g)} \longrightarrow C(graphite) + 2\,H_2(g)$	$\Delta_r H° = +75$ kJ/mol
4.	$\mathbf{2O_2(g)} \longrightarrow 2\,O_2(g)$	$\Delta_r H° = \ \ \ 0$ kJ/mol

$$\mathbf{CH_4(g) + 2O_2(g) \longrightarrow CO_2(g) + 2H_2O(g)} \qquad \Delta_r H° = -803 \ \ \text{kJ/mol}$$

(Oxygen gas is already in its standard state, so its equation is somewhat of a "nonevent.") Check for yourself that these four equations, when added, combine to give this overall equation. The numbers here are taken from our table and adjusted for doubling (2 and 4) and reversal (3 and 4).

We don't really have to go to the trouble of writing out all these equations. They are all "built into" the values given in a table of standard enthalpies of formation. All we have to do is to add up the standard enthalpies of formation of all the *products* and subtract the sum of the standard enthalpies of formation of all of the *reactants*, being careful to multiply by the appropriate number of moles in the equation, since the values in the table are all *per mole*. Mathematically, we write:

$$\Delta_r H° = \sum_{\text{products}} \Delta_f H° - \sum_{\text{reactants}} \Delta_f H° \tag{9.5}$$

Thus, in this case, we have

$$CH_4(g) + 2\,O_2(g) \longrightarrow CO_2(g) + 2\,H_2O(g)$$

and

	$CH_4(g)$	$O_2(g)$	$CO_2(g)$	$H_2O(g)$	
$\Delta_f H°$	-75	0	-393.5	-242	kJ/mol

$$\sum \Delta_f H°(reactants) = (1)\frac{-75\,\text{kJ}}{\text{mol}} + (2)\frac{0\,\text{kJ}}{\text{mol}} = -75\,\text{kJ/mol}$$

$$\sum \Delta_f H°(products) = (1)\frac{-393.5\,\text{kJ}}{\text{mol}} + (2)\frac{-242\,\text{kJ}}{\text{mol}} = -877.5\,\text{kJ/mol}$$

$$\Delta_r H° = -877.5\,\text{kJ/mol} - (-75\,\text{kJ/mol}) = -803\,\text{kJ/mol}$$

Just remember, *sum of products minus sum of reactants.* Be careful with those minus signs and to multiply by the number of moles of substance indicated in the equation! Be very careful when using these tables to check for phases. Don't use $\Delta_f H°[H_2O(g)]$ where $\Delta_f H°[H_2O(l)]$ is appropriate, or you will be disappointed. Figure 9.1 illustrates the relationship between $\Delta_r H°$ for a chemical reaction and standard enthalpies of formation for the reactants and products.

We now consider three examples of the use of standard enthalpies of formation. In each case, notice how the goal is more than just the arrival at a set of numbers. The goal is an explanation, an answer to the question, "Why?"

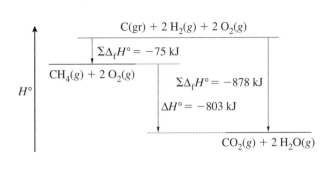

Figure 9.1
Using Hess's law to get $\Delta_r H^\circ$ based on standard molar enthalpies of formation. Note that to take the path from reactants to products through the standard reference state, one must reverse the arrow leading to the reactants. That is, one must *subtract* the sum of the reactant enthalpies of formation from the sum of the product enthalpies of formation.

Example 1: The Formation of Ozone

$$O_2(g) + O(g) \longrightarrow O_3(g)$$

We have

	$O_2(g)$	$O(g)$	$O_3(g)$	
$\Delta_f H^\circ$	0	249	143	kJ/mol

We calculate:

$$\sum \Delta_f H^\circ (products) = (1)\frac{143 \text{ kJ}}{\text{mol}} = 143 \text{ kJ/mol}$$

$$\sum \Delta_f H^\circ (reactants) = (1)\frac{0 \text{ kJ}}{\text{mol}} + (1)\frac{249 \text{ kJ}}{\text{mol}} = 249 \text{ kJ/mol}$$

$$\Delta_r H^\circ = 143 \text{ kJ/mol} - 249 \text{ kJ/mol} = -106 \text{ kJ/mol}$$

In this reaction, 106 kJ of energy is released per mole of O_3 formed. That's actually a fair amount of heat to be delivered to the surroundings.

In the stratosphere, ozone absorbs ultraviolet light from the Sun. This excites ozone into a high-energy electronic state and leads ultimately to the breaking of an O−O bond and the formation of O_2 and O. It is when these two fragments then recombine that the energy of the ultraviolet light is converted to heat. If ozone were not in the atmosphere, then that amount of energy would not be released as heat and would instead be transmitted to the surface of the Earth in the form of ultraviolet radiation. It might be *your* DNA that would finally absorb that energy, causing damage and the onset of cancer!

Example 2: The Evaporation of Water

$$H_2O(l) \longrightarrow H_2O(g)$$

One thing you have to be very careful about in using these tables is that you get the correct phase. We have

	$H_2O(l)$	$H_2O(g)$	
$\Delta_f H^\circ$	−286	−242	kJ/mol

and so

$$\sum \Delta_f H^\circ (products) = (1) \frac{-242 \text{ kJ}}{\text{mol}} = -242 \text{ kJ/mol}$$

$$\sum \Delta_f H^\circ (reactants) = (1) \frac{-286 \text{ kJ}}{\text{mol}} = -286 \text{ kJ/mol}$$

$$\Delta_r H^\circ = -242 \text{ kJ/mol} - (-286 \text{ kJ/mol}) = +44 \text{ kJ/mol}$$

(Watch those minus signs!) Thus, 44 kJ of energy is required to evaporate 1 mol of liquid water at 1 bar pressure and 298 K. Although this may not seem like much, it's only for 18 grams of water! Conversely, for every 18 g of water vapor that *condenses* to form liquid water, 44 kJ of energy is released into the atmosphere. In fact, it is the condensation of water vapor in the form of clouds and rain that provides much of the energy in a developing thunderstorm or hurricane.[3]

Example 3: Combustion of the Atmosphere

$$N_2(g) + 2\,O_2(g) \longrightarrow N_2O_4(g)$$

This reaction *is* the formation of $N_2O_4(g)$ from the elements in their standard state at 298 K. For $N_2O_4(g)$, $\Delta_f H^\circ = 9.16$ kJ/mol. Thus, we have simply

$$\Delta_r H^\circ = (1) \frac{9.16 \text{ kJ}}{\text{mol}} = 9.16 \text{ kJ/mol}$$

This is an extraordinarily small change in enthalpy. At the time of the testing of the first atomic bomb, there was great concern that this number might not be totally accurate. What if it were really -5 kJ/mol? What if these calculations for the standard change were not applicable to atmospheric conditions? What if there were some other related compound which was more stable than N_2? Might it not be possible that an atomic blast would provide the spark that would trigger this or some related reaction of the two major components of our atmosphere, leading to the end of all life on Earth? It was a legitimate concern. Would you have trusted the calculation if you were the president of the United States?

9.5 Enthalpy vs. Internal Energy

Looking back at how ΔH was derived, we find that the difference between a change in internal energy, ΔU, and a change in enthalpy, ΔH, is slight, amounting only to the work involved in a reaction:

$$\Delta H = \Delta U + P\Delta V = \Delta U - w \quad \text{(constant } P\text{)}$$

Work, w, is insignificant in all chemical reactions if no gases are involved. Even if gases are involved, work doesn't amount to much. If gases are present, if you *have* to make a distinction between ΔU and ΔH, then rather than actually working out the pressures and

---

[3] The process of evaporation also turns out to be a major source of energy for the British Isles. Heat from the tropical sun in the rainforests of South America is absorbed into the atmosphere in the form of water vapor. Strong upper atmosphere currents bring this air mass all the way across the Atlantic to the British Isles, where, of course, it rains. Rain is more than water, however. With every 18 grams of rainwater comes 44 kJ of heat energy. Thus, the flip side of rainy weather is an extra dose of warmth. (Just remember this on a cold, drizzly March day—if it weren't drizzling, it would be a whole lot colder outside!)

volumes, you can use the ideal-gas equation in the form $P\Delta V = (\Delta n_{gas})RT$ to get

$$\Delta H = \Delta U + P\Delta V = \Delta U + (\Delta n_{gas})RT \quad \text{(constant P and T)}$$

where Δn_{gas} is the change in number of moles of gas in the actual reaction or the difference in number of moles of gas in the equation as written (depending upon what you are doing). Be careful to check those phases—just gases here!

For example, consider these two very similar reactions:

	Δn_{gas}/mol	ΔH/kJ	ΔU/kJ
$2\,H_2(g) + O_2(g) \longrightarrow 2\,H_2O(g)$	-1	-483.6	-481.1
$2\,H_2(g) + O_2(g) \longrightarrow 2\,H_2O(l)$	-3	-571.7	-564.2

Can you spot the difference? The second one has H_2O as a *liquid* product. Even so, in both cases the difference between $\Delta H°$ and ΔU doesn't amount to much. Interestingly, using mean bond dissociation energies (Section 5.6) we would estimate that for *both* reactions $\Delta U = -509$ kJ. Remember, "mean" means "average." The values above are better.

If ΔH and ΔU are so similar, why not just estimate q_{sur} for a reaction using tables of mean bond dissociation energies? You can, but three major benefits of using tables of $\Delta_f H°$ include the following:

1. Tables of standard enthalpies of formation are considered some of the most accurate data ever compiled in chemistry. The values are not averages over many compounds the way mean bond dissociation energies are.
2. Tables of standard enthalpies of formation are very extensive, and are available on the Web at *http://webbook.nist.gov/chemistry*.
3. Using these tables, we don't have to draw Lewis structures, count bonds, or worry about detailed structural information such as which atom is connected to which.

Thus, tables of mean bond dissociation energies give us only a rough estimate of q_{sur}, while tables of $\Delta_f H°$ are much more accurate. Then again, if the standard enthalpies of formation of the compounds we are interested in are not known, then there is little we can do but either use tables of mean bond dissociation energies or (better, if available) do a computer-based computation (Section 5.9).

9.6 High Temperature Breaks Bonds

For the general reaction involving the breaking of bonds,

$$A-B(g) \longrightarrow A(g) + B(g)$$

we know that both the entropy of the system and the enthalpy of the system must increase. A bond is being broken, which means energy is required and $\Delta H > 0$. In addition, more gaseous particles are produced, which means $\Delta S > 0$ as well.

At any temperature, as heat is pulled into the system from the surroundings, the entropy of those surroundings must drop based on

$$\Delta S_{sur} = \frac{-\Delta H}{T} < 0 \quad \text{(in this case)}$$

This drop in entropy of the surroundings will be more important at low temperature and less important at high temperature (Figure 9.2). We can think of ΔH as relatively constant— basically its standard value. There isn't much we can do to change it, as it is tied strongly

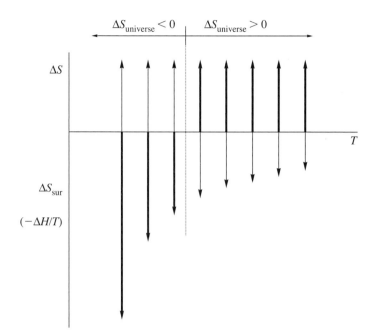

Figure 9.2
In this example, the reaction is endothermic, so that heat is absorbed from the surroundings, and the entropy of the surroundings must decrease (arrows pointing down). However, the reaction also involves an increase in entropy for the *system* (arrows pointing up). At each temperature, the dominant arrow is indicated in bold. The effect of temperature on $\Delta S_{sur} = q_{sur}/T$ is to make that term dominate at low temperature and become less important at high temperature. Thus, at low temperature it is the decrease in entropy of the surroundings that prevents the reaction from proceeding. At high temperature, the increase in entropy of the system drives the reaction forward.

to bond dissociation energies. It is the fact that T here is in the denominator that produces the effect. At low temperature the entropy change of the surroundings has a huge effect on the direction of a reaction. At high temperature, on the other hand, this term becomes less important, and the intrinsic changes in entropy in the system are what rule. At all temperatures, however, entropy rules.

We should not be surprised to find that systems with strong bonds are favored at low temperature, but a temperature can always be found that breaks those bonds. At the temperatures associated, for example, with the fireball of an atomic bomb, there are simply no molecules at all, only isolated electrons, nuclei, and a few atoms.

9.7 Summary

Entropy drives all observable phenomena. The key to figuring out whether a reaction will "go" lies in including the heat flow into the surroundings, q_{sur}. This heat coming *out* of the system $(-q)$ increases the entropy of the surroundings by jiggling particles in the surroundings up into their excited states. Thus, it is very possible to have a reaction proceed quite well, even if its own *system* entropy is decreasing, as long as bonds are being made and energy is being dumped out of the system into the surroundings.

Because the heat going into the surroundings must be along a reversible path—one that is essentially a Boltzmann distribution all along the way—it becomes a state function itself. Well, almost. In order to handle this, devious minds invented a new state function called *enthalpy, H*. Enthalpy by itself is not particularly informative. However, differences in enthalpy, ΔH, for chemical reactions indicate how much heat that reaction will require $(\Delta H > 0)$ or release $(\Delta H < 0)$ when overall pressure is kept constant.

Most importantly for the current discussion, when the pressure is kept constant, ΔH gets us the information we need to determine the entropy change in the surroundings. We have that under conditions of constant pressure (Equation 9.2),

$$\Delta H = q$$

and (Equation 9.3),

$$\Delta S_{sur} = \frac{-\Delta H}{T}$$

This, then, is the real importance of enthalpy: It gives us a handle on the change in entropy of the surroundings when a reaction occurs. This effect is highly temperature-dependent, being very important at low temperature and much less important at high temperature. Enthalpy allows us to write the Second Law of Thermodynamics in terms of the system only (Equation 9.4):

$$\Delta S_{universe} = \Delta S + \Delta S_{sur} = \Delta S - \frac{\Delta H}{T} > 0$$

Furthermore, since *H* is a state function, so is ΔH, and it is possible to use Hess's law to determine changes in enthalpy relative to a standard reference. This reference is "the elements in their standard states," which gives us tables of standard molar enthalpies of formation. From those we can calculate the *standard reaction enthalpy* based on Equation 9.5:

$$\Delta_r H^\circ = \sum_{products} \Delta_f H^\circ - \sum_{reactants} \Delta_f H^\circ$$

Although it may seem tedious, it is extremely important that you keep writing those little "o"s with standard values. Although for our purposes we will not distinguish between ΔH and ΔH°, there is a huge difference, a most important distinction, between ΔS and ΔS° that underlies all of the phenomena related to equilibrium.

One last state function, *Gibbs energy*, *G*, is left to be introduced, which combines the ideas of the entropy changes of the system and the entropy changes of the surroundings in one convenient form which can be depicted graphically. Hang in there!

Problems

The symbol $\boxed{A}$ indicates that the answer to the problem can be found in the "Answers to Selected Exercises" section at the back of the book.

9.1 $\boxed{A}$ (a) Would you expect the signs of $\Delta_r S°$ and $\Delta_r H°$ to be positive or negative for the evaporation of methane, written $CH_4(l) \longrightarrow CH_4(g)$? (b) Would you expect the signs of $\Delta_r S°$ and $\Delta_r H°$ to be positive or negative for the reverse reaction, condensation, written $CH_4(g) \longrightarrow CH_4(l)$?

9.2 (a) Would you expect the signs of $\Delta_r S°$ and $\Delta_r H°$ to be positive or negative for the evaporation of ethanol, written $C_2H_5OH(l) \longrightarrow C_2H_5OH(g)$? (b) Would you expect the signs of $\Delta_r S°$ and $\Delta_r H°$ to be positive or negative for the reverse reaction, condensation, written $C_2H_5OH(g) \longrightarrow C_2H_5OH(l)$?

9.3 $\boxed{A}$ (a) Would you expect the signs of $\Delta_r S°$ and $\Delta_r H°$ to be positive or negative for the sublimation of iodine, written $I_2(s) \longrightarrow I_2(g)$? (b) Would you expect the signs of $\Delta_r S°$ and $\Delta_r H°$ to be positive or negative for the reverse reaction, written $I_2(g) \longrightarrow I_2(s)$?

9.4 (a) Would you expect the signs of $\Delta_r S°$ and $\Delta_r H°$ to be positive or negative for the evaporation of acetone, written $CH_3COCH_3(l) \longrightarrow CH_3COCH_3(g)$? (b) Would you expect the signs of $\Delta_r S°$ and $\Delta_r H°$ to be positive or negative for the reverse reaction, condensation, written $CH_3COCH_3(g) \longrightarrow CH_3COCH_3(l)$?

9.5 $\boxed{A}$ For the melting of solid methane, written as a balanced equation, $CH_4(s) \longrightarrow CH_4(l)$, $\Delta_r H° = 0.9392$ kJ/mol, (a) How much heat is required to melt 18.0 g of methane? (b) Compare this to the amount of energy required to melt an equivalent mass of ice (6010 J). Why the difference?

9.6 Provided $\Delta_r H° = 4.94$ kJ/mol for the melting of solid ethanol, written as a balanced equation, $C_2H_5OH(s) \longrightarrow C_2H_5OH(l)$, how much heat is required to melt 25.0 g of ethanol?

9.7 $\boxed{A}$ Provided $\Delta_r H° = 29.1$ kJ/mol for the melting of solid bromine, written as a balanced equation, $Br_2(s) \longrightarrow Br_2(l)$, how much heat is required to melt 25.0 g of bromine?

9.8 Provided $\Delta_r H° = 5.77$ kJ/mol for the melting of solid acetone, written as a balanced equation, $CH_3COCH_3(s) \longrightarrow CH_3COCH_3(l)$, how much heat is required to melt 25.0 g of acetone?

9.9 $\boxed{A}$ Determine $\Delta_r H°$ for each of the following reactions from data given in Table 9.1 or Appendix D.

(a) $2 N_2H_2(l) + 5 O_2(g) \longrightarrow 4 NO_2(g) + 2 H_2O(g)$
(b) $CO_2(g) + 4 HCl(g) \longrightarrow CCl_4(g) + 2 H_2O(g)$
(c) $CCl_4(g) + 4 HCl(g) \longrightarrow CH_4(g) + 4 Cl_2(g)$

9.10 Determine $\Delta_r H°$ for each of the following reactions from data given in Table 9.1 or Appendix D.

(a) $4 Fe(s) + 3 O_2(g) \longrightarrow 2 Fe_2O_3(s)$
(b) $CH_4(g) + 2 O_2(g) \longrightarrow CO_2(g) + 2 H_2O(g)$
(c) $SiO_2(s) + 4 HF(g) \longrightarrow SiF_4(g) + 2 H_2O(g)$

9.11 $\boxed{A}$ Determine $\Delta_r H°$, in kJ/mol, for each of the following reactions as written, from data given in Table 9.1 or Appendix D:

(a) $Fe_2O_3(s) + 6 HCl(g) \longrightarrow 2 FeCl_3(s) + 3 H_2O(g)$
(b) $SiCl_4(g) + 2 H_2(g) \longrightarrow Si(s) + 4 HCl(g)$
(c) $CsCl(s) \longrightarrow Cs^+(g) + Cl^-(g)$

9.12 Using Table 9.1 or Appendix D, draw diagrams similar to Figure 9.1 for the following three reactions, as written, including numbers for $\Sigma \Delta_f H°$(reactants), $\Sigma \Delta_f H°$(products), and $\Delta_r H°$. In each case, indicate whether the reaction is *endothermic* or *exothermic*.

(a) $N_2(g) + 2 O_2(g) \longrightarrow N_2O_4(g)$
(b) $N_2(g) + 5/2 O_2(g) + H_2O(g) \longrightarrow 2 HNO_3(l)$
(c) $NO(g) + O(g) \longrightarrow NO_2(g)$

9.13 Using Table 9.1 or Appendix D, draw diagrams similar to Figure 9.1 for the following three reactions, including numbers for $\Sigma \Delta_f H°$(reactants), $\Sigma \Delta_f H°$(products), and $\Delta_r H°$. In each case, indicate whether the reaction is *endothermic* or *exothermic*.

(a) $Mg(OH)_2(s) \longrightarrow MgO(s) + H_2O(g)$
(b) $2 Mg(s) + O_2(g) \longrightarrow 2 MgO(s)$
(c) $MgCO_3(s) \longrightarrow CO_2(g) + MgO(s)$

9.14 Complete the table below for 25°C based on the relationship between ΔH and ΔU:

$$\Delta H = \Delta U + P \Delta V = \Delta U + (\Delta n_{gas}) RT$$

Reaction (number of moles as indicated)	Δn_{gas}	ΔH/kJ	ΔU/kJ
a. $B_2H_6(g) + 3 O_2(g) \longrightarrow$ $B_2O_3(s) + 3 H_2O(l)$	——	——	-2147.5
b. $TiCl_4(l) + 2 H_2O(l) \longrightarrow$ $TiO_2(s) + 4 HCl(g)$	——	67.0	——
c. $2 C_2H_5NO_2(l) + 13/2 O_2(g) \longrightarrow$ $4 CO_2(g) + 5 H_2O(g) + 2 NO_2(g)$	——	-2696	——

9.15 $\boxed{A}$ For each of the following reactions at 25°C as written, calculate Δn_{gas}, ΔH, and ΔU based on ΔH being approximately $\Delta H°$ and the relationship between ΔH and ΔU.

(a) $Cu_2O(s) + 1/2 O_2(g) \longrightarrow 2 CuO(s)$
(b) $H_2(g) + Br_2(g) \longrightarrow 2 HBr(g)$
(c) $H_2(g) + Br_2(l) \longrightarrow 2 HBr(g)$

9.16 For each of the following reactions at 25°C as written, calculate Δn_{gas}, ΔH, and ΔU based on ΔH being approximately $\Delta H°$ and the relationship between ΔH and ΔU.

(a) $Hg_2Cl_2(s) \longrightarrow 2 Hg(l) + Cl_2(g)$
(b) $C_3H_8(g) + 5 O_2(g) \longrightarrow 3 CO_2(g) + 4 H_2O(g)$
(c) $FeCl_2(s) \longrightarrow Fe(s) + Cl_2(g)$

9.17 [A] Limestone ($CaCO_3$) stalactites and stalagmites are formed by the reaction of Ca^{2+} in cave water with hydrogen carbonate according to the following equation:

$$Ca^{2+}(aq) + 2\,HCO_3^-(aq) \longrightarrow CaCO_3(s) + CO_2(g) + H_2O(l)$$

When 10.0 g of limestone forms at 298 K and 1 bar of pressure, 3.895 kJ of heat is absorbed. (a) Determine q and w for this reaction. [*Hint:* How many moles of gas are produced?] (b) For the equation as written, which is for 1 mol of $CaCO_3$, what are the values of $\Delta_r H$ and $\Delta_r U$?

9.18 Find $\Delta_f H^\circ$ of propane, propene, and propyne using the NIST Chemistry WebBook.[4] Then use those values to calculate $\Delta_r H^\circ$ for the following two reactions, as written:

(a) $H_2(g) + C_3H_6(g, \text{propene}) \longrightarrow C_3H_8(g, \text{propane})$
(b) $2\,H_2(g) + C_3H_4(g, \text{propyne}) \longrightarrow C_3H_8(g, \text{propane})$

9.19 Based on your answers to Problem 9.18(a) and (b), determine the amount of heat released in the reaction of 16 g of propyne with just enough hydrogen gas to give an equivalent molar amount of propene.

$CH_3CH_2CH_3$
propane

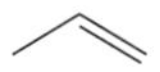
CH_3CHCH_2
propene

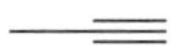
CH_3CCH
propyne

9.20 [A] Find $\Delta_r H^\circ$ for the following reactions as written using information from the NIST Chemistry WebBook (under "reaction thermochemistry data" of the appropriate fluorine- or chlorine-containing compound). Then use that information along with Hess's law and the known heats of formation of gaseous propane, HF, and HCl (from Appendix D) to determine $\Delta_f H^\circ$ of 2-fluoropropane and 2-chloropropane.

(a) $H_2(g) + C_3H_7F(g, \text{2-fluoropropane}) \longrightarrow C_3H_8(g, \text{propane}) + HF(g)$
(b) $H_2(g) + C_3H_7Cl(g, \text{2-chloropropane}) \longrightarrow C_3H_8(g, \text{propane}) + HCl(g)$

9.21 Look up the mean bond dissociation energies of F–F, C–F, Cl–Cl, and C–Cl σ bonds in Table 5.1 (page 100). Draw Lewis dot structures for 2-fluoropropane and 2-chloropropane. [Note that "2-" indicates that the F or Cl is attached at the second carbon from the end.] Explain why the heat of formation of 2-fluoropropane is so much more negative than the heat of formation of 2-chloropropane in terms of these energies.

9.22 [A] Find $\Delta_r H^\circ$ for the following reactions, as written:

(a) $Pb(s) + PbO_2(s) + 2\,SO_3(g) \longrightarrow 2\,PbSO_4(s)$
(b) $SO_3(g) + H_2O(l) \longrightarrow H_2SO_4(l)$

9.23 [A] Using your answer to the previous problem, determine $\Delta_r H^\circ$ for the reaction of lead sulfate with water as written:

$$2\,PbSO_4(s) + 2\,H_2O(l) \longrightarrow Pb(s) + PbO_2(s) + 2\,H_2SO_4(l)$$

9.24 Find $\Delta_r H^\circ$ for the following reactions, as written:

(a) $P_4(s,\text{white}) + 6\,Cl_2(g) \longrightarrow 4\,PCl_3(g)$
(b) $P_4(s,\text{white}) + 10\,Cl_2(g) \longrightarrow 4\,PCl_5(g)$

9.25 Using your answer to the previous problem, determine $\Delta_r H^\circ$ for the reaction

$$PCl_3(g) + Cl_2(g) \longrightarrow PCl_5(g)$$

9.26 Find $\Delta_f H^\circ$ for bromomethane (CH_3Br) and iodomethane (CH_3I) using the NIST Chemistry WebBook. Then use those values to calculate $\Delta_r H^\circ$ for the following two reactions:

(a) $HI(g) + CH_3I(g) \longrightarrow CH_4(g) + I_2(g)$
(b) $HBr(g) + CH_3Br(g) \longrightarrow CH_4(g) + Br_2(g)$

Brain Teasers

9.27 The ozone-depleting chemical dichlorodifluoromethane, CF_2Cl_2, has been proposed to be removed from the atmosphere by reaction with water vapor:

$$CF_2Cl_2(g) + 2\,H_2O(g) \longrightarrow CO_2(g) + 2\,HF(g) + 2\,HCl(g)$$

Argue for or against this proposal by calculating both $\Delta_r H^\circ$ and $\Delta_r S^\circ$ for this reaction, as written. Discuss the effect of temperature on the reaction along the lines of Figure 9.2.

9.28 Consider a thunderstorm that produced an inch of rain over a 200 square mile area in a period of one hour. (a) How much energy was released during this storm? (b) Power is a measure of rate of energy dispersed and is generally expressed in watts (W), where $1\text{ W} = 1\text{ J/s}$. How much power was produced by this storm? (c) Electric companies generally measure energy and bill their customers in kilowatt-hours (kWh). Show that this is a unit of energy, and then express the energy released during the storm in kWh. (d) Compare your answer to (c) to the 300–400 billion kWh released in the average hurricane, and to the 300 kWh used by a household in a month.

---

[4] *http://webbook.nist.gov/chemistry*. You should search for information based on a name, for example, *propane*, or a formula, such as C_3H_8, and select from the thermodynamic data options at least *gas phase* and *reactions*.

CHAPTER 10

Gibbs Energy (*G*)

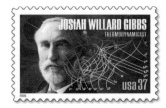

10.1 The Second Law Again, with a Twist

We really *don't* need anything more to explain all of equilibrium, but somehow chemists haven't been satisfied with the Second Law of Thermodynamics written as

$$\Delta S_{\text{universe}} = \Delta S + \Delta S_{\text{sur}} > 0$$

The "entropy of the surroundings" and the "entropy of the universe" are just a little too far out for us chemists. We need more "chemical" ideas which relate to bonding and energy. We want to focus on the *system*.

To start, let's think a little more about the melting reaction:

$$(\text{solid}) \longrightarrow (\text{liquid})$$

We know that for this reaction heat will be required to break the intermolecular bonds in the solid. This means that heat is going to go *into* the system, *out of* the surroundings, and the entropy of the surroundings will drop. Thus, for melting, $\Delta S_{\text{sur}} < 0$. In addition, we know that the entropy of the system itself will increase, because liquids can rotate, while solids cannot. For melting, $\Delta S > 0$. These aspects of melting will be true at all temperatures.

Figure 10.1 illustrates the effect of temperature on ΔS and ΔS_{sur}. At low temperature, since $\Delta S_{\text{sur}} = -q/T$, it is the change in entropy of the surroundings that dominates, and the melting reaction is not favored. As the temperature increases, though, we see that the entropy change of the surroundings becomes less important, and at some point the upward pointing arrow for the system is longer than the downward pointing arrow for the surroundings, and melting occurs.

The temperature at which these two arrows are both the same length is called the *melting point* of the substance. At that point $\Delta S_{\text{universe}} = 0$, and *neither* reaction is favored. That is, at the melting point, unless heat is put into or taken out of the system, the system will be stable and no changes will be observed. Indeed, a Thermos™ bottle is simply a device designed to not allow the transfer of heat. Ice water at 0°C will remain at 0°C in a Thermos bottle without changing because it is already in its most probable state. But if heat leaks in, the melting reaction will occur until all the ice has melted.

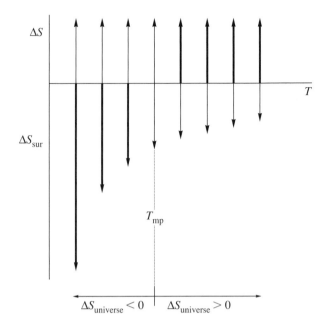

Figure 10.1
At any particular temperature, the longer (dominant) arrow is indicated in bold. It is the sum of the arrows for ΔS and ΔS_{sur} that gives $\Delta S_{universe}$. Below the melting point the longer arrow is the one for ΔS_{sur}. Above this temperature, the longer arrow is the one for ΔS. Melting occurs spontaneously only above the melting point, at temperatures for which the sum of the two arrows is positive.

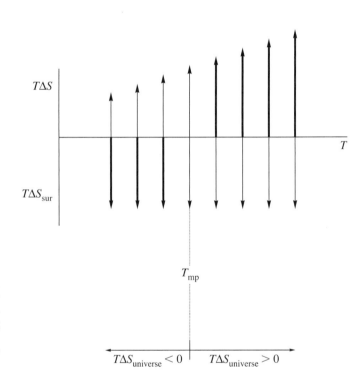

Figure 10.2
A slightly different way of showing the temperature effect on the melting reaction. Now it is the sum of the arrows for $T\Delta S$ and $T\Delta S_{sur}$ that must be positive, as that sum is $T\Delta S_{universe}$. Once again, below the melting point, the longer arrow is the one for ΔS_{sur}. Above this temperature, the longer arrow is the one for ΔS. Melting occurs spontaneously only above the melting point, at temperatures for which the sum of the two arrows is positive.

Now consider the Second Law in a form multiplied through by temperature, T (kelvin):

$$T\Delta S_{universe} = T\Delta S + T\Delta S_{sur} > 0$$

This switches from units of entropy (J/K) into units of energy (J). It's still the Second Law, and Figure 10.2 depicts the effect now of temperature on the same melting reaction, this time using arrows that include the temperature, $T\Delta S$ and $T\Delta S_{sur}$. Notice first that everything that was said regarding Figure 10.1 still applies. The entropy change in the surroundings dominates at low temperature, and the entropy change in the system dominates at high

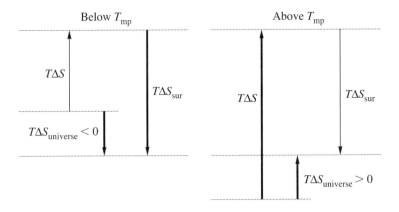

Figure 10.3
Vector graph for the same melting reaction as depicted in Figures 10.1 and 10.2, but now focusing on a temperature below T_{mp} (on the left) and a temperature higher than T_{mp} (on the right). To add vectors, simply put the tail of one ($T \Delta S_{sur}$) to the tip of the other ($T \Delta S$) and see where they lead ($T \Delta S_{universe}$). The Second Law is violated for melting at the lower temperature but not at higher temperature.

temperature. But now all the arrows for the surroundings are roughly the same length, and it is the arrows for the *system* that are changing length with temperature! Why is that?

In the case of the system, we are simply now multiplying by T, and the arrows, which were relatively constant, must now increase with temperature. In the case of the surroundings, since

$$\Delta S_{sur} = \frac{q_{sur}}{T} = \frac{-q}{T}$$

we have that $T \Delta S_{sur} = -q$, which is roughly constant with temperature. The same bonds must be broken at low temperature as at high temperature, after all.

Notice that these arrows add up because they represent relationships between state functions and thus must obey Hess's law. Arrows are handy, because they are *additive* and have *direction*. This is exactly like the mathematics we are working with. Throughout this book, you may have already noticed, an arrow goes up when the value it represents is positive and down when its value is negative. To add any two arrows, simply move the tail of one to the tip of another and see where they take you (Figure 10.3). In the case of melting, $T \Delta S$ is positive, while $T \Delta S_{sur}$ is negative. (Remember, ΔS_{sur} is negative whenever heat is extracted from the surroundings—that is, whenever $q > 0$.)

The Second Law in these terms simply says that **at any given temperature, the sum of the arrow for $T \Delta S$ and the arrow for $T \Delta S_{sur}$ must always be an arrow pointing up.** In this case the entropy of the system is going up, because the solid is turning into liquid (☺), while the entropy of the surroundings is going down (☹), because heat is being pulled out of the surroundings in order to break the bonds holding the solid together. Note that only above the melting point is the entropy of the universe found to be increasing ($\Delta S_{universe} > 0$), so melting only occurs spontaneously above that temperature.[1]

Now, it was shown in Chapter 9 that $q_{sur} = -q = -\Delta H$ for chemical reactions at constant pressure, and it can generally be assumed that ΔH does not depend (much) upon temperature.[2]

We have

$$T \Delta S_{sur} = -\Delta H$$

or

$$\Delta H = -T \Delta S_{sur}$$

---

[1] Melting also occurs *at* the melting point, but not *spontaneously*. Heat must be transferred from an external source, and that requires the temperature of the surroundings to be slightly higher than the temperature of the system.

[2] To be exact, both ΔS and ΔH do depend a little on temperature, and the only temperature we can really be sure about when using our tables is 298 K. But that is a subtlety we will ignore here. For an example of how ΔH and ΔS change with temperature, see Appendix E.

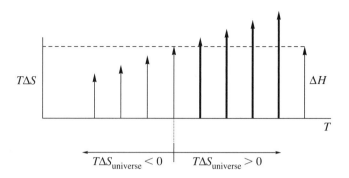

Figure 10.4
Vector graph for the same melting reaction as depicted in Figures 10.1 and 10.2, now including an arrow for $\Delta H = -T\Delta S_{sur}$. Now the second law can be stated, "$\Delta H < T\Delta S$."

This now suggests an even better way to depict the temperature effect on melting (Figure 10.4). Here, instead of showing all those constant $T\Delta S_{sur}$ arrows, we simply draw a line at the level of ΔH. Only when the arrow representing $T\Delta S$ is *above* this line is the reaction favored! Interestingly, a new form of the Second Law now becomes apparent, at least under the conditions of constant temperature and pressure:

$$\Delta H < T\Delta S \tag{10.1}$$

Note in Figure 10.4 that above the melting point, where the entropy of the universe is increasing, the ΔH arrow is shorter than the $T\Delta S$ arrow. That is, $\Delta H < T\Delta S$. Below the melting point, the reverse is true, and the Second Law is violated.

10.2 The Definition of Gibbs Energy: $G = H - TS$

If you think about it, this new form of the Second Law is really handy! It means that all we have to do to determine whether a reaction should proceed is to pick a temperature and then figure out (or estimate) ΔH and ΔS. Then, if $\Delta H < T\Delta S$, we are in business, and if it isn't, then we can forget it; the reaction won't go.

   By thinking about the reaction this way, we have accomplished what the chemist requires: We have focused on what is happening in the system instead of having to think about the surroundings all the time. In addition, we have switched to units more familiar to chemists—the units of energy. This suggests one last energy-like state function. Subtracting $T\Delta S$ from both sides of this equation and calling the difference "ΔG" we have

$$\Delta G = \Delta H - T\Delta S < 0 \tag{10.2}$$

Note that another way of thinking about ΔG is simply as $-T\Delta S_{universe}$, since

$$\Delta G = \Delta H - T\Delta S$$
$$= -T\Delta S_{sur} - T\Delta S$$
$$= -T\Delta S_{universe}$$

Thus, in Figure 10.3 an arrow for ΔG would just be the opposite of the arrow for $T\Delta S_{universe}$. Rewriting this last result, we have an important relationship:

$$\Delta S_{universe} = -\frac{\Delta G}{T} \tag{10.3}$$

For a reaction to be favorable, $\Delta S_{universe}$ must be positive—the arrow for $T\Delta S_{universe}$ must point up. In contrast, the arrow for ΔG must point down. **The value of ΔG for a chemical reaction must be negative for a reaction to be spontaneous.**

The state function G itself, without the Δ sign, is the *Gibbs energy*,[3] defined as

$$G = H - TS \qquad (10.4)$$

The relationship between ΔG, ΔH, and ΔS given in Equation 10.2 derives from finding changes in Gibbs energy under the condition of constant temperature:

$$\Delta G = \Delta H - \Delta(TS)$$
$$= \Delta H - T\Delta S \quad \text{(constant temperature only)} \qquad (10.5)$$

In chemistry, focusing as we do on chemical reactions, we will find that Gibbs energy is extraordinarily useful even if we don't stick to the conditions of constant temperature (so that $\Delta(TS) = T\Delta S$) and constant pressure (so that $\Delta H = q$). This is because all we will want to do is check to see if ΔG is zero (the system is at equilibrium) or if the *sign* of ΔG is positive or negative (telling us which way to go from here to get to equilibrium). What the sign of ΔG does is give us the *direction* of the change or the "potential" for change.

For example, say we have

$$A \longrightarrow 2\,B \qquad \Delta_r G = -100 \text{ kJ/mol}$$

Since we are just interested in the potential for change, we use the trick that the sign of ΔG (in kJ) for the reaction of *any* amount of reactants, no matter how small, must be the same as the sign of ΔG for the reaction of a *stoichiometric* amount of reactants (one mole of A in this case) reacting to form a *stoichiometric* amount of products (two moles of B)—what we call $\Delta_r G$ (in kJ/mol). Calculating that value, which is a property of the system *in its current state*—without regard to how that state might change in the future—is totally general.[4]

Note too, that if we were to find that

$$A \longrightarrow 2B \qquad \Delta_r G = 100 \text{ kJ/mol}$$

we are definitely NOT saying, "When one mole of A reacts to give two moles of B, the change in Gibbs energy is 100 kJ." What we mean to say is this: "The difference in Gibbs energy between having one mole of A and having two moles of B is 100 kJ." Can you detect the subtle difference? Because we are talking about just the *potential* for change, we suggest you read all "Δ_r" symbols as "differences in" when considering Gibbs energy.

It's like saying that there is a difference in altitude between Northfield and Denver of 4280 ft instead of saying that going from Northfield to Denver involves a change of 4280 ft. Both statements are true, but the first is more a statement of state, and the second is more focused on getting from one state to the other.

Make no mistake about it, if we say that $\Delta_r G = 100$ kJ/mol for $A \longrightarrow 2\,B$, we are *not* saying, "Going from A to B will require 100 kJ of Gibbs energy" No, indeed. We are saying that you are not going to go from A to B. Period. Whether $\Delta_r G = 100$ kJ/mol or $100\ \mu$J/mol makes no difference. As long as $\Delta_r G$ is positive, the reaction just isn't going to happen spontaneously. That would be like trying to get from Northfield to Denver in a canoe with no paddle. It can't be done. (Not without a lot of help, anyway!) Gibbs energy

---

[3] Gibbs energy was named after Josiah Gibbs, who was trained as an engineer. Gibbs was appointed Professor of Mathematical Physics at Yale University in 1871. He published this work during the years 1873–1878.

[4] The generality of $\Delta_r G$ is expressed in its other name, the *reaction potential*. It turns out that the value we get for the reaction potential under conditions of constant temperature and pressure, using Gibbs energies, is the same as we would get from, say, keeping the temperature and the volume constant. Thus, keeping the temperature and the pressure constant is just a convenient means of arriving at the reaction potential.

"flows" only downstream. The sign of $\Delta_r G$ is just an indicator of the probability of a reaction proceeding spontaneously in the direction written in a balanced chemical equation.

10.3 Plotting *G* vs. *T* (G–T Graphs)

You should now be quite comfortable with the idea of depicting a state function on the *y* axis of a graph. We did this for altitude *A* and internal energy *U* in Chapter 5 and for enthalpy *H* in Chapter 9. In all of these graphs, vertical arrows represent changes in a state function in going from one state to another. The horizontal axis is used just to set apart the different possible states, usually with reactants on the left and products on the right.

It is interesting, then, to expand this idea to Gibbs energy. Consider the definition of Gibbs energy as a function of *T*:

$$G = H - TS$$

Note that this equation is in the form $y = (m)x + (b)$, where we have

G	y	(dependent variable)
T	x	(independent variable)
−*S*	m	(slope)
H	b	(*y*-intercept)

A plot of *G* vs. *T*, where *G* is along the *y* axis and *T* is along the *x* axis, is shown in Figure 10.5. We will refer to these plots as G–T graphs.

Notice that the curve slopes downward, because *S* is positive, and so −*S*, which is the slope, is negative. Notice also that the curve is not a straight line. This is because *S* increases with temperature for all substances, making "−*S*" more and more negative with increasing temperature. You should note several aspects of this figure:

- At 0 K, $G = H$, because the *T* in $G = H - TS$ equals zero. That is, **the Gibbs energy at 0 K is simply the energy due to *bonding***. The more bonding a substance has, the lower its curve will start on this graph.
- As the temperature increases, *G* drops. This is because **entropy is positive for all real substances**, and the *slope* of the curve at any temperature is −*S*.
- Finally, the slope of *G* gets more and more negative with increasing temperature. This is because *S itself* **increases with temperature for all substances**, making the −*TS* term of $G = H - TS$ especially important at higher temperatures.

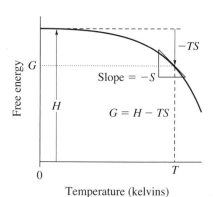

Figure 10.5
General G–T graph for any substance. The starting point for the curve is *H*, and the slope at any temperature equals −*S*.

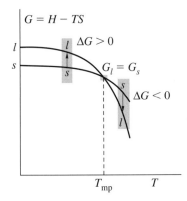

Figure 10.6
G–T graph depicting melting. Only above the melting point temperature T_{mp} is the reaction "$(s) \longrightarrow (l)$" favorable. Only when $\Delta G < 0$, at $T > T_{mp}$, is the Second Law satisfied.

10.4 Comparing Two or More Substances Using G–T Graphs

The most valuable aspect of G–T graphs, introduced by Gibbs in 1878, is that they allow the comparison of two states (Figure 10.6). The important thing is not the absolute positions of the curves, but instead the vertical *difference* between the curves, which is ΔG and will always be represented as an arrow. In this way, arrows and states are represented vertically very much the way they have been on other graphs we have used.

The difference here is that the horizontal axis, being T, allows us to see how the energies of the two states change with temperature *relative to each other*.

Realize that the situation depicted in Figure 10.6 is precisely the situation depicted in Figures 10.1–10.5. This is the same melting reaction, which happens spontaneously only above the melting point, T_{mp}. Remember, melting is

$$\text{solid} \longrightarrow \text{liquid}$$

Note that in Figure 10.6 the upward-directed arrow for this reaction at low temperature indicates that it isn't going to happen. A solid will not melt spontaneously below its melting point. In fact, below a temperature of T_{mp}, it is the *other* reaction, liquid-to-solid, which will occur instead.

Very importantly, if two states are compared, the state that is favored at any particular temperature is simply the one with the lower curve on a G–T graph (Figure 10.7). By "favored" here we mean *completely* favored. That is, the only substance which can exist *at all*. Thus, in Figure 10.7, wherever Curve B is below Curve A (at low temperature), the reaction

$$A \longrightarrow B$$

will continue until one of two things happen. Either

1. a limiting reactant is used up, or
2. somehow the curves *themselves* change until G_B is no longer lower than G_A.

Both of these are real possibilities that we will consider. For now, we consider only the first option.

In all G–T graphs, the actual curves are independent of the amounts of material actually present, just as are the coefficients in a balanced chemical equation. That is, these graphs, like the ones we have used to illustrate Hess's law, will always be based on some sort of balanced chemical equation. **One curve will always correspond to reactants and one will always correspond to products in a balanced chemical equation.** The amounts of reactants and products will simply be the stoichiometric molar amounts indicated in that balanced equation. Thus, at least for solids and liquids, as substances appear or disappear in a reaction there is no change in the diagram itself. The diagram simply relates to the stoichiometric amounts as always.

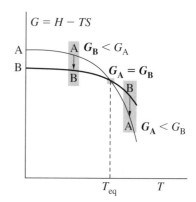

Figure 10.7
G–T graph depicting two states A and B. Below the temperature at which $G_A = G_B$ only B will exist in the most probable distribution; above this temperature only A will exist.

10.5 Equilibrium Is Where $\Delta_r G = 0$

Notice that in both Figures 10.6 and 10.7 the curves cross. The temperature corresponding to the crossing, T_{eq}, is extremely important. ***Two states can coexist in equilibrium with each other only at the temperature at which their G–T curves cross.*** Looking at Figure 10.6, you can see why this must be true. There is one point and only one point where the two curves cross. Only at that point is it true that $G_B = G_A$, that is, $\Delta_r G = 0$ for *both* of the following equations:

$$A \longrightarrow B \qquad \Delta_r G = G_B - G_A = 0$$
$$B \longrightarrow A \qquad \Delta_r G = G_A - G_B = 0$$

That is, for *both* reactions there is no difference in Gibbs energies of product and reactant. The "forward" reaction (A $\longrightarrow$ B) and the "reverse" reaction (B $\longrightarrow$ A) are both equally probable.

This is precisely what we mean by "equilibrium." We can more simply write:

$$A \rightleftharpoons B \qquad \Delta_r G = 0$$

The important thing to realize is that equilibrium is possible only at a specific temperature: that temperature for which the two G vs. T curves cross. Not every pair of curves will cross. Whether they do depends upon which one starts out higher (the state with the larger H and weaker bonding) and which one has the steepest descent (the state with the larger S and more disorder). Had the one that started out lower had the steeper slope, then the two curves would *never* have crossed, and the lower one would *always* be the one that is favored.

10.6 The "Low Enthalpy/High Entropy Rule"

A simple "Low Enthalpy/High Entropy rule" shows up here:

- At low temperature the TS term of $G = H - TS$ is of little significance, and the important term then is H, enthalpy. Thus, when two substances are compared at low temperature, their enthalpy difference, ΔH, is most important. **At low temperature, the substance with the lower enthalpy is favored**. This is simply another way of saying that it is at low temperature that entropy change in the *surroundings* is the important thing.
- At high temperature the TS term becomes more important, and it dominates the H term in $G = H - TS$. Thus, regardless of where the curves start, the one having the steeper negative slope (higher entropy, S) will at some temperature

become the lower of the two curves. Thus, **at high temperature, the substance with the higher entropy will be favored**. This is just another way of saying that at high temperature, the entropy change in the system is the important thing.

Thus we have

The state favored at . . .

. . . low temperature has . . .	. . . high temperature has . . .
. . . lower enthalpy	. . . higher entropy
. . . stronger bonds	. . . more ways of distributing its energy
. . . lower ground state	. . . more closely spaced energy levels

And the state favored at *any* temperature has the lower Gibbs energy!

10.7 A Quantitative Look at Melting Points: $0 = \Delta_{fus}H - T_{mp}\Delta_{fus}S$

G–T graphs clue us in to the general idea of which way a reaction is going to go at different temperatures. We want to determine, if nothing else, at exactly what temperature the two curves cross, the *equilibrium temperature*. As an example, consider the equilibrium between the solid and liquid states of water:

$$H_2O(s) \rightleftharpoons H_2O(l)$$

We are interested specifically in the place in figures such as Figures 10.6 and 10.7 where the two curves cross, where $\Delta_r G = 0$. This is the point where the two Gibbs energies are equal.

To analyze this situation more quantitatively, we consider the *forward* reaction:

$$H_2O(s) \longrightarrow H_2O(l)$$

We need to determine the difference in Gibbs energies for this equation and set that value equal to zero. To do this, we tabulate values of standard molar enthalpy and entropy and use the fact that $\Delta_r G = \Delta_r H - T\Delta_r S$.

The melting point, T_{mp}, can be determined if we know the actual values of $\Delta_r H$ and $\Delta_r S$ for the melting reaction. Since there are no gases here, we will be able to use all *standard* values. Thus, we want to know at what temperature the lines for liquid and solid cross, that is, where $\Delta_r G = 0$:

$$\Delta_r G = \Delta_r H - T\Delta_r S$$
$$0 = \Delta_{fus}H^\circ - T_{mp}\Delta_{fus}S^\circ$$
$$T_{mp} = \frac{\Delta_{fus}H^\circ}{\Delta_{fus}S^\circ}$$

(10.6)

In the case of the melting of water, it turns out that $\Delta_{fus}H^\circ$ is 6010 J/mol, while $\Delta_{fus}S^\circ$ is 22.0 J/mol·K, so we have

$$T_{mp} = \frac{\Delta_{fus}H^\circ}{\Delta_{fus}S^\circ} = \frac{6010 \text{ J/mol}}{22 \text{ J/mol·K}} = 273 \text{ K}$$

It is typical to find tables of melting points (T_{mp}) and *standard enthalpies of fusion*, $\Delta_{fus}H^\circ$. The standard enthalpy of fusion is the amount of energy required at the melting point to melt one mole of the solid. In that case we can use the above relationship to solve

Table 10.1
Thermodynamic data for the melting reaction

Substance	T_{mp}/kelvin	$\Delta_{fus}H°$/(kJ/mol)	$\Delta_{fus}S°$/(J/mol·K)
O_2	55	0.45	8.2
HCl	159	1.99	12.5
HI	222	2.87	12.9
CCl_4	250	2.51	10.0
H_2O	273	6.01	22.0
NaCl	801	30.2	37.7
NaF	992	29.3	29.5

for the entropy change on melting, $\Delta_{fus}S°$. Sometimes this can be informative. In Table 10.1 we see a comparison of these values for a variety of substances. Note that substances with higher melting points tend to have both larger enthalpy changes and larger entropy changes on going from solid to liquid. Ionic compounds such as NaCl and NaF tend to have very strong bonds as well as a lot of constraint in the solid state, both of which are lost in going to the liquid state. Water is intermediate between ionic salts and the other compounds listed.

10.8 The Gibbs Energy of a Gas Depends upon Its Pressure

Now consider the equilibrium between liquid and gaseous water:

$$H_2O(l) \rightleftharpoons H_2O(g)$$

In Figure 10.8 is shown the G–T graph for the water liquid/gas system at the "standard" vapor pressure of 1 bar. Note that the liquid curve starts below that for the gas, because liquid water has stronger bonding than gaseous water due to the presence in the liquid of weak intermolecular hydrogen bonds. In addition, the curve for gaseous water is much steeper than that for the liquid, because the gas naturally has much more entropy than the liquid. Below 373 K (100°C) liquid water has the lower Gibbs energy. Any gas present at 1 bar will start to condense *under these conditions*.

Recall, however, that the entropy of a gas depends upon its pressure (Chapter 8):

$$S_{gas} = S°_{gas} - R \ln(P_{gas}/\text{bar})$$

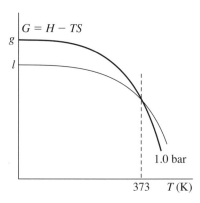

Figure 10.8
G–T graph for liquid H_2O (light line) and gaseous H_2O (heavy line) under the standard conditions of 1 bar partial pressure. The crossing point is 373 K (100°C). Below this temperature, any gas present will condense to form more liquid; above this temperature any liquid present will evaporate to form more gas.

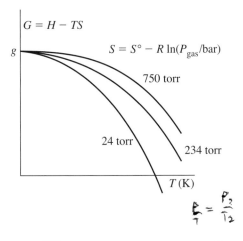

Figure 10.9
The shape of the curve for a gas on G–T graphs depends upon the pressure of the gas, because low partial pressure corresponds with high entropy (steeper negative slope).

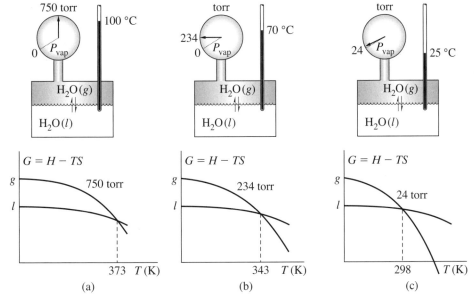

Figure 10.10
Three different equilibrium situations involving water liquid/gas are depicted. In (a) the temperature is 100°C (the "standard" boiling point of water), and the water vapor pressure is at the "standard" atmospheric pressure of 1 bar. In (b) the temperature is 70°C, and the corresponding gas pressure is only 234 torr (0.31 bar). Finally, in (c) at 25°C the pressure required to make the curve for the gas cross is only 24 torr (0.032 bar).

Since the entropy of the gas sets the slope of the curve, **the slope of the gas curve will depend upon the partial pressure of the gas.** Figure 10.9 depicts the differences to be expected for water vapor.

The standard condition of 1 bar pressure shown in Figure 10.8 is the top "750 torr" curve here. But as the partial pressure of the water vapor in the system drops, so, too, does the gas curve. This is because a lower pressure implies a larger molar volume and more entropy for a mole of the gas. Since the slope of the gas curve is $-S_{gas}$, as the gas entropy increases, its curve will become steeper at all temperatures. The immediate result of this is that the gas and liquid curves will cross at different temperatures depending upon the partial pressure of water vapor present (Figure 10.10).

All three of these conditions may be thought of as "equilibrium." In all cases the gas and liquid are in a dynamic state, exchanging particles and energy. In all cases, no observable phenomena occur. There is simply more gas present at higher temperature.

What if the system is not at equilibrium? In Figure 10.11 three scenarios are depicted for water at 70°C. In the first case, as depicted in Figure 10.11(a), the vapor pressure is too high, and the gas will be condensing, thus lowering its partial pressure. In (c) the vapor pressure is too low, and the liquid will be evaporating, thus increasing the pressure of the gas. Ultimately, both will end up at the same final state, depicted in (b).

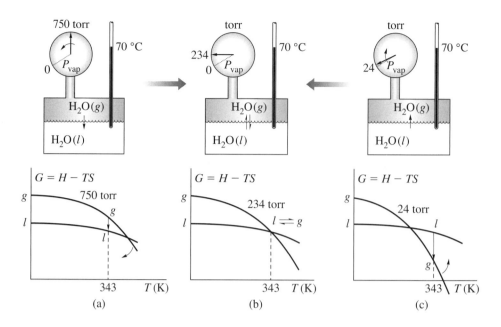

Figure 10.11
Three different situations all at 70°C. Only (b) is at equilibrium. In (a) the vapor pressure is too high, and gas is condensing to liquid. In (c) the vapor pressure is too low, and liquid is evaporating. As these processes occur, as long as the temperature of the system is maintained at 70°C by cooling in (a) or heating in (c), the vapor pressure of the gas will change, and the curve for the gas will change as well, until the vapor pressure of the gas reaches 234 torr, and the Gibbs energy situation is that depicted in (b).

The most important lesson here is that if a liquid is sealed in a container with any space above it at all, it will come to equilibrium with its vapor and increase the overall pressure in the container. ***This is why you should always be careful opening bottles of liquid chemicals, and why many liquid chemicals are refrigerated!*** Your concern should not be restricted to liquids. The equilibrium between a solid and its vapor is similar and is discussed in Section 12.3.

10.9 Vapor Pressure, Barometric Pressure, and Boiling

It is very important to realize that the "pressure" being referred to here is the partial pressure of water vapor—that part of the overall pressure of gas above the liquid that corresponds specifically to the H_2O present. If other air molecules are present (N_2, O_2, etc.), they have essentially no effect on this equilibrium between liquid water and its vapor. Where air gets involved is when the system is open to the environment. In that case, we have the possibility of boiling.

The total pressure of gas above a liquid open to the atmosphere is called "air pressure," "atmospheric pressure," or *barometric* pressure, P_{bar}, and is simply due to the weight of all the air above a liquid. Thus, the weight of the "standard atmosphere at sea level" is one "bar" which is equivalent to the weight of a 750-mm-high column of liquid mercury (750 mmHg, or 750 *torr*) or a 10,160-mm-high column of liquid water (10,160 mmH_2O).[5]

The barometric pressure determines the boiling point of a liquid open to the atmosphere, because that pressure is transmitted through the liquid to any potential bubble that might form in the process we call boiling.

The amount of pressure being transmitted to the bubble is always a little higher than atmospheric pressure, because the total pressure at any point under the surface of the liquid, where a bubble might form, is just the sum of the weight of all the air and water above that

---

[5] This number for one atmosphere's worth of liquid water, 10,160 mmH_2O, is based on the ratio of densities of water to mercury, 1 g/cm³ vs. 13.55 g/cm³. Thus, it takes a column of water 13.55 times higher than 750 mm to equal the weight of a column of mercury 750 mm high.

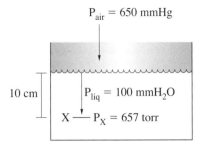

Figure 10.12
The pressure under the surface of a liquid, where a bubble might form, is just the sum of the weight of all the air and liquid above it.

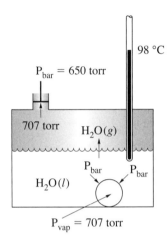

Figure 10.13
If the atmospheric pressure is low, the water will boil below 100°C, because the vapor pressure inside the bubble, P_{vap}, will be higher than the pressure being exerted throughout the liquid, just a bit more than P_{bar}. The bubble will grow and rise.

point, as shown in the calculation below and in Figure 10.12.

$$P_{total} = P_{air} + P_{liquid}$$

$$= 650 \text{ mmHg} + 10 \text{ cmH}_2\text{O}$$

$$= 650 \text{ mmHg} + 100 \text{ mmH}_2\text{O} \times \frac{750 \text{ mmHg}}{10160 \text{ mmH}_2\text{O}}$$

$$= 657 \text{ mmHg} = 657 \text{ torr}$$

On a typically beautiful day in Denver, Colorado, when the barometric pressure is 650 torr (Figure 10.13), water will boil at just 98°C, because the vapor pressure associated with equilibrium at that temperature is 707 torr.[6] As a bubble of gaseous water starts to form within the liquid, the pressure inside that bubble will be 707 torr, and the bubble will expand against the approximately 650 torr of pressure being exerted through the water and rise to the surface.

Figure 10.14 shows three different scenarios involving boiling. In (a), although the vapor pressure is 750 torr and the temperature is 100°C, still boiling does not occur, because there is also air in the space above the liquid, and the overall atmospheric pressure is therefore above 750 torr. The bubble will collapse instead of expand. In (b), the atmospheric pressure is less than 750 torr, and the bubble will expand. In (c), a vacuum pump has been attached to the system, and the total pressure above the water, P_{vac}, is less than the vapor pressure at the temperature indicated, 25°C. The bubble will expand, and boiling will occur.

Let's figure out exactly how the equilibrium vapor pressure and equilibrium temperature are related. Once again, because now we are interested in equilibrium, we set $\Delta G = 0$ and solve for the equilibrium temperature, which we call T_{eq}. Starting with the equation

$$\text{liquid} \longrightarrow \text{gas}$$

we write (Section 8.9, assuming pressure in bar):

$$\Delta_r G = \Delta_r H - T \Delta_r S$$
$$0 = \Delta_r H° - T_{eq}(\Delta_r S° - R \ln P_{gas}) \tag{10.7}$$

which can be solved for either T_{eq} or P_{gas} (in bars) directly using an equation-solving calculator as long as we know the values of $\Delta_r H°$ and $\Delta_r S°$.

---

[6] Note that the barometric pressure reported on the evening news is really barometric pressure "adjusted to sea level" based on the "standard atmosphere" which decreases in pressure roughly 1 inchHg per 1000 ft altitude. This is done so that lows and highs on weather maps are independent of local terrain. Thus, a reported pressure of "29.92 inchHg" (760 mmHg) in Denver is really only about 25 inchHg, or 635 mmHg actual pressure. "650 mmHg" pressure in Denver would be a "high pressure" day, generally associated with good weather.

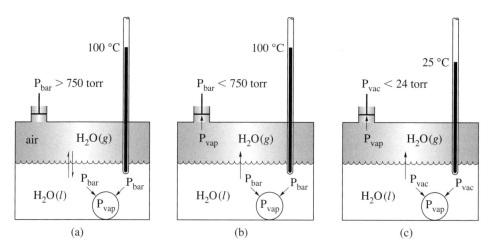

Figure 10.14

Three different *nonequilibrium* situations in which the system is open to the surroundings. In (a) a bubble will not form, because air is present, and the pressure being exerted throughout the liquid is higher than the equilibrium vapor pressure (750 torr). In (b) the atmospheric pressure is somewhat lower, and now the vapor pressure *inside the bubble* will be greater than the pressure of liquid. The bubble will expand and rise in the phenomenon we call *boiling*. This will continue as long as heat is applied and the piston is allowed to move upward. In (c) boiling is occurring at the unusually low temperature of 25°C because the pressure above the solution, P_{vac}, is at the unusually low value of 24 torr.

Alternatively, if no equation-solving calculator is available, this equation can be re-arranged to give either of Equations 10.8 or 10.9, from which T_{eq} or P_{gas} may be obtained.

$$T_{eq} = \frac{\Delta_r H^\circ}{\Delta_r S^\circ - R \ln P_{gas}} \tag{10.8}$$

$$\ln P_{gas} = \frac{-\Delta_r H^\circ}{R T_{eq}} + \frac{\Delta_r S^\circ}{R} \tag{10.9}$$

For example, using thermodynamic data for standard molar heats of formation ($\Delta_f H^\circ$) and standard molar entropies (S°) for water liquid and gas at 25°C, we get

	$H_2O(l)$	$H_2O(g)$	
$\Delta_f H^\circ$ (J/mol)	-286	-242	$\Delta_r H^\circ = +44$ kJ/mol
S° (J/mol-K)	70	189	$\Delta_r S^\circ = +119$ J/mol·K

$\Delta_r G = \Delta_r H - T \Delta_r S$

$0 = \Delta_r H^\circ - T_{eq}(\Delta_r S^\circ - R \ln P_{gas})$

$0 = 44000 \text{ J/mol} - T_{eq}[119 \text{ J/mol·K} - (8.31451 \text{ J/mol·K}) \ln P_{gas}]$

For the *standard* boiling point, we solve for T_{eq} using $P_{gas} = 1.0$ bar and get $T_{eq} = 370$ K.[7] Using $P_{gas} = 650$ torr, or 0.855 atm, we get that $T_{eq} = 366$ K, about 4°C lower.

---

[7] The fact that this is a bit low (373.15 K would be more accurate) indicates that our using values meant for 25°C introduces on the order of 1% error at 100°C in this case. For more accurate work beyond the level necessary here, one must adjust both ΔH° and ΔS° for the different populations of rotational

Note that if you are given P_{gas} in mmHg or torr, you must convert to bars first, because there is an implied "/bar" in the log term.

On the other hand, if you specify the temperature, T_{eq}, and you know $\Delta_r H°$ and $\Delta_r S°$, then you can solve this equation for the equilibrium pressure of the gas above the liquid, P_{gas}. Thus, for example, the vapor pressure of water at 25°C is calculated to be just 0.032 bar (24 torr). The actual value is 23.76 torr. That is, in a closed container holding pure liquid water at 25°C there will be 24 torr of water vapor in the airspace above the liquid. At 70°C we get 0.328 bar (246 torr), considerably higher. The actual value at this temperature is 233.7 torr. Our value of 246 torr is off a bit due to the fact that we are using numbers officially valid only for 25°C. Nonetheless, we are within a few percent. If we really want to be accurate, we would use the values for $\Delta_{vap} H°$ and $\Delta_{vap} S°$ given in Appendix E.

10.10 Summary

The Second Law of Thermodynamics states

$$\Delta S_{universe} = \Delta S + \Delta S_{sur} > 0$$

Using Gibbs energy, $G = H - TS$, this law can be restated simply as in Equation 10.2:

$$\Delta G = \Delta H - T\Delta S < 0$$

at least for constant pressure/temperature systems. The deltas in *all* that we do in this regard definitely mean *difference*, not *change*. We are never going to actually calculate the *change* in Gibbs energy for any real process. Instead, we will focus on reactions, add the subscript "r" and consider only the sign of $\Delta_r G$, which gives us a direct reading of which direction a reaction *under its current conditions* will proceed:

If $\Delta_r G < 0$, the reaction will proceed as written.
If $\Delta_r G > 0$, the *reverse* reaction will proceed.
If $\Delta_r G = 0$, we have equilibrium; both states are equally probable.

We must constantly strive to remember the origin and purpose of G. For our purposes now, at least, G is merely a tool with units of energy and only a sign that matters, pointing us in the direction of the most probable distribution *when the entropy of surroundings is included in the picture.*

The inclusion of the changes in entropy of the surroundings in the full treatment of the Second Law is the *sole* reason we are working with ΔH here, because ΔH is q at constant pressure, and $T\Delta S_{sur} = q_{sur} = -q$. Thus, $\Delta_r H$ simply gives us a handle on the effect the reaction would have on the entropy of the surroundings.

Plotting G vs. T for a chemical reaction is useful, even if in just a simple, qualitative way, with no real numbers. The starting point for these analyses is always 0 K, where we see that $G = H$, and thus $\Delta_r G = \Delta_r H$. This being the case, a quick rough comparison of the extent of bonding in products vs. reactants or knowledge about relative ground states gives us which state will start at a lower position (more bonding) along the vertical axis. This is precisely the same analysis we used previously to determine which of two substances, A or B, had the lower ground state on our Boltzmann energy-level diagrams.

A curve on a G–T graph bends downward because its slope, $-S$, becomes increasingly negative as the temperature rises. Thus, it is quite possible for one curve to cross another.

---

and vibrational excited states of reactants and products at different temperatures. Better values for the evaporation of water are given in Appendix E. In the case of the evaporation of water at 100°C, for example, we have $\Delta H° = 41.335$ kJ and $\Delta S° = 110.81$ J/K.

That temperature, where the two curves cross, and thus where $\Delta_r G = 0$, defines the *only* temperature at which equilibrium is achieved.

For solid/liquid equilibria, $\Delta_r G = 0$ only when $T = T_{mp}$, the "melting point" of the substance. At higher temperatures, the $-T\Delta_r S$ term of $\Delta_r H - T\Delta_r S$ is too negative, and $\Delta_r G$ itself goes negative. A negative $\Delta_r G$ always means that product (liquid in this case) will be favored.

Since we are looking only at the *sign* of $\Delta_r G$, there is no importance as to the actual amounts of liquid and solid present. Thus "favored" here means *completely* favored. When the temperature is above the melting point, T_{mp}, *any* solid present will melt. Likewise, when the temperature is below T_{mp}, $\Delta_r G$ will be positive for the melting reaction and it will not occur. Instead, reactant (solid) will be favored, and it is the *reverse* reaction, freezing, which is thermodynamically favorable ($\Delta_r G < 0$). Below the melting point, any liquid present will freeze.

The situation for liquids in equilibrium with their vapors is more interesting. The entropy of the gas is dependent upon its pressure. In a closed system, condensation or evaporation will occur until the Gibbs energy per mole of the gas equals the Gibbs energy per mole of the liquid—that is, until for both reactions,

$$\text{liquid} \longrightarrow \text{gas} \quad \text{and} \quad \text{gas} \longrightarrow \text{liquid}$$

we have the condition that $\Delta_r G = 0$. To satisfy this condition, we can apply what we learned in Chapter 8 and write that at equilibrium (assuming pressure in bars):

$$\Delta_r G = \Delta_r H - T\Delta_r S$$
$$0 = \Delta_r H^\circ - T_{eq}(\Delta_r S^\circ - R \ln P_{gas})$$

In Chapter 11 we see how this picture can be expanded now to include *all* types of equilibria. We will see that pressures of gases and concentrations of solutes will determine the slopes of curves, making them cross at different temperatures under different conditions. Most importantly, we will see how the fact that concentrations and pressures generally change as a reaction proceeds "animates" our graphs, allowing some of the curves to change until they *do* cross at a given temperature. We will see how the whole idea of the equilibrium constant arises from this effect of pressure and concentration on Gibbs energy, which ultimately derives from the effect of pressure and concentration on entropy.

We have now arrived at the full set of thermodynamic state functions we will need in this book. They are summarized in the endsheets of this book. For more details, see Appendix A.

Problems

The symbol Ⓐ *indicates that the answer to the problem can be found in the "Answers to Selected Exercises" section at the back of the book.*

Qualitative Understanding

10.1 Draw a vector graph similar to Figure 10.3 for each of the following situations. Which of these situations violate(s) the Second Law of Thermodynamics?

(a) The entropy change in the system is small and positive, while the entropy change in the surroundings is large and positive.

(b) The entropy change in the system is large and positive, while the entropy change in the surroundings is small and negative.

(c) The entropy change in the system is small and negative, while the entropy change in the surroundings is also small and negative.

(d) The entropy change in the system is small and positive, while the entropy change in the surroundings is negative and exactly the same magnitude as the entropy change in the system.

10.2 Ⓐ Explain in terms of particles, energy levels, and temperature, why curves on G–T graphs generally get steeper as T increases.

10.3 Ⓐ Sketch G–T graphs for the following descriptions of systems involving substances **A** and **B**:

(a) **A** has more bonding, but **B** has more ways to distribute energy.

(b) **A** has the lower ground state and the more closely spaced energy levels.

(c) For **A**⟶**B**, $\Delta_r H = -10$ kJ and $\Delta_r S = -40$ J/K.

10.4 Sketch G–T graphs for the following descriptions of systems involving substances **A** and **B**:

(a) **A** has a lower ground state energy, and **B** has more closely spaced energy levels.

(b) For **A** ⟶ **B**, the change in entropy is positive and the change in enthalpy is negative.

(c) **A** is a gas and **B** is a solid.

10.5 Ⓐ Sketch G–T graphs for the following descriptions of systems involving substances **A** and **B**:

(a) **A** has a higher ground state energy, and **B** has more closely spaced energy levels.

(b) For **A**⟶**B**, the change in entropy is negative and the change in enthalpy is negative.

(c) **A** is a liquid and **B** is a solid.

10.6 Sketch G–T graphs for the following descriptions of systems involving substances **A** and **B**:

(a) **A** has a lower ground state energy, and more closely spaced energy levels.

(b) For **A**⟶**B**, the change in entropy is negative and the change in enthalpy is positive.

(c) **A** is a solid and **B** is a gas.

10.7 Match each of the G–T graphs shown below with the corresponding energy level system also shown.

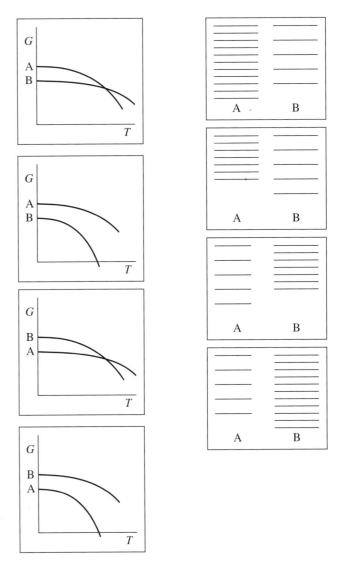

10.8 Ⓐ Based on data in Table 9.1 or Appendix D, sketch G–T graphs for each system below. Indicate the crossing temperatures in each case that the curves cross. Discuss each in terms of bonding, constraints, and the favored substance at low and high temperatures.

(a) $I_2(s)/I_2(g)$
(b) $Br_2(g)/Br_2(l)$
(c) C(graphite)/C(diamond)
(d) $P_4(s, \text{red})/P_4(s, \text{white})$

10.9 Based on data in Table 9.1 or Appendix D, sketch G–T graphs for each system below. Indicate the crossing temperatures in each case that the curves cross. Discuss each in terms of bonding, constraints, and the favored substance at low and high temperatures.

(a) $SO_3(g)/SO_3(l)$
(b) $Sn(s, white)/Sn(s, gray)$
(c) $TiCl_4(g)/TiCl_4(l)$
(d) $S(s, rhombic)/S(g)$

10.10   Based on data in Table 9.1 or Appendix D, sketch G–T graphs for each system below. Indicate the crossing temperatures in each case that the curves cross. Discuss each in terms of bonding, constraints, and the favored substance at low and high temperatures.

(a) $Pb(g)/Pb(s)$
(b) $ZnS(s, wurtzite)/ZnS(s, sphalerite)$
(c) $N_2O_4(l)/N_2O_4(g)$
(d) $HgS(s, red)/HgS(s, black)$

10.11   For the $H_2O(l)/H_2O(g)$ system, draw a G–T graph and discuss it fully. Why do the curves start where they do? Why are they curved and not straight? Why are they curving the way they do? If they cross, at what temperature exactly do they cross? Indicate the favored reaction with a vertical arrow at each of the four temperatures mentioned in Problem 10.16.

10.12   Ⓐ Create a table for the reaction of one mole of gaseous I_2 going to one mole of solid I_2. Calculate T(K), q(J), q_{sur}(J), ΔS(J/K), ΔS_{sur}(J/K) and $\Delta S_{universe}$ for this reaction at the following temperatures: (a) 25°C (b) 125°C (c) 159°C (d) 175°C.

10.13   It is known that freezing one mole of methanol (CH_3OH) releases 3.176 kJ of energy. Using the value 1.117 J/K for the standard molar entropy of $CH_3OH(s)$, create a table for the reaction of a mole of CH_3OH going to a mole of $CH_3OH(l)$. Calculate T(K), q(J), q_{sur}(J), $\Delta_r S$(J/K), ΔS_{sur}(J/K), and $\Delta S_{universe}$ for the melting of 1.0 g of methanol at the following temperatures: (a) −125°C (b) −95°C (c) −50°C (d) 0°C.

Applications

10.14   For the melting of ammonia (NH_3), $\Delta_{fus}H° = 5.65$ kJ/mol and $\Delta_{fus}S° = 28.9$ J/mol·K. (a) What is the melting point of ammonia? (b) How much energy is required to melt 1.0 g of NH_3? (c) Draw a G–T graph for the $NH_3(s)/NH_3(l)$ system and discuss it.

10.15   Ⓐ Complete the following table, which gives standard molar values for the *vaporization* reaction, $X(l) \longrightarrow X(g)$:

	$\Delta_{vap}H°/(kJ/mol)$	$\Delta_{vap}S°/(J/mol·K)$	T_{eq}/K
ethane, C_2H_6	14.70	_____	184.1
propane, C_3H_8	_____	81.26	231.0
butane, C_4H_{10}	22.39	82.30	_____
pentane, C_5H_{12}	25.76	_____	309.2
hexane, C_6H_{14}	28.85	_____	341.9

Discuss the trends in $\Delta_{vap}H°$, $\Delta_{vap}S°$, and T_{eq} based on what you know about bonding and entropy.

10.16   For the vaporization of *just 1.0 gram* of water according to the equation

$$H_2O(l) \longrightarrow H_2O(g)$$

fill in the table below, using Appendix E for the values of $\Delta_{vap}H°$ and $\Delta_{vap}S°$ at the temperatures listed. At what temperature is $\Delta S_{universe} = 0$?

	(a)	(b)	(c)	(d)
$T/°C$	0	25	50	120
T/K	_____	_____	_____	_____
q/J	_____	_____	_____	_____
q_{sur}/J	_____	_____	_____	_____
$\Delta S/(J/K)$	_____	_____	_____	_____
$\Delta S_{sur}/(J/K)$	_____	_____	_____	_____
$\Delta S_{universe}/(J/K)$	_____	_____	_____	_____

10.17   Based on information in Table 10.1, answer the following questions: (a) Is the difference in melting points between water and sodium chloride due *primarily* to differences in bonding or entropy? (b) What about these solids (shape, complexity, bonding, etc.) might explain the difference in melting points?

10.18   Ⓐ Methane, CH_4, and water have almost the same mass, yet methane boils at −161°C, while water boils at 100°C. Given for *vaporization*:

$$CH_4(l) \longrightarrow CH_4(g) \qquad \Delta_{vap}H° = 8.17 \text{ kJ/mol}$$

(a) Determine $\Delta_{vap}S°$ for methane. (b) Compare $\Delta_{vap}H°$ and $\Delta_{vap}S°$ for methane to those given for water in Section 10.9. Which explains best the very large difference in boiling point between these two substances, the entropy change or the enthalpy change, upon going from liquid to gas? (d) To what, in terms of bonding or disorder, could this be attributed?

10.19   Using the values of $\Delta_r H°$ and $\Delta_r S°$ for the vaporization of water given in Section 10.9, calculate the vapor pressure of water in torr at the following temperatures: 90°C, 99°C, 100°C, and 110°C. Compare these values to the values given in Appendix E, which takes into account the fact that both $\Delta_{vap}H°$ and $\Delta_{vap}S°$ are temperature-dependent.

Brain Teasers

10.20   Using the more accurate values of $\Delta_{vap}H°$ and $\Delta_{vap}S°$ given in Appendix E, calculate the boiling point of water at the following vapor pressures: (a) 0.01 bar (b) 100 torr (c) 28.85 inHg (d) 1.5 bar. [*Hint:* Estimate the temperature first using Appendix E. Then use the values of $\Delta_{vap}H°$ and $\Delta_{vap}S°$ given in Appendix E for pressure (interpolating if necessary) to complete the calculation.]

10.21 Using algebra, show that for a liquid/gas system,

$$\text{liquid} \longrightarrow \text{gas}$$

the vapor pressure of the gas at equilibrium is given by

$$P_{\text{gas}} = e^{\Delta_r S^\circ_{\text{universe}}/R}$$

where $\Delta_r S^\circ_{\text{universe}}$ is defined as the difference in entropy of the universe between having 1 mole of liquid and having 1 mole of gas at 1 bar partial pressure.

10.22 A system has the energy level diagram shown below. At a certain temperature particles initially put in the ground state of A completely transfer to B, leaving not a trace in A. Yet at another temperature any particles in B completely transfer to A. What is going on?

A   B

CHAPTER 11

The Equilibrium Constant (K)

11.1 Introduction

In Chapter 8, we concluded that the entropy of a gas is dependent upon its pressure as given in Equation 8.1:

$$S_x = S_x^\circ - R \ln(P_x/\text{bar})$$

and for a solute in solution (Equation 8.2):

$$S_x = S_x^\circ - R \ln([X]/M)$$

This led to the idea of a *reaction quotient*, Q, as the adjustment to the standard entropy change in a chemical reaction (Equation 8.3):

$$\Delta_r S = \Delta_r S^\circ - R \ln Q$$

Q is formed by taking pressures (for gases, in bar) and concentrations (for solutes, in mol/L) of products over reactants all to their stoichiometric powers (Section 8.9). For example, we have

$$H_2O(l) \longrightarrow H_2O(g) \qquad Q = P_{H_2O(g)}$$

$$N_2O_4(g) \longrightarrow 2\,NO_2(g) \qquad Q = \frac{P_{NO_2}^2}{P_{N_2O_4}}$$

$$HF(aq) \longrightarrow H^+(aq) + F^-(aq) \qquad Q = \frac{[H^+][F^-]}{[HF]}$$

$$CO_2(g) + OH^-(aq) \longrightarrow HCO_3^-(aq) \qquad Q = \frac{[HCO_3^-]}{P_{CO_2}[OH^-]}$$

In each case, the change in entropy of the system when the reaction occurs is going to depend upon the exact pressures of gases and concentrations of solutes during the reaction. Admittedly, this sounds a bit fishy. After all, don't the pressures and concentrations of reactants go down in a chemical reaction? And don't the pressures and concentrations of products go up? The answer to that may sound even more fishy: Maybe. In any case, *we don't care!*

Remember, all we are really after is the answer to a simpler question: **Will this reaction go *at all*?** That is, we are just interested in the *hypothetical* conversion of a stoichiometric number of moles of reactant into a stoichiometric number of moles of product in the situation involving the pressures and concentrations specified in the "reaction conditions." It is simply *convenient* to work with the stoichiometric number of moles of reactants and products. The idea is that if the reaction of a mole isn't favorable, then neither will be the reaction of a micromicromicromole. Working with moles instead of micromicromicromoles is simply easier.

The answer to our simpler question is the answer to any of the following three:

1. Is this in the direction of the most probable distribution?
2. Would the entropy of the universe increase?
3. Is the reaction Gibbs energy negative?

Gibbs energy is simply a trick used by chemists to answer this question quantitatively. We will observe a reaction if and only if $\Delta_r G < 0$. If that is *not* the case, then there are only two possibilities:

$\Delta_r G > 0$ In this case, we will observe a reaction, but it will be the *reverse* of what we have written for the chemical equation.

$\Delta_r G = 0$ Neither the forward nor the reverse reaction will be observed, yet both will be occurring simultaneously.

The implications of this second possibility are explored in this chapter. The idea is that when $\Delta_r G = 0$ the reaction is at the end of its road. We want to know what that end looks like. Perhaps it is already there, but more likely it is not. By knowing what that end should look like (what the value Q should be), we might be able to figure out how to get there.

Thus, ultimately, we are going to be interested in more than just hypothetical reactions. We will be very much interested in knowing whether a reaction will go, which way it will go, and finally how far it will go. The basic questions we want to ask are depicted in Figure 11.1.

We focus on equilibrium simply because that is the end of the road for all chemical reactions. Equilibrium is represented by the most probable distribution; it is the point where no reaction can decrease the Gibbs energy of the system any further.

There are two approaches to this problem. One is quantitative and involves serious calculations. It is the typical approach taken in all standard physical chemistry texts. The other, involving G–T graphs, is qualitative and gives us a much better basic understanding of what is going on, why the reaction is happening in the first place, and what to expect. We will do a little of both of these approaches in this chapter, starting with some easy calculations. The first step is to figure out what Q should be at equilibrium.

Figure 11.1
The basic questions one may ask about a chemical reaction. Will it go under these conditions of temperature and pressure? (Is $\Delta_r G \neq 0$?) Will it go the way we expect it to? (Is $\Delta_r G < 0$?) What does the end of the road look like? (What is Q_{final}?) How far will the reaction go? (What is x?)

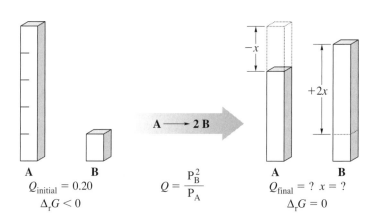

$$A \longrightarrow 2\,B$$

$$Q_{\text{initial}} = 0.20$$
$$\Delta_r G < 0$$

$$Q = \frac{P_B^2}{P_A}$$

$$Q_{\text{final}} = ?\quad x = ?$$
$$\Delta_r G = 0$$

11.2 The Equilibrium Constant

Consider again the dissociation of N_2O_4, where now we are particularly interested in learning about the end of the road, equilibrium:

$$N_2O_4(g) \rightleftharpoons 2\,NO_2(g)$$

We start right out with definitions for the enthalpy change and entropy change for the hypothetical reaction of one mole of N_2O_4 to give two moles of NO_2 (the stoichiometric amounts):

$$\Delta_r H = \Delta_r H^\circ$$

$$\Delta_r S = \Delta_r S^\circ - R \ln Q \quad \text{where } Q = \frac{P^2_{NO_2}}{P_{N_2O_4}}$$

Remember, those pressures must be in bars! Since for equilibrium we require that $\Delta_r G = 0$, we have

$$\Delta_r G = \Delta_r H - T\Delta_r S$$

$$0 = \Delta_r H^\circ - T_{eq}[\Delta_r S^\circ - R \ln Q] \quad \text{where } Q = \frac{P^2_{NO_2}}{P_{N_2O_4}}$$

Now, the really interesting thing about this equation is that at a given equilibrium temperature, everything in this equation is a constant! Remember that $\Delta_r H^\circ$ and $\Delta_r S^\circ$ are constants because they are the "standard" values. The temperature is basically a constant because we will specify it. A little algebra isolates Q on the left and all the other constants on the right:

$$0 = \Delta_r H^\circ - T_{eq}[\Delta_r S^\circ - R \ln Q]$$

$$0 = \Delta_r H^\circ - T_{eq}\Delta_r S^\circ + RT_{eq} \ln Q$$

$$-RT_{eq} \ln Q = \Delta_r H^\circ - T_{eq}\Delta_r S^\circ$$

and

$$\ln Q = \frac{-(\Delta_r H^\circ - T_{eq}\Delta_r S^\circ)}{RT_{eq}}$$

$$Q = e^{\frac{-(\Delta_r H^\circ - T_{eq}\Delta_r S^\circ)}{RT_{eq}}} \equiv K_{eq}$$

Thus, *for a given equilibrium temperature*, the reaction quotient, Q, will end up a very specific number when the system reaches its equilibrium destination. We call it "K_{eq}." As for the *value* of K_{eq}, no problem! We can easily calculate the *standard* reaction enthalpy, $\Delta_r H^\circ$, using a table of standard enthalpies of formation. Likewise, we can calculate the *standard* reaction entropy, $\Delta_r S^\circ$, using tables of standard molar entropies. At 25°C these values are right on; at other temperatures they will have to serve as approximations.

For the reaction of one mole of N_2O_4 to give two moles of NO_2 at 25°C, we calculate from tables that

$$\Delta_r H^\circ = (2)\frac{33.18\,\text{kJ}}{\text{mol}} - (1)\frac{9.16\,\text{kJ}}{\text{mol}} = 57.20\,\text{kJ/mol}$$

$$\Delta_r S^\circ = (2)\frac{240.06\,\text{J}}{\text{mol}\cdot\text{K}} - (1)\frac{304.29\,\text{J}}{\text{mol}\cdot\text{K}} = 175.83\,\text{J/mol}\cdot\text{K}$$

and from that we can get the value of K_{eq} at 25°C:

$$K_{eq} = e^{\frac{-(\Delta_r H^\circ - T_{eq}\Delta_r S^\circ)}{RT_{eq}}}$$

$$= e^{\frac{-(57200\,\text{J/mol}) - (298\,\text{K})(175.83\,\text{J/mol}\cdot\text{K})}{(8.31451\,\text{J/mol}\cdot\text{K})(298\,\text{K})}}$$

$$= e^{-1.938}$$

$$= 0.144$$

Our conclusion: The end of the road for the reaction of N_2O_4 to give NO_2 is when

$$Q = \frac{P_{NO_2}^2}{P_{N_2O_4}} = 0.144$$

(Note that if we started with just N_2O_4, that wouldn't be a very long road!)

The same thing goes for solutions, only here you would put concentrations in for the pressures. For example, for the dissociation of water:

$$H_2O(l) \rightleftharpoons H^+(aq) + OH^-(aq) \qquad Q = [H^+][OH^-]$$

we get a constant we can write as K_{eq} or the more familiar "K_w":

$$[H^+][OH^-] = e^{\frac{-(\Delta_r H^\circ - T_{eq}\Delta_r S^\circ)}{RT_{eq}}} \equiv K_w$$

The value 1.0×10^{-14} for K_w is simply what you get when you put 55,900 J in for $\Delta_r H^\circ$, -80.54 J/K in for $\Delta_r S^\circ$, and 298 K in for the temperature.

Note that this value for K_w is only valid at 25°C. At higher temperatures, there is more dissociation, because in this case $\Delta_r H^\circ > 0$; strong bonds are being broken. For example, at 100°C, $K_w \approx 1.0 \times 10^{-12}$. Thus, at 100°C the pH of neutral water is *not* 7.00. In neutral water, where $[H^+] = [OH^-]$, at 100°C we have

$$[H^+][OH^-] = 1.0 \times 10^{-12}$$

$$[H^+]^2 = 1.0 \times 10^{-12}$$

$$[H^+] = 1.0 \times 10^{-6}$$

$$pH = -\log[H^+] = 6.00$$

In fact, both the pH and the pOH of boiling water are about 6.0!

The above analysis suggests that for *any* situation we can write

$$\Delta_r G = \Delta_r H^\circ - T[\Delta_r S^\circ - R \ln Q] \qquad (11.1)$$

At equilibrium, when $\Delta_r G = 0$ and $T = T_{eq}$, we can write a general expression for the reaction quotient Q, which is a constant we call "K_{eq}":

$$Q = K_{eq} = e^{\frac{-(\Delta_r H^\circ - T_{eq}\Delta_r S^\circ)}{RT_{eq}}} \qquad (11.2)$$

But we have to be careful here. The reaction quotient, Q, involving concentrations and/or pressures of reactants and products must follow certain rules. Specifically, we have these rules:

	Rule	Reason
1.	Gases appear in the reaction quotient as partial pressures, in bars; dissolved species appear as concentrations, in moles per liter. Pure solids and pure liquids do not appear in equilibrium expressions.	Only gases and solutes have molar entropies that depend upon concentration; pressure is simply a measure of concentration; the standard for gases is 1 bar; the standard for solutes is $1\ M$.
2.	Products appear in the numerator; reactants appear in the denominator. Each is raised to the power equal to its coefficient in the balanced chemical equation.	The entropy effect is logarithmic and taken as *products minus reactants*: $(c \ln C + d \ln D) - (a \ln A + b \ln B)$ $= \ln[C^c D^d / A^a B^b]$
3.	Equilibrium constants are temperature-dependent. Adding heat shifts the equilibrium so as to absorb some of the added heat.	K_{eq} is related to temperature by Equation 11.2, involving the *standard reaction* quantities $\Delta_r H°$ and $\Delta_r S°$. (Note, however, that it is really just the sign of $\Delta_r H°$ that determines the effect of temperature.)

Thus, we have derived three important "rules" for writing equilibrium expressions learned in an introductory chemistry course.

11.3 Determining the Values of $\Delta_r H°$ and $\Delta_r S°$ Experimentally

Taking the logarithm of both sides of Equation 11.2, we have a fully general case now for the relationship of K_{eq} to T_{eq}:

$$\ln K_{eq} = \left(\frac{-\Delta_r H°}{R} \right) \frac{1}{T_{eq}} + \frac{\Delta_r S°}{R} \qquad (11.3)$$

In and of itself, this isn't particularly useful, especially since with an equation-solving calculator we can just solve Equation 11.1 for any of the variables directly. The usefulness of Equation 11.3 comes when we do not know either $\Delta_r H°$ or $\Delta_r S°$. It is then that Equation 11.3 allows us, given the value of K_{eq} at two or more temperatures, to solve for both $\Delta_r H°$ or $\Delta_r S°$ simultaneously. If two (or preferably many more) measurements of K_{eq} are available, then what we can do is to graph our data. Plotting $\ln K_{eq}$ vs. $1/T_{eq}$, we should get a straight line with slope of $-\Delta_r H°/R$ and intercept $\Delta_r S°/R$. (Figure 11.2). Depending upon the sign of $\Delta_r H°$, this line may slope upwards or downwards.

Notice that the high temperature limit, either seen from inspection of Equation 11.3 or from the y-intercept in Figure 11.2, is

$$\ln K_{eq} = \frac{\Delta_r S°}{R}$$

$$K_{eq} = e^{\Delta_r S°/R} \quad \text{(high-temperature limit)}$$

Once again we conclude that at high temperature it is the entropy of the *system* that determines the position of equilibrium, and in fact the limit of K_{eq} is simply e taken to the power of $\Delta_r S°/R$.

Importantly, by determining $\Delta_r H°$ and $\Delta_r S°$ experimentally, we might learn valuable information about the reaction. A positive $\Delta_r H°$ implies that the reactant has the lower ground state and the stronger bonding. A positive $\Delta_r S°$ indicates that the entropy of the product is higher, and its energy levels are more closely spaced.

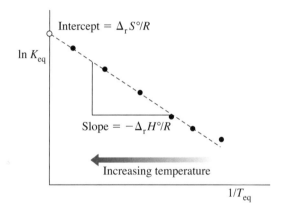

Figure 11.2
Determining the values of $\Delta_r H^\circ$ and $\Delta_r S^\circ$ experimentally involves measuring K_{eq} at two or more temperatures. The slope of a graph of $\ln K_{eq}$ vs. $1/T_{eq}$ should give a straight line with slope $-\Delta_r H^\circ/R$ and y-intercept $\Delta_r S^\circ/R$. Higher temperature is toward the left.

11.4 The Effect of Temperature on K_{eq}

This now also suggests a simple way of determining K_{eq} at one temperature given K_{eq} at another temperature as long as $\Delta_r H^\circ$ is known, even if $\Delta_r S^\circ$ is not known. Subtracting Equation 11.3 for T_1, with K_1, from a similar equation for T_2, with K_2, the entropy term drops out and we get:

$$\ln K_2 - \ln K_1 = \left(\frac{-\Delta_r H^\circ}{R}\right)\left(\frac{1}{T_2} - \frac{1}{T_1}\right)$$

and

$$\ln \frac{K_2}{K_1} = \frac{\Delta_r H^\circ}{R}\left(\frac{1}{T_1} - \frac{1}{T_2}\right) \tag{11.4}$$

(Note that T_1 and T_2 are switched because the minus sign in front of $\Delta_r H^\circ$ was incorporated into the expression.) This result is one of the forms of the *Clausius–Clapeyron equation*, and shows that if K increases with T (left side positive when term in parentheses positive) then it must be true that the reaction absorbs heat ($\Delta_r H^\circ > 0$).

Equation 11.4 is useful for several purposes. Say we know K and T for two temperatures. Substituting these values into Equation 11.4 gives us $\Delta_r H^\circ$ for the reaction. Perhaps we know $\Delta_r H^\circ$, T_1, T_2, and K_1, and want to know K_2. Again, no problem with an equation-solving calculator.

For example, say we wanted to know the boiling point of water in Denver, Colorado, on a nice day where the atmospheric pressure is 0.8 bar. The system we are dealing with is the evaporation of liquid water:

$$H_2O(l) \rightleftharpoons H_2O(g) \qquad K_{eq} = P_{H_2O(g)} = P_{vap}$$

At the boiling point, the vapor pressure above the liquid equals the atmospheric pressure, so $P_{vap} = 0.80$ bar. Here T_1 is the normal boiling point of water, 373 K, K_1 is 1.0 bar (the pressure for the normal boiling point), and $K_2 = 0.8$. Using the best value of $\Delta_r H^\circ$ we can get (Appendix E, 100°C), we calculate

$$\ln \frac{K_2}{K_1} = \frac{\Delta_r H^\circ}{R}\left(\frac{1}{T_1} - \frac{1}{T_2}\right)$$

$$\ln \frac{0.8}{1} = \frac{41300\,\text{J/mol}}{8.31451\,\text{J/mol}\cdot\text{K}}\left(\frac{1}{373} - \frac{1}{T_2}\right)$$

Solving this for T_2 gives 367 K (94°C), a full 6°C lower than normal. No wonder it takes so long to boil an egg in Denver!

11.5 A Qualitative Picture of the Approach to Equilibrium

In most cases when gases and solutes are involved in a chemical reaction, the trip toward equilibrium is *not* at constant pressure or concentration for any of the reactants or products involved. In fact, the temperature is not likely to remain constant either. How then are we to use Gibbs energy, which is specifically designed to give information only in situations of constant pressure and temperature, to any avail?

G–T graphs to the rescue! We can consider a G–T graph to be a snapshot of the Gibbs energy situation anywhere along the reaction route. It is when the curve for the reactants crosses the curve for the products at the temperature of the system that we announce, "Done! The reaction is over." Until then, though, something is going to be going on. Either reactants are going to be turning into products, or products are going to be turning into reactants.

For example, consider again the dissociation of N_2O_4:

$$N_2O_4(g) \longrightarrow 2\,NO_2(g) \qquad Q = \frac{P_{NO_2}^2}{P_{N_2O_4}}$$

Presume for a moment that the partial pressures of both N_2O_4 and NO_2 are 1 bar. This situation is depicted in Figure 11.3. This graph is constructed with just the minimal amount of information. As in Section 11.2, from tables we have that (at least at 25°C, which will be just fine for this qualitative picture)

$$\Delta_r H^\circ = \left(\frac{2\,mol\,NO_2}{mol\,rxn}\right)\frac{33.18\,kJ}{mol\,NO_2} - \left(\frac{1\,mol\,N_2O_4}{mol\,rxn}\right)\frac{9.16\,kJ}{mol\,N_2O_4} = 57.20\,kJ/mol$$

$$\Delta_r S^\circ = \left(\frac{2\,mol\,NO_2}{mol\,rxn}\right)\frac{240.06\,J/K}{mol\,NO_2} - \left(\frac{1\,mol\,N_2O_4}{mol\,rxn}\right)\frac{304.29\,J/K}{mol\,N_2O_4} = 175.83\,J/mol \cdot K$$

Thus, energy is required ($\Delta_r H^\circ > 0$) and the product curve (2 NO_2) starts out above the reactant curve (N_2O_4). In addition, for the standard conditions of 1 bar of each reactant and product, the entropy will increase ($\Delta_r S^\circ > 0$). This means that the entropy of the products (2 NO_2) is greater than the entropy of the reactants (N_2O_4), and the "2 NO_2" curve will fall faster than the N_2O_4 curve. In fact, we can quickly calculate the crossing point. For these "standard" conditions, $Q = 1$, so $\ln Q = 0$, and we have

$$0 = \Delta_r H^\circ - T_{eq}\Delta_r S^\circ$$

$$T_{eq} = \frac{\Delta_r H^\circ}{\Delta_r S^\circ} = \frac{57200\,J/mol}{175.83\,J/mol \cdot K} = 325\,K$$

You should always do this calculation when you set up a G–T graph for any real system. Start with the standard state and determine the crossing temperature T_{eq}. In this case the curves would cross somewhat above room temperature.

Now consider what would happen if you really had a pressure of 1 bar of N_2O_4 and a pressure of 1 bar of NO_2 in a flask at 25°C. Would the system be at equilibrium? Definitely not! Just a glance at Figure 11.3 indicates that at any real temperature the reaction is going to go in reverse under these conditions. There is way too much NO_2 in the system.

Another way of looking at this is that $Q = 1$, whereas K_{eq} was calculated above to be only 0.144 at 25°C. "$Q > K$" always means the reaction will go backward. Consider, though, what will happen when the reaction does go backward: The pressure of NO_2 will go down while the pressure of N_2O_4 will rise. *This will change their curves* (Figure 11.4).

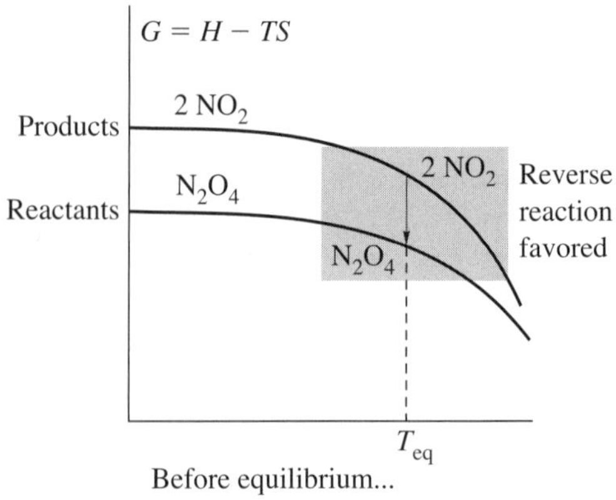

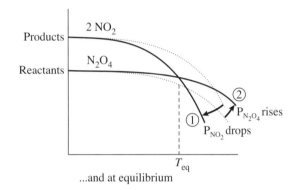

Figure 11.3
G–T graph for the N_2O_4/2 NO_2 system prior to equilibrium. In this depiction, both gases are at the standard condition of 1 bar partial pressure for each gas. The reverse reaction will be favored.

Figure 11.4
G–T graph for the N_2O_4/2 NO_2 system at equilibrium. The partial pressure of NO_2 has dropped and the partial pressure of N_2O_4 has risen. The effect on the curves is for them to move closer and closer together until finally, reaching equilibrium at T_{eq}, they cross.

Basically, *both* curves will adjust until they cross at T_{eq}.[1] In this particular case there is too much NO_2 in the system. (The favored reaction is the *reverse* reaction.) As the pressure of NO_2 diminishes, its curve drops, because lower pressure means higher molar entropy, and entropy is what is determining the slope of this curve.

However, at the same time the pressure of N_2O_4 must *increase*, because the only way to decrease the pressure of NO_2 is to *react it* to form N_2O_4. The curve for N_2O_4 "comes up to meet" the falling NO_2 curve! Thus, the situation is a bit more complex than before, but other than that it should sound quite reasonable.

11.6 Le Châtelier's Principle Revisited

In Section 1.14, for the isotope exchange reaction involving H_2, D_2, and HD, we saw that when the system in its most probable state (with $Q = 4$) was disturbed, it returned to a new most probable distribution with, again, $Q = 4$.

What must happen if the system at equilibrium in Figure 11.4 is suddenly disturbed by the addition of reactant, N_2O_4? Probably by now your experience would tell you that some of the added N_2O_4 must be used up, and NO_2 will be formed. Let's see if our G–T graphs bear this out.

---

[1] In fact, even the temperature can change, and we can still see what is going to happen, but we'll just consider the temperature constant in this case. If the reaction really were carried out, we know that heat will be required. Either we are going to have to add energy to the system to maintain its original temperature or the temperature will drop as the reaction sucks heat out of the surroundings.

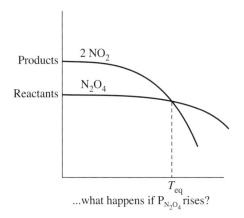

...what happens if $P_{N_2O_4}$ rises?

Figure 11.5
The same situation as depicted after equilibrium is achieved in Figure 11.4 prior to the addition of more N_2O_4.

Shown in Figure 11.5 is the equilibrium situation at the end of the road in the previous scenario. Basically, the scenario we now imagine looks something like this:

1. The N_2O_4 pressure instantly rises due to the addition of N_2O_4 to the system, decreasing the molar entropy of the N_2O_4 and flattening its curve.
2. The favored state is now "2 NO_2" and the forward reaction occurs, reducing the excess pressure of N_2O_4, and making its curve start to drop back toward its original position.
3. Simultaneously, because NO_2 is also a gas, its curve rises to meet the falling N_2O_4 curve.
4. A new crossing point develops, and the system is once more at equilibrium.

The system is back in equilibrium, and since the temperature hasn't changed (presumably), Q will again be back to its original equilibrium value.[2]

The new equilibrium situation still must satisfy exactly the same equations involving T, $\Delta_r H°$, and $\Delta_r S°$, and so must end up with exactly the same equilibrium constant, even though the final pressures and concentrations will certainly be different. This is the essence of Le Châtelier's principle (Section 1.14), which states that a system that is pushed away from equilibrium will adjust so as to reestablish a new equilibrium. If reactant is added, then the forward reaction will take place; if products are added, the reverse reaction will take place to get rid of some of the added product instead.

Adding a reactant that is a gas or solute simply increases its concentration, decreases its molar entropy, and, on these graphs, leads to a flatter curve. Being flatter, the curve for the reactants is now higher than that of the products at T_{eq}. The imbalance leads to the *forward* reaction having a $\Delta_r G < 0$ at T_{eq}. That forward reaction proceeds, steepening the curvature of the reactant curve (and flattening the product curve if any of the products are gases or solutes).

Similarly, adding a product that is a gas or solute simply flattens *its* curve. Being flatter, the curve for the *products* is now higher than that of the reactants at T_{eq}. The imbalance leads to the *reverse* reaction having a $\Delta_r G < 0$ at this temperature.

Furthermore, if *heat* is added so as to increase the temperature of the system, then the reaction observed will be the one which can *absorb* some of that heat. Consider Figure 11.5

---

[2] You should make sure that in your mind this scenario of causes and effects makes sense. Be certain you can reproduce the steps in your mind or on paper using simple G–T graphs.

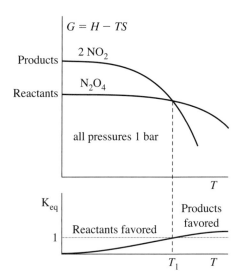

Figure 11.6
Correlation between a G–T graph depicting standard conditions and a graph of K_{eq} vs. T. The temperature at which the two curves cross on the G–T graph corresponds to the temperature where $K_{eq} = 1$.

again. Can you figure out which one of these reactions:

$$N_2O_4 \longrightarrow 2\,NO_2 \text{ (forward)}$$

or

$$2\,NO_2 \longrightarrow N_2O_4 \text{ (reverse)}$$

absorbs heat when it occurs?[3]

Now, which reaction should occur ($\Delta_r G < 0$) if the equilibrium depicted in Figure 11.5 is disturbed by moving T_{eq} to the right? Isn't this the very same reaction? This should sound familiar.

In fact, there is a direct correlation between the sorts of graphs of K vs. T we used in Chapter 6 and these G–T graphs. Figure 11.6 introduces a corresponding graph of K vs. T below the diagram of Figure 11.5, which is for the standard conditions where each partial pressure is 1 bar.

Note that above T_1, this reaction will go forward from the standard state to reach equilibrium. Therefore K_{eq} above T_1 must be greater than 1. The opposite is true below T_1, where reaching equilibrium will involve product becoming reactant; below T_1, K_{eq} will be less than 1.

In a nutshell, wherever the product curve is below the reactant curve on a G–T graph depicting the standard state, $K_{eq} > 1$, and wherever the reactant curve is below the product curve, $K_{eq} < 1$.

Note that only gas and solute curves move as the reaction proceeds, because only they have pressure- or concentration-dependent molar entropies; curves for solids and liquids remain stationary during chemical reactions.

11.7 Determining Equilibrium Pressures and Concentrations

To actually come up with the final equilibrium concentrations, we have to do a little algebra. The result depends upon how the reaction is done: for example, in a sealed flask at constant

---

[3] Look at where the curves start, at 0 K. At that point $\Delta_r G = \Delta_r H$. For which direction of reaction is $\Delta_r H° > 0$? The answer is *forward*. Remember, all dissociation reactions require heat to break the bond. This will be true at all temperatures, not just 0 K.

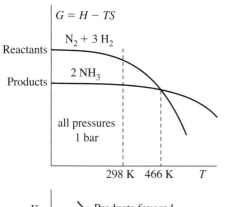

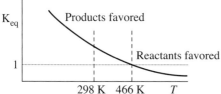

Figure 11.7
The situation for the standard conditions
in the ammonia synthesis system.

volume, or in some sort of a system that can adjust its volume to keep the external pressure
constant. Let's take a close look at the synthesis of ammonia from nitrogen and hydrogen
gas as an example:

$$N_2 + 3\,H_2 \rightleftharpoons 2NH_3 \qquad \text{(all gases)}$$

First we need some data for the *standard* differences in enthalpy and entropy for the reaction
as written:

	$N_2(g)$	$H_2(g)$	$NH_3(g)$
$\Delta_f H°$ (kJ/mol)	0	0	−46.19
$S°$ (J/mol·K)	191.50	130.58	192.5

From these values, we derive that for the equation as written, $\Delta_r H° = -92.38\,\text{kJ/mol}$ and
$\Delta_r S° = -198.24\,\text{J/mol·K}$. Now, just looking at these values, we see that the situation is
something like that in Figure 11.7. Reactants have both the higher ground state and the higher
entropy. The crossing temperature from $\Delta_r H°/\Delta_r S°$ is calculated to be 466 K (193°C).

The equilibrium-constant expression in this case is

$$\frac{P_{NH_3}^2}{P_{N_2}P_{H_2}^3} = e^{\frac{-(\Delta_r H° - T\Delta_r S°)}{RT}}$$

which evaluates to 6.88×10^5 at 298 K. So, our depiction in Figure 11.7 is quite representative. At 298 K, the forward reaction will be favored.

Let's try some real numbers. We'll start with 0.5 bar of N_2 and 0.5 bar of H_2 at 298 K
and see what happens. First we make a table summarizing the changes:

	N_2 +	$3\,H_2$ ⇌	$2\,NH_3$ (all gases)
P_{initial}	0.5	0.5	0
Δ	−x	−3x	+2x
P_{eq}	0.5 − x	0.5 − 3x	2x

We then substitute these equilibrium pressures into the constant for 25°C:

$$\frac{(2x)^2}{(0.5 - x)(0.5 - 3x)^3} = 6.88 \times 10^5$$

For x we get 0.164 using an equation-solving calculator, and from that we get our equilibrium pressures:

$$N_2 \quad + \quad 3\,H_2 \quad \rightleftharpoons \quad 2\,NH_3 \quad \text{(all gases)}$$

P_{eq}	$0.5 - x$	$0.5 - 3x$	$2x$	
$=$	0.336	0.0078	0.328	(bar)

Clearly H_2 was the limiting reactant here, and it has just about run out.

The theoretical yield (in the perhaps unusual units of bar) and percent yield are calculated from the fact that we are starting with 0.5 bar of the limiting reactant, H_2:

$$(0.5\text{ bar}H_2)\frac{2\text{ bar }NH_3}{3\text{ bar }H_2} = 0.333\text{ bar }NH_3 \text{ theoretically}$$

$$\frac{0.328\text{ bar }NH_3\text{ actually}}{0.333\text{ bar }NH_3\text{ theoretically}} \times 100\% = 98\% \text{ yield}$$

Unfortunately, no catalyst has been discovered that allows this reaction to be run at room temperature. Instead, we must go to around 400°C to get this reaction to go at any appreciable rate. At 400°C (673 K) we have roughly[4]

$$K_{eq} = e^{-\frac{(\Delta_r H^\circ - T\Delta_r S^\circ)}{RT}} = 6.53 \times 10^{-4}$$

Ah, now this is going to be a problem! We are on the other side of that crossing point in Figure 11.7, and $K_{eq} < 1$. Nonetheless, we have to make the best of it. Solving for x in

$$\frac{(2x)^2}{(0.5 - x)(0.5 - 3x)^3} = 6.53 \times 10^{-4}$$

we get $x = 0.0031$. Almost no reaction occurs at all!

This won't do! Let's increase the pressure and see what happens. Going to 5 bar each for the starting pressures of both N_2 and H_2 (10 bar total pressure) we get

$$\frac{(2x)^2}{(5 - x)(5 - 3x)^3} = 6.53 \times 10^{-4}$$

Solving for x, we get $x = 0.245$. That's better. Substituting, we get our final pressures:

$$N_2 \quad + \quad 3\,H_2 \quad \rightleftharpoons \quad 2\,NH_3 \quad \text{(all gases)}$$

P_{eq}	$5 - x$	$5 - 3x$	$2x$	
$=$	4.755	4.26	0.49	(bar)

So, indeed, increasing the pressure overall has increased the amount of product, although the percent yield is only 15% based on the limiting reactant still being H_2:

$$(5\text{ bar}H_2)\frac{2\text{ bar }NH_3}{3\text{ bar }H_2} = 3.33\text{ bar }NH_3 \text{ theoretically}$$

$$\frac{0.49\text{ bar }NH_3\text{ actually}}{3.33\text{ bar }NH_3\text{ theoretically}} \times 100\% = 15\% \text{ yield}$$

---

[4] To do this correctly, we need $\Delta_r H^\circ$ and $\Delta_r S^\circ$ valid for 673 K, giving $K_{eq} = 2.0 \times 10^{-4}$.

At least we are going in the right direction. Let's try 50 bar each:

$$\frac{(2x)^2}{(50-x)(50-3x)^3} = 6.53 \times 10^{-4}$$

Solving this time for x, we get $x = 9.0$ bar. Substituting, we get our final pressures:

	N₂	+	3 H₂	⇌	2 NH₃	(all gases)
P_{eq}	$50 - x$		$50 - 3x$		$2x$	
=	41		23		18	(bar)

and we are up to 54% yield:

$$(50 \text{ bar H}_2)\frac{2 \text{ bar NH}_3}{3 \text{ bar H}_2} = 33.3 \text{ bar NH}_3 \text{ theoretically}$$

$$\frac{18 \text{ bar NH}_3 \text{ actually}}{33.3 \text{ bar NH}_3 \text{ theoretically}} \times 100\% = 54\% \text{ yield}$$

Not bad. It will have to do. The final pressure in the system is $(41 + 23 + 18) = 82$ bar, so the final percent ammonia in the mixture is $18/82 \times 100\% = 22\%$.

This exercise is not so far off from what real chemists do when they are trying to optimize the yield of their reactions. The idea is to understand enough about the reaction you are dealing with that you can predict the outcome of different possible alterations in the system. If we increase the temperature, will the yield go up? How about the overall pressure? How about adding more of reactant X or reactant Y? Ideas relating to relative ground states, bonding, $\Delta_r H^\circ$, and $\Delta_r S^\circ$ can all be used to help pinpoint the best conditions for a chemical reaction.

11.8 Equilibration at Constant Pressure (optional)

Note that increasing the overall pressure in the last calculation works because there are 4 moles of gas on the left and 2 moles on the right, so the equilibrium responds to an increase in pressure by shifting to the right. At the same time, however, the overall pressure has dropped from 100 bar initially to 82 bar at equilibrium.

What if we did this reaction in some sort of a chamber with a piston that allowed us to keep that pressure at 100 bar while the reaction proceeded? As H₂ and N₂ are consumed then, the volume would decrease to keep the pressure at 100 bar. How does this change the analysis? What would happen?

The answer is that we will get a higher yield. But the calculation is pretty tricky. The only change is to add a parameter α that adjusts the final equilibrium pressures so that, in the end, the overall pressure is still 100 bar:

	N₂	+	3 H₂	⇌	2 NH₃	(all gases)
$P_{initial}$	50		50		0	
Δ	$-x$		$-3x$		$+2x$	
P_1	$50 - x$		$50 - 3x$		$2x$	
P_{eq}	$\alpha(50-x)$		$\alpha(50-3x)$		$\alpha(2x)$	

P_1 refers to the (imaginary) pressure "after" the chemical reaction but "before" the volume change bringing us to the desired final pressure. In actuality, these two steps are happening simultaneously, but we are free to separate a change in state into two steps if it makes it easier (which it does!).

Thus, we imagine letting the reaction change in two steps such that *in the end* we have the desired values. Evaluating α is not too difficult. We just equate the total equilibrium pressure to 100 bar:

$$P_{eq}(\text{total}) = \alpha(50 - x) + \alpha(50 - 3x) + \alpha(2x)$$

$$100 = \alpha(100 - 2x)$$

$$\alpha = \frac{100}{100 - 2x}$$

Now, substituting *these* equilibrium pressures into our equilibrium expression, we get a rather nasty beast:

$$\frac{[\alpha(2x)]^2}{[\alpha(50 - x)][\alpha(50 - 3x)]^3} = \frac{(2x)^2}{\alpha^2(50 - x)(50 - 3x)^3}$$

$$= \frac{(2x)^2}{\left(\frac{100}{100-2x}\right)^2 (50 - x)(50 - 3x)^3} = 6.53 \times 10^{-4}$$

Believe it or not, using an equation-solving calculator, this is not too difficult to solve. We get $x = 9.67$. This means $\alpha = 1.24$ and we have

	N$_2$	+	3 H$_2$	$\rightleftharpoons$	2 NH$_3$	(all gases)
P_{eq}	$\alpha(50 - x)$		$\alpha(50 - 3x)$		$\alpha(2x)$	
=	50		26		24	(bar)

Note that, sure enough, the final pressures add up to 100 bar. Our product mixture is now a little more enriched in ammonia: 24% NH$_3$.

11.9 Standard Reaction Gibbs Energies, $\Delta_r G_T^\circ$

In the case of standard conditions, when $Q = 1$ and $\ln Q = 0$, Equation 11.1 reduces to

$$\Delta_r G_T^\circ = \Delta_r H^\circ - T\Delta_r S^\circ \qquad (11.5)$$

(The subscript "T" is added to help remind us that this value may be quite sensitive to temperature.)

Using this relationship in Equation 11.2 gives

$$K_{eq} = e^{-\frac{(\Delta_r H^\circ - T_{eq}\Delta_r S^\circ)}{RT_{eq}}} = e^{-\Delta_r G_T^\circ / RT_{eq}} \qquad (11.6)$$

which can be rearranged to give

$$\Delta_r G_T^\circ = -RT \ln K_{eq} \qquad (11.7)$$

The equilibrium constant, K_{eq}, and the *standard reaction Gibbs energy*, $\Delta_r G_T^\circ$, of a reaction are intimately related. If we know K_{eq}, we can get $\Delta_r G_T^\circ$, and vice-versa. If we have either (a) both of $\Delta_r H^\circ$ and $\Delta_r S^\circ$ or (b) just $\Delta_r G_T^\circ$, then we know K_{eq} at the specified temperature. Notice that both of these quantities depend intimately on exactly how the associated balanced chemical equation is written.

This explains the presence of a column for $\Delta_f G_{298}^\circ$, or *standard Gibbs energies of formation,* in the table of thermodynamic data in Appendix D along with $\Delta_f H_{298}^\circ$ and S_{298}°. The idea is to help you out at 298 K and avoid a calculation. However, "298" here is extremely important. It's not too bad to use $\Delta_f H_{298}^\circ$ at temperatures a little off from 298—you'll be pretty close—but you're looking for trouble if you use $\Delta_f G_{298}^\circ$ at any other temperature if there is any significant change in standard molar entropy for the reaction. Hiding in that value is $-(298 \text{ K})\Delta_r S^\circ$.

If you *happen* to be concerned specifically at 298 K and aren't interested in any other temperature, and you want to get the value of K_{eq}, then, by all means, use the table value for $\Delta_f G^\circ_{298}$. Otherwise, that column should be avoided. It's a trap.

Finally, substituting the definition of $\Delta_f G^\circ_T$ into Equation 11.1 but allowing Q to not be 1, we get

$$\Delta_r G = \Delta_r G^\circ_T + RT \ln Q \qquad (11.8)$$

This says that the reaction quotient can be thought of as a correction to the standard reaction Gibbs energy, similarly to the case for entropy (Equation 8.3):

$$\Delta_r S = \Delta_r S^\circ_T - R \ln Q$$

11.10 The Potential for Change in Entropy of the Universe is $R \ln K/Q$

An interesting relationship between the reaction quotient, Q, and the equilibrium constant, K, involves the hypothetical change in entropy of the universe if a stoichiometric amount of reactants were to convert to products under the specified current reaction conditions. Combining Equations 11.7 and 11.8, we get

$$\Delta_r G = -RT \ln K + RT \ln Q = -RT \ln \frac{K}{Q}$$

Using Equation 10.3 as a guide, we could define the potential for entropy change in the universe due to a chemical reaction to be $\Delta_r S_{universe}$:

$$\Delta_r S_{universe} = -\Delta_r G/T = R \ln \frac{K}{Q} \qquad (11.9)$$

Three conditions are possible:

If $Q < K$, we have that $\Delta_r S_{universe} > 0$. The forward reaction will be observed.
If $Q = K$, we have that $\Delta_r S_{universe} = 0$. No reaction of any type will be observed.
If $Q > K$, we have that $\Delta_r S_{universe} < 0$. The reverse reaction will be observed.

The most important thing you can learn from this study is that everything we know about equilibrium can be interpreted in terms of entropy, and entropy is just a measure of thermodynamic probability.

Furthermore, when we have the standard conditions, then $Q = 1$, and the real "identity" of the equilibrium constant K is revealed. A little rearrangement of the above equation indicates that K is a direct measure of the change in entropy of the universe that would occur upon conversion of a stoichiometric amount of reactants to products under the standard conditions of 1 bar partial pressure of each gas and 1 M concentration for each solute:

$$K = e^{\Delta_r S^\circ_{universe}/R} \qquad (11.10)$$

We assert:

It is the potential for the distribution of matter and energy (as measured by entropy) that is the real driving force behind all chemical reactions.

and

Probability rules all natural processes.

This is truly an awesome finding.

11.11 Beyond Ideality: "Activity"

Up to this final point we have concerned ourselves only with "ideal" substances. As with all theories, our conceptualization of entropy, enthalpy, system, and surroundings is an approximation, a model. The question remains: How does adding nonideality change the model? Some of the greatest minds in chemistry have contemplated this issue, and it is still an active area of research in theoretical chemistry. It might be interesting to at least take a peek at the sort of thinking that goes into extending the model to more "real" situations, just to give us some idea of how that might be done. In this section we show just one way to adjust the model.

First, consider where the "ideal" has shown up. One place was in our using the ideal-gas equation, $PV = nRT$, in converting from ratios of volumes to ratios of pressures when we considered the effect of concentration and pressure on entropy (Chapter 8). Another, more subtle, application of ideality was way back in Chapter 3 when we assumed that all of the energy in a system can be attributable to individual particles. Both of these applications of ideality are only approximations to real systems, because they assume that individual particles are independent. But this is not really true.

The particles of real gases are not actually independent. They attract or repel each other with what we call van der Waals forces. These forces become especially important as the gas gets concentrated. So, first off, what we can say about our model is that we expect it to break down when we consider gases at high pressure.

Similarly, real species in solution interact with each other and the solvent in complex ways. Ions in particular exhibit strong attractions and repulsions. These forces are expected to be especially strong at higher concentration. It turns out, though, that even at just 0.1 M concentrations, our ideal model breaks down and loses much of its predictive power.

There are any number of ways to extend the ideal model. We could, of course, throw out the whole thing and start over. But a simple practical extension is really all we need, and that extension is this: We've considered the effect of concentration and pressure on entropy. We know that these real interparticle forces should show up in the form of noncovalent bonding. Let's now propose that there is some sort of effect of concentration and pressure on *enthalpy* as well, because it is enthalpy that relates to bonding in thermodynamics.

The simplest way to deal with this is to propose that the enthalpy of a species has a logarithmic concentration dependence, just like the one involving entropy. We have already:

$$S_x = S_x^\circ - R \ln(P_x/\text{bar}) \quad (8.1)$$

$$S_x = S_x^\circ - R \ln([\text{X}]/M) \quad (8.2)$$

We add:

$$H_x = H_x^\circ + RT \ln \gamma_x \quad (11.11)$$

The γ_x (gamma-x) here just expresses the idea that we don't know exactly how much effect this is going to be. In the case of gases, the value of γ_x might depend upon the partial pressure of the gas, or it might depend upon the *total* pressure of all the gases in the system. In the case of ions in solution, γ_x might depend upon the concentration of x, the total ion concentration, the charges of the ions, and, perhaps, the *dielectric constant* of the solvent (a measure of how well the solvent screens the forces between ions). In any case, γ also might depend upon the temperature. The RT factor keeps us in units of energy. The plus sign is there rather than a minus sign because this is the way Gilbert Lewis did it in 1907, and everyone has since then followed his lead.

It's a nice exercise to follow through what the effect of this extension of the model will be in calculating $\Delta_r G$ for a chemical reaction as written:

$$\text{A}(aq) + \text{B}(aq) \longrightarrow 2\text{C}(g)$$

If you do this yourself you will see that we get the very same expression as before:

$$\Delta_r G = \Delta_r H^\circ - T[\Delta_r S^\circ - R \ln Q]$$

only now Q has a slightly different definition:

$$Q = \frac{(\gamma_C \, P_C)^2}{\gamma_A [A] \gamma_B [B]}$$

The terms $\gamma_x P_x$ and $\gamma_x [X]$ take the name *activities* and are generally written just a_x. The factor γ_x itself is called the *activity coefficient*. Basically our "ideal" model is simply the assumption that $\gamma_x = 1$. Since 1907, people have discovered all sorts of interesting ways of defining the activity coefficient. The end result is better predictions for the equilibrium pressures of gases and concentrations of solutes.

For this introduction, the exact definition of γ_x isn't particularly important. You can read all about it if you continue on in the area of *physical chemistry*. The point we want to make here is simply that this sort of scientific research—starting with a simple model and then extending it in different creative ways—is standard practice in chemistry.

11.12 Summary

Starting with the following four concepts:

1. reaction enthalpies ($\Delta_r H$) deviate little from their standard values ($\Delta_r H^\circ$),
2. reaction entropies ($\Delta_r S$) deviate significantly from their standard values ($\Delta_r S^\circ$) by $-RT \ln Q$,
3. reaction Gibbs energy, $\Delta_r G$, is $\Delta_r H - T \Delta_r S$, and
4. $\Delta_r G = 0$ at equilibrium,

we have arrived at the general equation governing all situations (Equation 11.1):

$$\Delta_r G = \Delta_r H^\circ - T[\Delta_r S^\circ - R \ln Q]$$

In many ways this is the most important equation in this book. It is the "master equation" from which many others will be derived. At equilibrium, $\Delta_r G = 0$, and this equation becomes

$$0 = \Delta_r H^\circ - T_{eq}[\Delta_r S^\circ - R \ln Q]$$

Rearrangement now gives Equation 11.2:

$$Q = e^{\frac{-(\Delta_r H^\circ - T_{eq} \Delta_r S^\circ)}{R T_{eq}}} = K_{eq}$$

The reaction quotient Q is constant at a given equilibrium temperature. The value of that constant, which we call K, can be determined from $\Delta_r H^\circ$ and $\Delta_r S^\circ$ in this way.

In addition, rearrangement of this equation gives Equation 11.3:

$$\ln K_{eq} = \left(\frac{-\Delta_r H^\circ}{R} \right) \frac{1}{T_{eq}} + \frac{\Delta_r S^\circ}{R}$$

This gives us a handy way to determine the *standard* values of $\Delta_r H^\circ$ and $\Delta_r S^\circ$ for any reaction, as long as we can determine equilibrium concentrations and/or pressures (and, thus, the value of K_{eq}) at various equilibrium temperatures. The idea is that rearranging the equation this way allows us to plot $\ln K_{eq}$ vs. $1/T_{eq}$ and expect to find a straight line with slope $-\Delta_r H^\circ / R$ and y-intercept $\Delta_r S^\circ / R$.

And if we just want to compare equilibrium constants at two different temperatures, we find that only the change in enthalpy, not entropy, is involved (Equation 11.4):

$$\ln \frac{K_2}{K_1} = \frac{\Delta_r H^\circ}{R} \left(\frac{1}{T_1} - \frac{1}{T_2} \right)$$

We have seen in this chapter that G–T graphs are useful in depicting the overall qualitative picture of what is going on as a reaction approaches equilibrium, and why it is headed in the direction it is in the first place.

In the end, we have found that the relationship among the reaction quotient, the equilibrium constant, and the potential for entropy change in the universe due to a chemical reaction is extraordinarily simple (Equation 11.9):

$$\Delta_r S_{universe} = R \ln \frac{K}{Q}$$

Throughout this book we have used a very simple model involving ideal gases, ideal solutions, and the idea that individual molecules absorb and release countable numbers of energy "quanta." Of course, nature is not "ideal" in this sense. A more sophisticated treatment of the subject of thermodynamics would take additional considerations into account. The concept of chemical activity is just one way that the model can be extended.

In this book we have seen that every aspect of chemical reactivity—the approach to equilibrium—really boils down to probability. Equilibrium is the end of the road, and this chapter is really the end of this book.[5] Does God play dice? Only God knows for sure.

---

[5] Chapters 12 and 13 simply apply our model to phase changes and electrochemistry.

Problems

The symbol ⏍ *indicates that the answer to the problem can be found in the "Answers to Selected Exercises" section at the back of the book.*

Calculating K_{eq}

11.1 ⏍ In each of the following cases, determine $\Delta_r H°$ and $\Delta_r S°$ for the forward reaction at 298 K, and use that information to predict the equilibrium constant K_{eq} at 298 K for the equilibrium as written.

(a) $2\,NO_2(g) \rightleftharpoons N_2O_4(g)$
(b) $CO(g) + Cl_2(g) \rightleftharpoons COCl_2(g)$
(c) $2\,H_2(g) + CO(g) \rightleftharpoons CH_3OH(l)$

11.2 In each of the following cases, determine $\Delta_r H°$ and $\Delta_r S°$ for the forward reaction at 298 K, and use that information to predict the equilibrium constant K_{eq} at 298 K for the equilibrium as written.

(a) $N_2H_4(l) + 3\,O_2(g) \rightleftharpoons 2\,NO_2(g) + 2\,H_2O(g)$
(b) $CO_2(g) + 4\,HCl(g) \rightleftharpoons CCl_4(g) + 2\,H_2O(g)$
(c) $CCl_4(g) + 4\,HCl(g) \rightleftharpoons CH_4(g) + 4\,Cl_2(g)$

11.3 In each of the following cases, determine $\Delta_r H°$ and $\Delta_r S°$ for the forward reaction at 298 K, and use that information to predict the equilibrium constant K_{eq} at 298 K for the equilibrium as written.

(a) $4\,Fe(s) + 3\,O_2(g) \rightleftharpoons 2\,Fe_2O_3(s)$
(b) $CO_2(g) + 2\,H_2O(g) \rightleftharpoons CH_4(g) + 2\,O_2(g)$
(c) $SiO_2(s) + 4\,HF(g) \rightleftharpoons SiF_4(g) + 2\,H_2O(g)$

11.4 ⏍ Predict the numerical value of K_w for water at 0.0°C, 25.0°C, 60.0°C, and 100.0°C based on the following information:

$$2\,H_2O \longrightarrow H_3O^+ + OH^-$$

$$\Delta_r H° = +55.9 \text{ kJ/mol}$$

$$\Delta_r S° = -80.54 \text{ J/mol·K}$$

11.5 Acetic acid can be manufactured from methanol and carbon monoxide in a carbonylation reaction, which is an equilibrium:

$$CH_3OH(l) + CO(g) \rightleftharpoons CH_3COOH(l)$$

For acetic acid, $\Delta_f H° = -484$ kJ/mol, $S°_{298} = 158$ J/mol·K. (a) Predict K_{eq} for this reaction at 25°C and 50°C. (b) Will the yield of acetic acid be improved if the temperature is increased or decreased? Why?

11.6 Determine the vapor pressure of Br_2 at (a) 10°C, (b) 25°C, and (c) 55°C.

Determining K_{eq} Experimentally

11.7 ⏍ Shown below are several observations of vapor pressures of gaseous lead over molten (liquid) lead at various temperatures. Make a plot of ln P_{vap} vs. $1/T$ and use it to determine $\Delta_{vap} S°$, $\Delta_{vap} H°$, and the normal boiling point of lead. [*Hint:* Careful with those units!]

T/K	P_{vap}/mmHg
1500	19.72
1600	48.48
1700	107.2
1800	217.7
1900	408.2

11.8 Determine $\Delta_r H°$ and $\Delta_r S°$ for the forward reaction of

$$C(s) + CO_2(g) \rightleftharpoons 2\,CO(g)$$

based on the data shown below. Discuss your results in terms of the relative ground state energies and energy level spacings of products vs. reactants.

$T/°C$	K_{eq}
850	1428
950	7484
1050	27500
1200	170000

11.9 ⏍ (a) Given the data shown below, without doing any calculation, determine the *signs* of $\Delta_r H°$ and $\Delta_r S°$, or explain why that is not possible.

$T/°C$	K_{eq}
450	2.20
500	1.20
550	0.74
600	0.50
650	0.36

(b) Using these data, determine $\Delta_r H°$ and $\Delta_r S°$ for the forward reaction as written:

$$2\,C_3H_6(g) \rightleftharpoons C_2H_4(g) + C_4H_8(g)$$

11.10 (a) Given the data shown below, without doing any calculation, determine the *signs* of $\Delta_r H°$ and $\Delta_r S°$, or explain why that is not possible.

$T/°C$	K_{eq}
1000	1.63×10^{-5}
1250	8.36×10^{-5}
1500	2.74×10^{-4}
1750	6.79×10^{-4}
2000	1.39×10^{-4}

(b) Using these data, determine $\Delta_r H°$ and $\Delta_r S°$ for the forward reaction as written:

$$2\,HBr(g) \rightleftharpoons H_2(g) + Br_2(g)$$

11.11 Ⓐ Calculate $\Delta_r H°$ for the following reactions based on data in Appendix D, and use that information along with the value of K_{eq} at the specified temperature to predict K_{eq} for the reaction at 298 K.

(a) $2\,SO_2(g) + O_2(g) \rightleftharpoons 2\,SO_3(g)$ $K_{eq} = 0.05$ at 700 K
(b) $Br_2(l) \rightleftharpoons Br_2(g)$ $K_{eq} = 1.0$ at 331 K

11.12 Determine $\Delta_r H°$ and $\Delta_r S°$ for the following isomerization reactions. In each case also predict the limiting value of K_{eq} at high temperature. Which substance has the closer spaced energy levels, reactant or product in each case? Sketch a graph of K vs. T in each case showing clearly the limits at high and low temperature. For (a) use $K_{eq} = 4.4$ at 300 K and 0.52 at 400 K; for (b) use $K_{eq} = 0.27$ at 300 K and 0.55 at 1000 K.

(a) $C_4H_{10}(g, \text{$n$-butane}) \longrightarrow C_4H_{10}(g, \text{isobutane})$
(b) $C_8H_{10}(l, \text{$m$-xylene}) \longrightarrow C_8H_{10}(l, \text{o-xylene})$

11.13 In each of the following cases, determine $\Delta_r H°$ and $\Delta_r S°$ for the forward reaction at 298 K, and use that information to predict the equilibrium constant K_{eq} at 298 K for the equilibrium as written. (For Hg(g): $\Delta_f H° = 61.4$ kJ/mol and $S° = 175.0$ J/mol·K.)

(a) $PCl_3(g) + Cl_2(g) \rightleftharpoons PCl_5(g)$ $K_{eq} = 0.0408$ at 500 K
(b) $H_2(g) + I_2(g) \rightleftharpoons 2\,HI(g)$ $K_{eq} = 51.1$ at 750 K
(c) $HgO(s, \text{red}) \rightleftharpoons Hg(g) + \frac{1}{2}\,O_2(g)$ $K_{eq} = 0.065$ at 400 K

G–T Graphs

11.14 Ⓐ Sketch G–T graphs for the following reactions under standard conditions. Determine T_{eq} and indicate that temperature on your graph.

(a) $I_2(s) \longrightarrow I_2(g)$
(b) $Br_2(l) \longrightarrow Br_2(g)$
(c) $Sn(\text{gray}) \longrightarrow Sn(\text{white})$

11.15 Sketch G–T graphs for the following reactions under standard conditions. Determine T_{eq} and indicate that temperature on your graph.

(a) $SO_3(l) \longrightarrow SO_3(g)$
(b) $TiCl_4(g) \longrightarrow TiCl_4(l)$
(c) $S(s, \text{rhombic}) \longrightarrow S(g)$

11.16 Sketch G–T graphs for the following reactions under standard conditions. Determine T_{eq} and indicate that temperature on your graph.

(a) $Pb(g) \longrightarrow Pb(s)$
(b) $ZnS(s, \text{wurtzite}) \longrightarrow ZnS(s, \text{sphalerite})$
(c) $N_2O_4(l) \longrightarrow N_2O_4(g)$

11.17 Ⓐ Based on the information given below, draw a G–T graph for the dissolving of carbon dioxide in water under standard conditions. Use your graph to argue that it is better to keep a bottle of champagne cold, not hot, to prevent it from going flat. [*Hint*: What is the effect of raising the temperature on the concentration of $CO_2(aq)$?]

$$CO_2(g) \rightleftharpoons CO_2(aq)$$

with

	$\Delta_f H°_{298}$/ (kJ/mol)	$S°_{298}$/ (J/mol K)
$CO_2(g)$	−393.78	213.8
$CO_2(aq)$	−413.2	120

11.18 Using a G–T graph, depict the situation in neutral water, where $[H_3O^+] = [OH^-] = 1 \times 10^{-7}$ (equilibrium) at 25°C assuming

$$2\,H_2O \longrightarrow H_3O^+ + OH^-$$
$$\Delta_r H° = +55.9\text{ kJ/mol}$$
$$\Delta_r S° = -80.54\text{ J/mol·K}$$

Determine the pH of water at 0°C.

11.19 Consider the decomposition of solid Ag_2O to form $Ag(s)$ and $O_2(g)$. (a) Write a balanced equation for this reaction. (b) Determine $\Delta_r H°$ and $\Delta_r S°$ for the reaction based on the equation you have written. Sketch a G–T graph for the reaction in its standard state, clearly indicating the crossing temperature (if any). (c) Discuss how the addition or removal of $O_2(g)$ from the standard state will affect the curves on your graph. How will the addition or removal of $O_2(g)$ affect the yield of $Ag(s)$ at 30°C? (d) Will increasing the temperature increase or decrease the yield of $Ag(s)$? Why?

Brain Teasers

11.20 The conversion of methane, CH_4, into compounds that contain more than one carbon atom, such as ethane, C_2H_6, is an important industrial process. (a) Using Appendix D, calculate $\Delta_r H°$, $\Delta_r S°$, and $\Delta_r G°$ at 25°C for the reaction

$$2\,CH_4(g) \longrightarrow C_2H_6(g) + H_2(g)$$

(b) Is this reaction spontaneous at 25°C when all gases are at 1 bar pressure? (c) In practice, the conversion of methane to ethane is carried out industrially using a slightly different reaction:

$$2\,CH_4(g) + 1/2\,O_2(g) \longrightarrow C_2H_6(g) + H_2O(g)$$

Why might this reaction be preferable? [*Hint*: Calculate $\Delta_r H°$, $\Delta_r S°$, and $\Delta_r G°$ at 25°C for this reaction.] Discuss the differences between the two reactions. Is the primary difference related to bonding or disorder?

11.21 $[K^+]$ in blood plasma is about $5.0 \times 10^{-3}M$, but $[K^+]$ in muscle cell fluid is much higher at about $0.15\ M$. The plasma and the intracellular fluid are separated by the cell membrane. From this information and a G–T graph, argue that $\Delta_r G$ for the reaction

$$K^+(\text{plasma}) \longrightarrow K^+(\text{muscle})$$

must be positive. How might a cell make this reaction happen? [Think Hess's law.] How might the body use the difference in concentration outside and inside the cell to "drive" other reactions? [Think Hess's law again!]

CHAPTER 12

Applications of Gibbs Energy: Phase Changes

This chapter, unlike the others, may be read in essentially any order. It is a collection of the following applications, which will be discussed in turn using G–T graphs. Some of this material has been discussed in earlier sections, but it is brought together here to emphasize the similarities and differences in the different situations:

12.1 Review

As review, consider Figure 12.1, which summarizes the finding of Section 10.4.

Bonding differences are seen at the intersection with the y-axis at 0 kelvin. Thus, curves for substances with stronger bonding start lower on the graph. Curves generally slope downward because the slope at any point on the curve is $-S$, the absolute entropy, and absolute entropy for substances is always positive, at least at 1 bar pressure. Curves for solids and liquids are "fixed" and relatively flat; curves for gases and solutes are "flexible" and are steeper for lower pressure, flatter for higher pressure.

In this chapter, each curve belongs not only to a single state such as "reactants" or "products" but also to a specific substance or phase, such as "liquid water" or "solid CO_2."

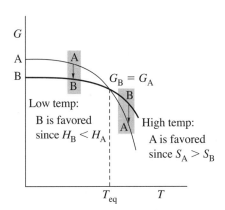

Figure 12.1
General G–T graph depicting two states, each of which is a single substance. Different curves represent different *phases* such as solid, liquid, gas, or aqueous (dissolved). The stable phase at any given temperature will always be the lowest curve at that point.

The chemical reactions are really *phase changes*. Our guiding reaction will always be

$$X(\textit{phase 1}) \longrightarrow X(\textit{phase 2})$$

Note that these curves refer to stoichiometric quantities, which in this case just means one mole of the substance in either the "reactant" phase or the "product" phase.

These, then, are all the ideas we need to understand what is going on in a wide variety of real-world situations. Although some math is presented here, this chapter is mostly designed to help you picture how bonding and the distribution of energy ultimately determine the course of events in physical changes familiar to you from your own experience.

Application: Melting and Freezing

A quantitative treatment of melting and freezing was introduced in Section 10.7. There it was pointed out that, since neither the entropy of the solid nor the entropy of the liquid depend substantially on pressure, their molar Gibbs energy curves on a G–T graph (Figure 12.2) can be considered fixed. This results in only a single temperature at which $\Delta_r G = 0$, and a characteristic melting point, T_{mp}, for every pure substance. Below this temperature, all liquid present will freeze; above this temperature, all solid will melt. The value of T_{mp} is easily derived from the definition of Gibbs energy by setting $\Delta G = 0$ and $\Delta_r H$ and $\Delta_r S$ to their standard values for melting, $\Delta_{fus} H^\circ$ and $\Delta_{fus} S^\circ$:

$$\Delta_r G = \Delta_r H - T \Delta_r S$$
$$0 = \Delta_{fus} H^\circ - T_{mp} \Delta_{fus} S^\circ$$
$$T_{mp} = \frac{\Delta_{fus} H^\circ}{\Delta_{fus} S^\circ}$$

(12.1)

So, for example, for water, since $\Delta_{fus} H^\circ = 6010$ J/mol and $\Delta_{fus} S^\circ = 22$ J/mol·K, we have

$$T_{mp} = \frac{6010 \text{ J/mol}}{22 \text{ J/mol·K}} = 273 \text{ K}$$

12.2 Evaporation and Boiling

If you seal a liquid in a container, it will come to equilibrium with its vapor (Section 10.8). That is, if you had a sensitive pressure meter connected to the container, you would see its reading rise as the number of molecules in the "head space" above the liquid increased over time and then stabilized.

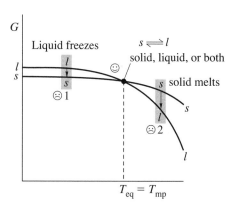

Figure 12.2
G–T graph for a typical solid/liquid system. Only at the melting point, T_{mp}, is equilibrium achieved.

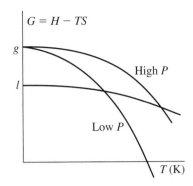

Figure 12.3
G–T graph for a general liquid/gas system. Due to the fact that the entropy of the gas is dependent upon its vapor pressure, its curve has variable slope and can intercept the liquid at any temperature.

This is the phenomenon of evaporation, depicted in Figure 12.3. The difference here with melting and freezing is that in this case the gas curve is sensitive to the vapor pressure of the gas. At higher vapor pressure, the gas is effectively compressed, leading to a lower molar entropy and, as a result, a flatter curve. At low pressure, the gas is effectively expanded, leading to a much higher molar entropy for the gas and a much steeper downward sloping curve.

The relationship between temperature and vapor pressure is easily derived (Section 10.9):

$$\Delta_r G = \Delta_r H - T \Delta_r S$$
$$0 = \Delta_{vap}H^\circ - T_{eq}(\Delta_{vap}S^\circ - R \ln P_{vap})$$
$$T_{eq} = \frac{\Delta_{vap}H^\circ}{\Delta_{vap}S^\circ - R \ln P_{vap}}$$

(12.2)

where P_{vap} is the vapor pressure of the gas in bars (dimensionless). If the overall atmospheric pressure above the liquid is less than this vapor pressure, then the liquid will be observed to boil. The standard boiling point of a liquid is defined simply as the temperature at which its gas has an equilibrium vapor pressure of 1 bar. Thus, for the standard boiling point, we have $\ln P_{vap} = 0$, and the $\ln P_{vap}$ term drops out. We are left with an expression which looks very similar to that for melting (Equation 12.1):

$$T_{bp} = \frac{\Delta_{vap}H^\circ}{\Delta_{vap}S^\circ}$$

(12.3)

If we focus not on the standard boiling point but on *any* temperature, then we find that we can have three possible situations (Figures 10.10 and 12.4). Either the molar Gibbs energy of the gas is higher than that for the liquid (too high vapor pressure, 12.4a), the Gibbs energies are equal (equilibrium vapor pressure, 12.4b), or the molar Gibbs energy of the gas is less than that of the liquid (low vapor pressure, 12.4c). In the first case the favored reaction is gas-to-liquid, and we see condensation; in the second case, nothing is observed, because the system is at equilibrium; in the third case, the favored reaction is liquid-to-gas, and we see evaporation.

Application: Humidity and Dew Point

This explains why water in a sidewalk puddle evaporates even though it never boils and the temperature never reaches 100°C. *Relative humidity* in the atmosphere is related to the vapor pressure of water in the air. *100% humidity* means that the gas and liquid curves are crossing at the current temperature (Figure 12.4b), and no more water vapor can be held by the air at this temperature. The system is at equilibrium, and depending upon whether it is being cooled (as at night) or being heated (as when the sun rises) fog or dew will be

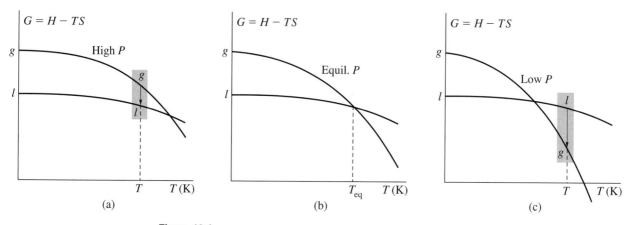

Figure 12.4
Three possible situations for a liquid/gas system. In (a) the vapor pressure is too high for the system temperature, and gas will condense. In (b) the liquid and vapor are in equilibrium. In (c) the vapor pressure is too low, and liquid is evaporating.

Figure 12.5
(a) G–T graph for the evaporation of water from a puddle at 30% humidity. The vapor pressure is 30% of the equilibrium pressure determined for temperature T. (b) The *dew point*, T_{DP}, is the temperature for which the current vapor pressure of water in the air would be the equilibrium vapor pressure.

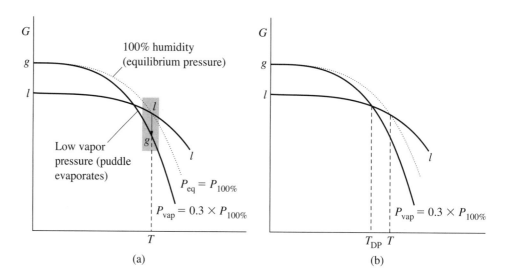

either forming or clearing. It is even possible for the air to get supersaturated and achieve, at least temporarily, a state where the vapor pressure of water is higher than the equilibrium amount (Figure 12.4a). In this case, rain or fog will be observed.

Generally, though, the actual vapor pressure of water in the air is only a fraction of what is necessary to maintain equilibrium with liquid water in a puddle (Figure 12.4c and Figure 12.5a). Thus P_{vap} is generally on the low-pressure side of equilibrium for the temperature of a puddle, and the water in a puddle evaporates. What we call "30% relative humidity" is the condition such that the actual vapor pressure of water in the air, P_{vap}, is only 30% of the *equilibrium* vapor pressure for water for a specific temperature. In this case, although the gas curve starts to rise, it never makes it up to the liquid curve, since it is impossible for the small amount of water evaporating from a puddle to significantly change the vapor pressure of the whole *atmosphere*. The puddle simply continues to evaporate until it is dry.

On the other hand, if night falls and the temperature plunges, it is quite possible to reach the *dew point*—the temperature at which the current actual-pressure gas curve crosses the liquid curve (Figure 12.5b). The dew point is often reported on the evening news because it is a good guess for how low the temperature might get overnight. If the temperature is falling and the dew point is reached, then any further attempt to lower the temperature will result in a gas-to-liquid reaction (dew or fog formation), releasing heat (q_{sur}, from $\Delta_{vap}H°$) into the atmosphere and countering the drop in temperature in precisely the same way that cooling liquid water past 0°C is impossible until all the water has frozen. The dew point thus serves as a rough estimate of the expected nighttime low temperature.

Calculating the dew point if the temperature is 20°C and the relative humidity is 30% using an equation-solving calculator is a snap. One simply considers the equation for the evaporation of water:

$$H_2O(l) \longrightarrow H_2O(g)$$

for which $\Delta_{vap}H° = 44$ kJ/mol and $\Delta_{vap}S° = 119$ J/mol·K. Here $Q = P_{vap}$, the vapor pressure. We apply the general equation relating Gibbs energy and reaction quotient, Q:

$$\Delta_r G = \Delta_r H° - T(\Delta_r S° - R \ln Q)$$

1. First we calculate the equilibrium vapor pressure at 20°C by setting $\Delta_r G = 0$ and $T = 293$ K. Solving for Q, we get 0.0235. "100% humidity" at 20°C corresponds to a vapor pressure of water of 0.0235 bar.
2. Since the *actual* pressure is 30% of this *equilibrium* pressure, we change Q to 0.30 × 0.0235 = 0.00706 and then solve for T, the temperature at which this lower vapor pressure is the equilibrium pressure. We get $T = 275$ K (2°C). This is our dew point. Of course, this might be off a little. Could be a frost tonight!

12.3 Sublimation and Vapor Deposition

The same analysis can be carried out for a gas in relation to its solid phase. Specifically, below the melting point we know that solid will be favored over liquid. However, just as for the liquid above the melting point, it is also true that the solid will be in equilibrium with its gas at all temperatures below the melting point. Thus, the gas curve can simply swing down even further to cross the solid instead of the liquid (Figure 12.6). Note that below T_{mp} (0°C) a new equilibrium we hadn't considered is possible:

$$H_2O(s) \rightleftharpoons H_2O(g)$$

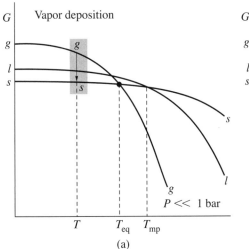
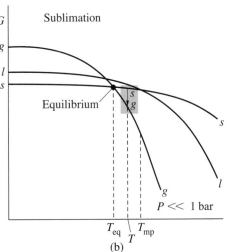

Figure 12.6
G–T graph in the case of (a) vapor deposition and (b) sublimation. In both cases the temperature is below the melting point and the vapor pressure is low. Temperatures below the equilibrium temperature T_{eq} relating the solid to the vapor result in vapor deposition. Temperatures between T_{eq} and the melting point result in sublimation.

The forward reaction here is called *sublimation* and is to a solid what evaporation is to a liquid. The reverse is called *vapor deposition*. Sublimation occurs when the temperature is above T_{eq} but below T_{mp}. Vapor deposition occurs at all temperatures below T_{eq}.

Application: Frost and Refrigerators

During the winter in Minnesota, for example, there isn't much water vapor in the air. If the sun comes out, raising the temperature just a bit, then patches of ice on sidewalks will sublime and disappear without ever melting. The reverse reaction, *frost formation*, occurs when the temperature falls to T_{eq} (still called the dew point). In that case the vapor pressure of water above a cold surface (a leaf, tree limb, or car window) becomes the equilibrium vapor pressure for that temperature. Further cooling will not lower the temperature. Instead, frost will form. This generally happens at night, when the temperature drops faster than the humidity.

But frost will not form if the vapor pressure of water in the air is too low. A frost-free refrigerator is simply one designed to keep the vapor pressure of water inside the freezer compartment so low that frost cannot form and sublimation occurs instead. Thus, the down side of a frost-free refrigerator is that continual sublimation causes ice to disappear and frozen foods to dry out.

12.4 Triple Points

Since the gas curve can change with pressure, it should be easy to imagine that there is a specific temperature and vapor pressure where the solid, liquid, and gas curves all cross at the same point (Figure 12.7). This is called the *triple point*, and it arises when a liquid/solid mixture is sealed in a container so that its vapor can come to equilibrium with it.

In fact, ice water in a sealed insulated container protected from air is *not* at 0°C. Rather, it is at 0.0098°C. Its vapor pressure will stabilize at 4.58 torr. Attempting to raise the temperature any further will simply melt the ice.

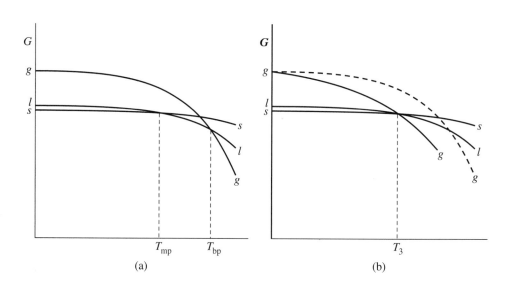

Figure 12.7
G–T graphs illustrating (a) the normal situation where $T_{bp} > T_{mp}$ with the vapor pressure of the gas at 1 bar. At most two phases may coexist. At the exact pressure and temperature of the *triple point*, T_3, (b) all three phases may coexist and $T_{bp} = T_{mp}$.

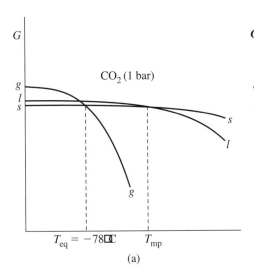
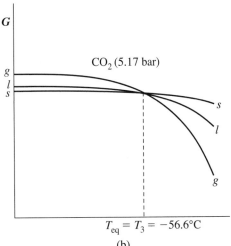

Figure 12.8
G–T graphs illustrating the situation for carbon dioxide. The gas curve starts much closer to the liquid and solid than in the case of water, as there is no hydrogen bonding in the case of CO_2. Thus, under standard conditions (a) "dry ice" sublimes above $-78°C$ without ever turning liquid. However, if the vapor pressure is allowed to increase to 5.17 bar, the triple point will be reached at $-56.6°C$, and liquid CO_2 can be observed.

Application: Liquid Carbon Dioxide

The strange effect of the triple point can be observed easily with carbon dioxide. The G–T graph for the CO_2 system is shown in Figure 12.8. Notice that under normal conditions, "dry ice" sublimes to the gas without ever melting. This is largely due to the fact that there is not much in the first place holding the liquid together the way there is in water with its hydrogen bonds. Thus, it takes only a little heat to turn solid CO_2 to liquid and liquid CO_2 to gas. Still, the gas has much higher entropy than either the liquid or solid, so its curve drops precipitously even at 1 bar pressure. Notice that the liquid form of CO_2 is never favored at 1 bar pressure.

However, if you fill a 5-mL disposable polyethylene pipette with powdered dry ice, quickly clamp it shut with a pair of pliers, and hold it tightly **behind a blast shield while wearing goggles,** you will soon become one of the few people to have ever observed liquid carbon dioxide first-hand.[1] The temperature will start out at $-78°C$, but as the dry ice sublimes, and the vapor pressure inside the pipette increases, the temperature, T_{eq}, will increase, approaching the melting point temperature. At some point the temperature and pressure will stabilize at the triple point values of $-56.6°C$ and 5.17 bar. A first milky then clear liquid phase will half-fill the pipette. Solid CO_2 will fall to the bottom of the mixture.

These conditions will hold until all the solid CO_2 has melted. Continued heating after the solid has melted allows the temperature and pressure to increase further until the pipette can hold no more, and it will burst at a seam. As the pressure returns to 1 bar and the gas expands, the temperature inside the burst pipette will plummet, and you will observe, once again, only dry ice in the pipette! **Do not do this demonstration without a blast shield!**

12.5 Critical Points and Phase Diagrams

As the vapor pressure of a gas increases, its molar entropy approaches that of the liquid form. This decrease in entropy of the gas results in a flattening of its curve, and its intersection with the liquid curve moves to higher temperature. (To be sure, the entropy of the liquid phase is changing as well, but not as quickly as for the gas.) If this continues, then a point

---

[1] The authors thank Bob Becker for this demonstration discovery.

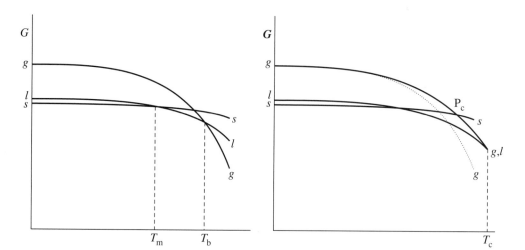

Figure 12.9
As the vapor pressure of the gas increases, its curve starts to flatten. At some point its curvature matches that of the liquid, and at that *critical point* of temperature and pressure, the distinction between gas and liquid vanishes.

is reached where the curvature of the gas curve (S_{gas}) matches the curvature of the liquid curve (S_{liquid}). Once this *critical point* is reached (Figure 12.9), the gas and liquid curves effectively merge, and there is no longer any distinction that can be made between gas and liquid.

At the critical point, the familiar meniscus disappears, and the *supercritical fluid* takes on properties which are both similar to and different from both gases and liquids. In particular, the fluid starts to flow very smoothly and to scatter light in what is called *critical opalescence*. In 1910, Einstein showed that this scattering was the very same scattering phenomenon that causes the sky to appear blue and is due to small differences in density of the fluid resulting from tiny fluctuations in temperature throughout the fluid. For water, the critical point is at 374°C and 221 bar. For carbon dioxide, though, the critical point is only 31°C—just above room temperature—and 74 bar. Thus, if the pipette in the demonstration above were strong enough and we were able to get the system to just a few degrees above room temperature, we would see the liquid "disappear" as its meniscus is lost and no further distinction between gas and liquid would be discernable.

If one tracks the crossing points of the gas curve with first the solid, then the liquid curve and makes a plot of pressure vs. crossing temperature, we have what is called a *phase diagram* (Figure 12.10).

All of the aforementioned special temperatures and pressures, including the standard melting point, the standard boiling point, the triple point, and the critical point are easily identified on the phase diagram. Three regions are shown, one for solid, one for liquid, and one for gas. In any of these regions, the specified phase has the lowest Gibbs energy and will be present exclusively.

Along any line, we have an equilibrium between two phases ($\Delta_r G = 0$). The standard melting and boiling points are the temperatures where the horizontal 1-bar pressure meets the solid–liquid and liquid–gas lines, respectively. The triple point is where all three lines intersect. The critical point is the "end of the line" for the liquid–gas boundary.

Application: Ice Skating and Ice IX

Subtle pressure effects upon the entropy of solids and liquids can also be depicted on these diagrams, and appear as a near-vertical line. Usually, as indicated in Figure 12.10, that solid/liquid line slopes upward, toward the right, indicating that increasing the pressure at any given temperature ultimately leads to the formation of solid exclusively. However, the phase diagram for water, depicted in Figure 12.11 (wildly exaggerated), shows that for water, the solid/liquid line slopes the "wrong" way, indicating that pressurizing ice, at least between 0.01 and −0.01°C, should lead to melting. For many years this fact was thought

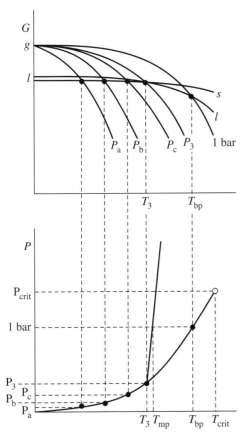

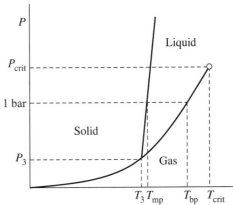

Figure 12.10
Construction of a phase diagram from a G–T graph involves tracking the effect of pressure on the crossing points of the gas–solid, gas–liquid, and even the solid–liquid curves. The finished diagram as it usually appears in textbooks is shown on the right.

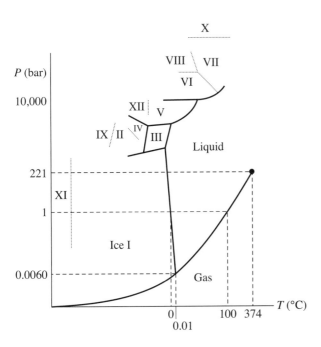

Figure 12.11
A wildly exaggerated phase diagram for water. Although there are twelve known crystalline forms of ice, fortunately, there is no ice IX at room temperature and pressure. Some of the phase boundaries, indicated as dotted lines, are still only roughly known.

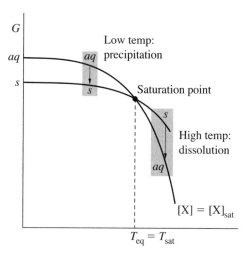

Figure 12.12
G–T graph for an aqueous solution for which
$\Delta H° > 0$ for dissolving. $\Delta G = 0$ at saturation.
Note that in this case, higher temperature
should lead to more dissolution.

to be the explanation of how ice skates slide on ice under the weight of a skater, but this conjecture has recently been disproved[2] and was rather unlikely in any event, considering how below about $-2°C$, compression of ice simply leads to a new form of solid ice.

In fact, there are actually at least twelve forms of ice: ice I, ice II, etc. Prior to its discovery, "Ice IX" was only in the imagination of science fiction writer Kurt Vonnegut, who speculated about a form of ice that somehow would be more stable than liquid water at atmospheric pressure, were it ever produced.[3] This doomsday ice would presumably solidify all of the liquid water on the planet and lead to the end of all life. Rest assured, there is no room on the actual phase diagram for any form of ice which is stable anywhere near room temperature and below a pressure of about 200 bar except for regular old ice I.

12.6 Solubility: $0 = \Delta_r H° - T(\Delta_r S° - R \ln[X]_{sat})$

Generally the reaction of a solid dissolving in water requires the breaking of weak inter-atomic interactions in the solid and the making of weaker ones with solvent water. Thus, $\Delta_r H°$ is positive for the equation

$$X(s) \longrightarrow X(aq)$$

and we could draw a G–T graph something like Figure 12.12.

A solution containing a solute in equilibrium with its undissolved solid form is said to be *saturated*. The solute's concentration when saturated, $[X]_{sat}$, is called the *solubility* of X. The forward reaction, in which the solid dissolves, is called *dissolution*; the reverse reaction, where solid is formed, is *precipitation*.

For a given concentration of X, the crossing will occur, once again, at a specific temperature, T_{sat}. That means that at this particular temperature the solution is saturated with X. If the temperature is raised even just a little bit, more solid will dissolve (Figure 12.13a). If the temperature is lowered even a fraction of a degree, then solid will precipitate from the solution (Figure 12.13b).

Note that the aqueous curve will change shape as the reaction proceeds, since during the reaction the concentration of X changes. If the temperature is raised, then dissolution occurs and [X] increases. As [X] increases, the molar entropy of X in solution decreases (because a mole now takes up less volume), and the aqueous curve gets flatter. This pulls the saturation point to the right, toward the new, higher, temperature.

[2] See *http://www.amasci.com/miscon/ice.txt* or James White, *The Physics Teacher*, *30*, 495, **1992**.
[3] In *Cat's Cradle*. But see *http://www.amsci.org/Amsci/issues/Sciobs96/Sciobs96-09Ice.html*.

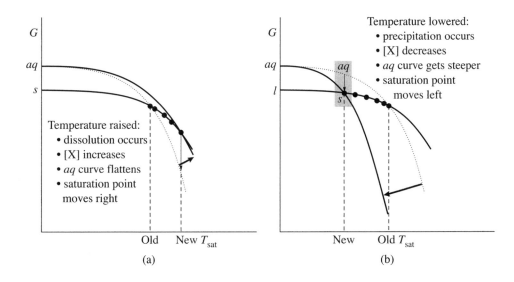

Figure 12.13
G–T graphs showing the effect on solubility of (a) raising the temperature and (b) lowering the temperature when $\Delta_r H^\circ > 0$ for dissolving. In each case the aqueous curve shifts due to the change in [X] with reaction.

Eventually, either (1) the curves cross again at the new temperature, and the solution becomes saturated again, or (2) all the solid dissolves.

The reverse occurs when the temperature is lowered (Figure 12.13b). In that case X in solution precipitates, lowering [X] and making the *aq* curve steeper. This pulls the saturation point toward the new (lower) temperature.

If you want to know exactly how much of a substance will dissolve at a particular temperature T_{eq}, then you need $[X]_{sat}$. To get it, you consider, again, the differences in molar enthalpy and entropy for the two substances involved. Write the general equation, determine $\Delta_r H^\circ$ and $\Delta_r S^\circ$ for the dissolving reaction, set $\Delta_r G = 0$, and solve for $[X]_{sat}$. Just the setup and result are shown here.

$$X(s) \rightleftharpoons X(aq)$$

We have:

$$\Delta_r G = \Delta_r H - T \Delta_r S$$
$$0 = \Delta_r H^\circ - T_{eq}(\Delta_r S^\circ - R \ln[X]_{sat})$$

and

$$\ln[X]_{sat} = -\left(\frac{\Delta H^\circ}{R}\right)\frac{1}{T_{eq}} + \frac{\Delta S^\circ}{R}$$

If $\Delta_r H^\circ$ and $\Delta_r S^\circ$ are not known, you can go to the lab and measure $[X]_{sat}$ at various temperatures. If you then plot $\ln[X]_{sat}$ vs. $1/T$, you should get a straight line with slope of $-\Delta_r H^\circ/R$ and intercept $\Delta_r S^\circ/R$.

Continuing on as we did for gases, we arrive once again with an equilibrium expression:

$$[X]_{sat} = e^{-\frac{\Delta_r H^\circ - T_{eq}\Delta_r S^\circ}{RT_{eq}}} = \left\{ \begin{array}{c} \text{constant} \\ \text{for a given} \\ \text{temperature} \end{array} \right\} \equiv K_{sp}$$

where K_{sp} is the equilibrium constant at the particular temperature of interest (also called the *solubility product*) for the equilibrium:

$$X(s) \rightleftharpoons X(aq) \qquad K_{sp} = [X]_{sat}$$

Note that this time the *solid* does not appear in the equilibrium expression since its molar entropy is always its standard value.

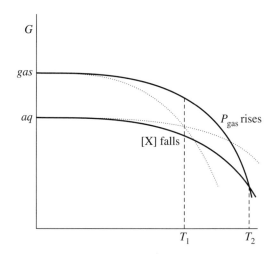

Figure 12.14
When a saturated solution of a gas (at T_1) is heated (to T_2), then the [gas] *decreases* and P_{gas} *increases*. Thus, the solution is "degassed."

Application: Degassing of Liquids

Interestingly, not all solids absorb heat when they dissolve. Some release heat, and virtually all gases do, too. In those cases, we get the interesting phenomenon that increasing the temperature actually *decreases* the solubility (Figure 12.14). Heating a solution thus provides a simple and effective way to *degas* it. This also explains why ice made from hot water is so crystal-clear. Cold water has dissolved air, which, when the ice is made, comes out of solution as little bubbles that cause the ice to appear cloudy. Hot water, on the other hand, has almost no air in it, so as it freezes, no air comes out. Its ice is clear.

In the case of gases dissolving, the equilibrium expression contains both pressures and concentrations:

$$X(g) \rightleftharpoons X(aq) \qquad K_{sp} = [X]_{sat}/P_{gas} \text{ (unitless)}$$

The expression in this case is called *Henry's law*. The constant in this case is called *Henry's law constant*, but it is just an equilibrium constant, the same as all others. For N_2 in water at 25°C, the constant K_{sp} is 6.8×10^{-4}. Thus, the solubility of N_2 gas in water at its standard atmospheric partial pressure is

$$[N_2]_{sat} = K_{sp}P_{N_2} = (6.8 \times 10^{-4})0.78 = 5.3 \times 10^{-4}$$

The units for gas pressure, of course, must be bar, and the resultant units of concentration are mol/L.

Scuba divers are well aware of the extent to which vapor pressure and dissolved gas concentration are related. At depth, the pressure is so great that a tremendous amount of N_2 can be dissolved in blood. If decompression occurs too fast, the diver's blood will release this gas into the bloodstream, causing great pain and possible death. Only a slow ascent allows this gas to come out at a rate that can be removed safely from the bloodstream.

Application: Atmospheric CO_2 and Coral Reefs

Coral reefs are composed primarily of calcium carbonate, $CaCO_3$, which is produced by corals and other *calcifying* organisms. In order to produce more reef, these organisms take advantage of the fact that the ocean water around a reef is *supersaturated* in $CaCO_3$. That is, there is more Ca^{2+} and/or CO_3^{2-} in ocean water than that allowed by the equilibrium:

$$CaCO_3(s) \rightleftharpoons Ca^{2+}(aq) + CO_3^{2-}(aq) \qquad K_{sp} = [Ca^{2+}]_{sat}[CO_3^{2-}]_{sat}$$

The state of supersaturation is characterized by $Q > K_{sp}$ for the forward reaction:

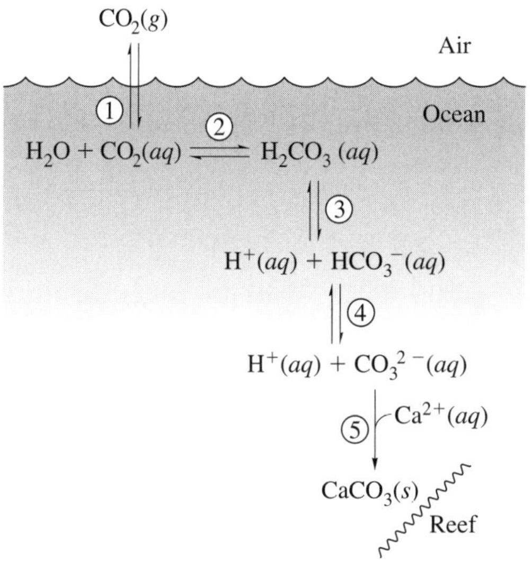

Figure 12.15
Five equilibria leading from atmospheric carbon dioxide to coral reef calcium carbonate. The situation is complicated by the presence of H^+. The last step, involving the precipitation of $CaCO_3$, is irreversible because the ocean waters are supersaturated in $CaCO_3$.

$$Q = [Ca^{2+}][CO_3^{2-}] > [Ca^{2+}]_{\text{sat}}[CO_3^{2-}]_{\text{sat}} = K_{\text{sp}}$$

This reaction will spontaneously go *backward* in normal ocean water around coral reefs, precipitating $CaCO_3$. Apparently, reef-building organisms take advantage of the fact that the reverse reaction is "free" to maintain their existence and build their home, the reef. In fact, it has recently been argued that Q has to be at least three times K_{sp} for reef to form, or at least to form fast enough to keep up with erosion.[4]

Where does the CO_3^{2-} necessary to form coral reefs come from? The seemingly obvious answer is atmospheric carbon dioxide (Figure 12.15). Thus, we might think of the calcium carbonate of reefs as being at the "end of the line" of a complex set of equilibria starting with carbon dioxide in the atmosphere. Atmospheric CO_2 dissolves in ocean water (1), reacting to form carbonic acid, H_2CO_3, which is a weak diprotic acid (2). This carbonic acid dissociates (3) to give bicarbonate, HCO_3^-, which further dissociates (4) to give carbonate, CO_3^{2-}. Finally, this carbonate is what combines with calcium ions to give calcium carbonate (5). This last reaction is largely *irreversible* due to the fact that the ocean is supersaturated in $CaCO_3$.

At first glance, one might imagine that due to Le Châtelier's principle, increasing the partial pressure of CO_2 in the atmosphere might increase the concentration of CO_3^{2-} in the ocean, increasing Q and helping these organisms make the reef. Unfortunately, that analysis is an improper use of Le Châtelier's principle. A thought experiment shows why.

Consider the following general set of equilibria linking substance A with substance D:

$$A \rightleftharpoons B \rightleftharpoons C \rightleftharpoons D$$

Surely if the concentration of A is increased, Le Châtelier's principle argues for a "ripple" effect through the system: The concentration of B will increase somewhat, which will cause the concentration of C to increase a little, which will finally cause the concentration of D to increase a tiny amount. Thus, an increase in the concentration of A will ultimately result in an increase in *all* of the other species in the equilibrium, including D.

---

[4] See J. Kleypus et al., *Science*, vol. 284, page 118, April 2, 1999.

However, the situation with regard to carbonic acid in the ocean is slightly different. In this case we have for equilibria (3) and (4), above:

$$(3)\ \ A \rightleftharpoons H + B \qquad K_3 = \frac{[H][B]}{[A]}$$

$$(4)\ \ B \rightleftharpoons H + C \qquad K_4 = \frac{[H][C]}{[B]}$$

The presence of this extra product, H, changes the whole picture and makes the effect of adding substance A much less obvious. If the concentration of A increases, then the concentrations of both H and B must increase as well due to equilibrium (3). But what now for equilibrium (4)? We are increasing the concentrations of both B and H, a reactant and a product! What effect will this disturbance have on the concentration of C?

The answer turns out to be that the effect of adding A is to increase the concentration of C only when [C] (in units of mol/L) is less than K_4. That is, the concentration of C will increase only when it is low. In the case when [C] = K_4, adding A will have no effect on the concentration of C, and when [C] > K_4, adding A will actually cause a *decrease* in the concentration of C![5]

In this case, $K_4 \approx 5.6 \times 10^{-11}$. Thus, simply pumping CO_2 into pure water cannot raise the concentration of CO_3^{2-} above $5.6 \times 10^{-11} M$. However, the concentration of CO_3^{2-} in the present-day ocean is known to be about 0.0001 M. This high value simply indicates that the ocean is not pure water. The pH of the ocean is around 8.2 (so that [H^+] is about $6 \times 10^{-9} M$). Effectively, this low concentration of H^+ shifts both equilibria (3) and (4) to the right, increasing the concentration of CO_3^{2-} to a level much, much higher than K_4.

The high concentration of CO_3^{2-} in the ocean demands that the effect of an increase in atmospheric CO_2 is a *decrease* in the concentration of CO_3^{2-}. This decrease in [CO_3^{2-}] lowers Q for the dissolving of calcium carbonate and endangers the health of coral reefs worldwide. Model calculations indicate that many coral reefs may be severely threatened by 2065, by which time the atmospheric CO_2 level is expected to be double its pre-industrial level, and Q/K_{sp} is calculated to drop below 3 in many areas of the ocean.

12.7 Impure Liquids: $S = S^\circ - R \ln x$

Generally we have said that only the entropies of gases and solutes are sensitive to pressure or concentration. But it is also possible to change the entropy of a pure liquid by mixing it with another substance. The entropy of an *impure* liquid can be understood precisely along the terms discussed in Chapter 8 relating to volumes and expansions. However, in this case we think of the mix-up not in terms of volume, per se, but instead in terms of number of moles of substances.

Take liquid water as our example. When a substance is dissolved, the entropy of the liquid water itself must go up, because it has gotten mixed up with the molecules (or ions)

---

[5] When [C] > K_4, it must also be true that [H] < [B] since

$$\frac{[H][C]}{[B]} = K_4 \quad \Rightarrow \quad \frac{[H]}{[B]} = \frac{K_4}{[C]} < 1$$

Note that it is the *ratio* of [H]/[B] that determines whether [C] will increase. For example, say [H] = 0.01 and [B] = 1. Then [H]/[B] = 0.01, and [C] = 100K_4. Now if [H] and [B] both increase by 0.01 as a result of the addition of A, then [H]/[B] = 0.02/1.01 ≈ 0.02. That means [C] would now have to equal 50K_4 at equilibrium, which is much less than 100K_4. The addition of A has resulted in a *decrease* in [C].

added. Thus, the entropy *per mole* of the liquid water is increased when substance A is added as follows:

$$S_{\text{liq}} = S_{\text{liq}}^{\circ} + R \ln \left(\frac{n_{H_2O} + n_A}{n_{H_2O}} \right)$$

$$= S_{\text{liq}}^{\circ} + R \ln \left(\frac{1}{x_{H_2O}} \right) \qquad (12.4)$$

$$= S_{\text{liq}}^{\circ} - R \ln x_{H_2O}$$

We ignored this effect in Chapter 8 because it did not contribute significantly to the overall picture of solutes. After all, the molarity of water is 55.5 mol/L, so a few tenths of a mol/L of solute really has very little effect on the mole fraction of water in the mix, x_{H_2O}. For example, in 0.1 *M* sucrose solution, the mole fraction of water is calculated to be

$$x_{H_2O} = \frac{55.5}{55.5 + 0.1} = 0.998$$

Nonetheless, the effect of this slight increase in entropy of liquid water is in fact detectable.

Application: Freezing Point Depression

A very interesting effect (Figure 12.16) arises if the liquid in a solid/liquid system is increased in entropy this way. The effect of the extra entropy for the liquid is to lower its Gibbs energy, making its curve extra steep. The result is that there is a new melting point, T_{mp}^{*}, which is depressed relative to the standard melting point, T_{mp}. The effect is called *freezing point depression*. Let's check out the effect quantitatively based on the equation:

$$X(s) \longrightarrow X(l)$$

for which we have:

$$\Delta_r S = S_{\text{liq}} - S_{\text{solid}} = (S_{\text{liq}}^{\circ} - R \ln x_{H_2O}) - S_{\text{solid}}^{\circ}$$

$$= \Delta_r S^{\circ} - R \ln x_{H_2O}$$

and

$$\Delta_r G = \Delta_r H - T \Delta_r S$$

$$0 = \Delta_r H^{\circ} - T_{\text{mp}}^{*} (\Delta_r S^{\circ} - R \ln x_{H_2O}) \qquad (12.5)$$

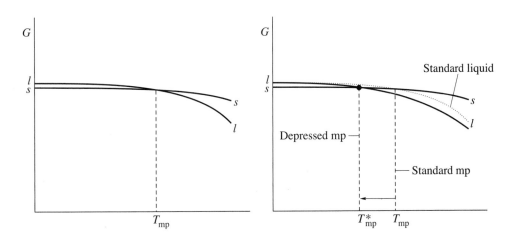

Figure 12.16
On the left is the situation for a typical solid/liquid system; on the right is the same system, now with the liquid phase having increased entropy. The result is a depressed melting point.

where in the last step we have put in all the values relating to equilibrium at the new melting point, T_{mp}^*. Notice that this looks an awful lot like all of the other equations we have gotten for equilibrium. The logarithmic term could even be incorporated into the concept of a reaction quotient by making the new rule: **Impure liquids are included in the reaction quotient as their mole fraction.**

You can do some very interesting things with this relationship. For example, say you wanted to know the molar mass of an unknown substance X. An easy way to estimate it would be to dissolve 375 g of the substance in 1000 g of water and check the melting point. Say you get $-10°C$. Then, substituting values into Equation 12.5 for what we know, specifically that $\Delta_r H° = 6010$ J, $\Delta_r S° = 22$ J/K, and $T_{mp}^* = 263$ K, we get $x_{H_2O} = 0.8985$. That is, for every 0.8985 mol of water we have $(1 - 0.8985) = 0.1015$ mol of substance X.

We can use this information to calculate for the molar mass of substance X:

$$\frac{375 \text{ g X}}{1000 \text{ g H}_2\text{O}} \times \frac{18.0 \text{ g H}_2\text{O}}{1 \text{ mol H}_2\text{O}} \times \frac{\mathbf{0.8985} \text{ mol H}_2\text{O}}{\mathbf{0.1015} \text{ mol X}} = 59.7 \text{ g/mol X}$$

How about that!

Note that if the substance dissolving is a strong electrolyte, then you must account for all the moles of all the ions in "mol X". For example, how much NaCl should we have to add to 100 g of water to lower its melting point to $-5°C$? We have in this case $\Delta_r H° = 6010$ J/mol, $\Delta_r S° = 22$ J/mol·K, $T_{mp}^* = 268$ K. Solving for x_{H_2O}, we get 0.946. Thus, for every 0.946 mol of H_2O, we need $(1 - 0.946) = 0.054$ mol of $(Na^+ + Cl^-)$, or just 0.027 mol of NaCl, and

$$100 \text{ g H}_2\text{O} \times \frac{1 \text{ mol H}_2\text{O}}{18 \text{ g H}_2\text{O}} \times \frac{\mathbf{0.027} \text{ mol NaCl}}{\mathbf{0.946} \text{ mol H}_2\text{O}} \times \frac{58.5 \text{ g NaCl}}{\text{mol NaCl}} = 9.3 \text{ g NaCl}$$

Let's do it! (Just add salt to ice water and see how cold you can get it.) The mathematical trick here is to convert mole fraction into a mole ratio, which can then be used as a conversion factor.

There's an approximation to Equation 12.5 which is rather tricky to derive but easier to use. It is historically important (the freezing point depression effect has been known since at least 1721, when Daniel Fahrenheit made his first mercury thermometer and assigned "0°F" to be the coldest he could get adding salt to ice water), and it is still taught even today. This approximation reads

$$\Delta T_f \approx k_f \, m_X \tag{12.6}$$

where we now define the *freezing point depression*, $\Delta T_f = T_{mp} - T_{mp}^*$. The parameter k_f is a constant for a given solvent, and m_X is the *molality of X*, calculated as "moles of X per kg of solvent." The value of k_f for water is 1.85 K·kg solvent/mol solute.

So in the above case of unknown substance X, where the melting point dropped 10 K, we would get

$$10 \text{ K} \approx \frac{1.85 \text{ K} \cdot \text{kg H}_2\text{O}}{\text{mol X}} m_X$$

$$m_X \approx 5.40 \frac{\text{mol X}}{\text{kg H}_2\text{O}}$$

Working with molality is not too difficult if you just consider it another useful sort of conversion factor. Since we had 375 g of X per 1000 g of H_2O, we could write:

$$\frac{375 \text{ g X}}{1000 \text{ g H}_2\text{O}} \times \frac{1000 \text{ g}}{\text{kg}} \times \frac{\mathbf{1} \text{ kg H}_2\text{O}}{\mathbf{5.40} \text{ mol X}} = 69 \text{ g X/mol X}$$

This result is a little off from 60, obtained using $\Delta_{\text{fus}}H°$ and $\Delta_{\text{fus}}S°$, but it is close enough. It's just a quick calculation, not intended to be tremendously accurate. It just gives a ball-park value that can be used mainly to distinguish among possible empirical formulas such as CH_2O or $C_2H_4O_2$ or $C_3H_6O_3$. In this case, we conclude that substance X is probably $C_2H_4O_2$, since the molar mass of $C_2H_4O_2$ is 60, while the molar masses of the others are 30 and 90.

It turns out that with a few mathematical tricks, k_f can be shown to be

$$k_f \approx \frac{RT_{\text{mp}}}{\Delta_{\text{fus}}S°(\text{mol}_{\text{liq}}/\text{kg}_{\text{liq}})} \tag{12.7}$$

Here T_{mp} is the standard melting point of the liquid. There are roughly 55.5 mol of H_2O per kg of H_2O, and $\Delta_{\text{fus}}S° = 22$ J/mol·K for water, so all of these constants end up evaluating to the known value for k_f for water, 1.85 K·kg/mol.

Application: Boiling Point Elevation

Similarly, increasing the entropy of the liquid also affects the boiling point. The new boiling point, T_{bp}^*, will be higher than the standard boiling point, T_{bp}, resulting in the phenomenon called *boiling point elevation* (Figure 12.17). Running through the same analysis as for freezing point depression, this time using the equation

$$X(l) \longrightarrow X(g)$$

we use 1 bar for the vapor pressure of the gas at its standard boiling point and arrive at

$$0 = \Delta_r H° − T_{\text{bp}}^* \left(\Delta_r S° − R \ln \frac{1}{x_{H_2O}} \right) \tag{12.8}$$

Note that the mole fraction is now in the denominator because the impure liquid is the *reactant* instead of the *product* this time. Once again, an approximation can be made, giving

$$\Delta T_b \approx k_b\, m_X \tag{12.9}$$

where we define the *boiling point elevation*, $\Delta T_b = T_{\text{bp}}^* − T_{\text{bp}}$. The parameter k_b is a constant for a given solvent, and m_X is once again the molality of X, in units of moles of X per

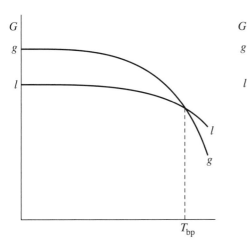

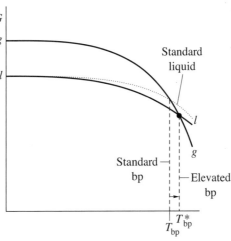

Figure 12.17
On the left is the situation for a typical liquid/gas system; on the right is the same system, now with the liquid phase having increased entropy. The result is an elevated boiling point.

kg of solvent. The value of k_b for water is 0.52 K·kg solvent/mol solute and comes from evaluating:

$$k_b \approx \frac{RT_{bp}}{\Delta_{vap}S°(mol_{liq}/kg_{liq})} \quad (12.10)$$

Perhaps we could use this to our advantage. How much salt do you have to add to a pot of spaghetti water to make it boil at 101°C instead of 100°C so that the spaghetti cooks a little faster and we can save some time? We calculate

$$1\,K \approx \frac{0.52\,K \cdot kg\,H_2O}{mol\,X}\,m_X$$

$$1.92\,\frac{mol\,X}{kg\,H_2O} \approx m_X$$

The fact that NaCl breaks up into Na^+ and Cl^- is to our advantage here. We need only 0.96 mol of NaCl per kg of water to accomplish this feat. Thus, for the whole pot of spaghetti, which might hold 1.5 kg of water, we will need

$$1.5\,kg\,H_2O \times \frac{\textbf{0.96}\,\textbf{mol NaCl}}{\textbf{1}\,\textbf{kg}\,\textbf{H}_2\textbf{O}} \times \frac{58.44\,g\,NaCl}{mol\,NaCl} = 84\,g\,NaCl$$

Blaauch! Too salty! OK, so maybe this is not a good way to get spaghetti to cook faster.

Application: Raoult's Law, $P = x_{H_2O}P^o$

If we look back at increasing the entropy of the liquid and focus now on the effect on vapor pressure at a specific temperature, we find a clear expectation (Figure 12.18). Since the liquid curve is lower, the vapor curve will have to be lower as well, so that they can still cross at the temperature of interest.

To evaluate the extent of this vapor pressure lowering, we simply extend the analysis we did for boiling point elevation, considering the same chemical equation:

$$X(l) \longrightarrow X(g)$$

including now the non-standard vapor pressure, P_{vap} (in bars), in the calculation:

$$0 = \Delta_{vap}H° - T(\Delta_{vap}S° - R\ln Q) \quad \text{where } Q = \frac{P_{vap}}{x_{H_2O}} \quad (12.11)$$

In effect, we are saying, once again, that if a liquid is impure, then its mole fraction should be included in the reaction quotient, Q. With an equation-solving calculator, there is no real problem solving for P_{vap}. For example, for the vapor pressure over a 1.0 *molal* sucrose

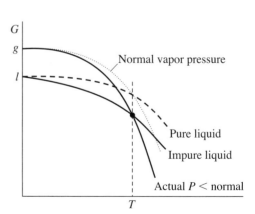

Figure 12.18
When a liquid is impure, its increased entropy causes its Gibbs energy to drop, and the corresponding vapor pressure curve must be for a lower vapor pressure.

solution at 25°C, we calculate based on the idea that 1 kg of water contains 55.5 mol of water:

$$x_{H_2O} = \frac{55.5 \text{ mol } H_2O}{56.5 \text{ mol total}} = 0.982$$

Using $\Delta_{vap}H^\circ = 44.0$ kJ, $\Delta_{vap}S^\circ = 119$ J/K, and $T = 298$ K, we solve for $Q = 0.0318$. Thus,

$$\frac{P_{vap}}{0.982} = 0.0318$$

and $P_{vap} = 0.0312$ bar $= 24$ mmHg.

If you compare Equation 12.11 with what we would get for the pure liquid:

$$0 = \Delta_{vap}H^\circ - T(\Delta_{vap}S^\circ - R \ln P^\circ)$$

it should be obvious that what we were solving for in solving for Q was just the equilibrium vapor pressure over the pure liquid, which we call here P°. Thus we have that

$$P^\circ = \frac{P_{vap}}{x_{H_2O}}$$

or

$$P_{vap} = x_{H_2O}\, P^\circ \qquad (12.12)$$

This result, known as *Raoult's Law,* simply says that the vapor pressure over a solution is the mole fraction of the solvent times the equilibrium vapor pressure over the pure solvent at that temperature. Of course, this is for an "ideal" solution (Section 8.2), and many solutions deviate from this law due to weak interactions between solvent and solute.

12.8 Summary

In this chapter several applications have been introduced. In every case we can start with the basic equation for the effect of concentration on the change in entropy for a chemical reaction:

$$\Delta_r S = \Delta_r S^\circ - R \ln Q$$

which leads at equilibrium, where $\Delta_r G = 0$, to

$$0 = \Delta_r H^\circ - T_{eq}(\Delta_r S^\circ - R \ln Q) \qquad (12.13)$$

This then is the "master equation" from which all equilibrium phenomena can be analyzed. In order to treat impure liquids, we simply add to our rules for writing the reaction quotient the following one additional rule: **Impure liquids are included in the reaction quotient as their mole fraction, x.**

Beyond that, the relationships governing freezing point depression and boiling point elevation are simply approximations that were made historically prior to the development of thermodynamics. At one point in the history of chemistry these were the principal ways of learning the identity of an unknown substance. Still, even today as rough approximations they are valuable and see use in a wide variety of industrial settings.

Mostly, it is enough to be able to explain these phenomena qualitatively. With a simple G–T graph and a few basic principles such as the idea that liquids have stronger bonding and less entropy than gases, we can easily make sense of a whole variety of phenomena and at least make some general predictions that will be correct. It is this predictability arising, as it does, from the chaos of energy distributed in a quantized fashion in real chemical systems that makes molecular thermodynamics so useful.

Problems

The symbol Ⓐ *indicates that the answer to the problem can be found in the "Answers to Selected Exercises" section at the back of the book.*

Note: If you have a graphing calculator, be sure to take a look at http://www.stolaf.edu/depts/chemistry/imt/js/tinycalc /ti-calc.htm before approaching these problems. The following single equation put into the solver can be used to do the calculations for many of these problems.

```
drG = drHo - T*(drSo - Rc*ln(Q))
```

All of these problems involve equilibrium conditions, so $\Delta_r G = 0$. *For melting,* $Q = 1$; *for boiling and sublimation,* $Q = P_{vap}/bars$. *Just set* drG = 0, *specify the known values for any three of* drHo, T, drSo, *and* Q, *then solve for the one you don't know. Be sure to use joules (not kilojoules) for* $\Delta_r H°$ *and* $\Delta_r S°$. *Use 8.3145 J/mol·K for* Rc.

Melting, Evaporation, and Boiling

12.1   Explain using a G–T graph why pure substances have single sharp melting points.

12.2 Ⓐ Show using a G–T graph why at least at 1 bar pressure graphite and diamond can *never* be in equilibrium.

12.3 Ⓐ Given that for ammonia $\Delta_{fus}H° = 5.65$ kJ/mol and $\Delta_{fus}S° = 28.9$ J/mol·K, determine the melting point of ammonia.

12.4   Given that for hydrogen peroxide, H_2O_2, $\Delta_{fus}H° = 12.5$ kJ/mol and $\Delta_{fus}S° = 45.8$ J/mol·K, determine the melting point of hydrogen peroxide.

12.5 Ⓐ Given that for chloromethane, CH_3Cl, $\Delta_{fus}H° = 6.43$ kJ/mol and $\Delta_{fus}S° = 36.6$ J/mol·K, determine the melting point of chloromethane.

12.6   Given that for bromomethane, CH_3Br, $\Delta_{fus}H° = 5.98$ kJ/mol and $\Delta_{fus}S° = 33.3$ J/mol·K, determine the melting point of bromomethane.

12.7 Ⓐ Br_2 melts at -7.2°C. If you found that it takes 9.43 J of heat to melt 0.140 g of Br_2, what are $\Delta_{fus}H°$ and $\Delta_{fus}S°$ for Br_2? (*Careful:* remember, $\Delta_{fus}H°$ and $\Delta_{fus}S°$ are *per mole*, not for 0.140 g, of Br_2.)

12.8   Acetic acid melts at -97.7°C. If you found that it takes 16.64 J of heat to melt 0.2560 g of $CH_3COOH(s)$, what are $\Delta_{fus}H°$ and $\Delta_{fus}S°$ for acetic acid?

12.9   Chloroacetic acid melts at 63°C. If you found that it takes 48 J of heat to melt 0.3690 g of $CH_2ClCOOH(s)$, what are $\Delta_{fus}H°$ and $\Delta_{fus}S°$ for chloroacetic acid?

12.10 Ⓐ Using values for $\Delta_f H°$ and $S°$ at 298 kelvins, estimate the standard boiling point of (a) carbon tetrachloride, CCl_4, and (b) benzene, C_6H_6. For gaseous benzene, $\Delta_f H° = 82.93$ kJ/mol and $S° = 269$ J/mol·K.

12.11   Using values for $\Delta_{fus}H°$ and $\Delta_{fus}S°$ at 298 kelvins, estimate the standard boiling point of mercury, $\Delta_f H° = 61.4$ kJ/mol and $S° = 175.0$ J/mol·K for $Hg(g)$. Determine the vapor pressure of mercury at 25°C.

12.12   Determine the vapor pressures of both carbon tetrachloride and benzene at 25.0°C. These compounds are suspected of being carcinogenic. (That is, they cause cancer.) Based on their vapor pressures, what might you put on the warning label of a bottle containing them? For gaseous benzene, $\Delta_f H° = 82.93$ kJ/mol and $S° = 269$ J/mol·K.

12.13 Ⓐ Estimate the standard boiling point of the following substances based on values in Appendix D: (a) H_2O, (b) HNO_3.

12.14   Estimate the standard boiling point of the following substances based on values in Appendix D: (a) C_2H_5OH, (b) CH_3OH.

12.15 Ⓐ Estimate the standard boiling point of the following substances based on values in Appendix D: (a) $SnCl_4$, (b) $TiCl_4$.

12.16   Estimate the standard boiling point of the following substances based on values in Appendix D: (a) N_2O_4, (b) SO_3.

12.17 Ⓐ Explain, using a G–T graph, why it takes longer to hard boil an egg in Denver, Colorado (altitude 5280 ft), than it does in Northfield, Minnesota (altitude 1010 ft).

12.18   Explain, using a G–T graph, why food steamed in a pressure cooker cooks faster than in a simple pot with a loose lid. If the valve is set to blow at 30 psi (2 bar), how hot can the steam in the pot get?

12.19   Explain, using a G–T graph, what "dew point" means, and why, on a night that the air is stable, the temperature rarely falls below the dew point.

12.20 Ⓐ Why will the water in a puddle eventually evaporate, instead of establishing the equilibrium vapor pressure?

12.21   When cells of the onion plant, *Allium cepa*, are damaged, cytoplasmic enzymes mix with sulfur-containing amino acids from the cells to produce the active volatile component in onions, (Z)-propanethial S-oxide, shown below. This compound is a *lacrimator*, evaporating into the air and causing you to cry when cutting onions. Explain using a G–T graph why refrigerating onions before cutting them is easier on the eyes.

Sublimation and Vapor Deposition

12.22   Explain, using a G–T graph, how frost forms on a still night, when the air is stable, the humidity is fairly low (but not too low), and the temperature slowly drops.

12.23 Ⓐ If one evening, the temperature is 40°F and the humidity is 80%, what's the dew point? Should farmers be worried about a frost that night?

12.24   If one evening, the temperature is 35°F and the humidity is 50%, what's the dew point? Should farmers be worried about a frost that night?

12.25 Show, using a G–T graph, that the vapor pressure above *supercooled* liquid water at $-20°C$ is higher than the vapor pressure above ice at the same temperature.

12.26 For the sublimation of ice, $\Delta_r H° = 11.95$ kcal/mol and $\Delta_r S° = 34.2$ cal/mol·K. Use this information to find the temperature at which the standard $H_2O(g)$ curve crosses the standard $H_2O(s)$ curve on a G–T graph. Adding a curve for $H_2O(l)$, show that above $0°C$, the vapor pressure over ice is higher than the vapor pressure over liquid water.

Triple Points, Critical Points, and Phase Diagrams

12.27 Ⓐ Explain why liquid CO_2 is not stable at pressures less than 5 bar.

12.28 Describe how one can observe the triple point of water.

12.29 Ⓐ Describe the phenomenon of a critical point.

12.30 Draw the phase diagram for water below 200 bar pressure and describe it.

12.31 Sketch the phase diagram for O_2 given the following data: triple point 54.3 K and 1.14 mmHg; critical point 154.6 K and 50 bar; standard melting point 54.75 K; standard boiling point 90.25 K.

12.32 Sketch the phase diagram for benzene, C_6H_6, given the following data: triple point 278.680 K and 35.89 mmHg; critical point 561.75 K and 48.7575 bar; standard melting point 278.68 K; standard boiling point 353.240 K.

12.33 Ⓐ Given the following information on the phase changes of ethanol (C_2H_5OH): (a) Construct a G–T graph showing three curves, one for each of the phases of ethanol. Be sure to indicate the temperature at any point where two curves cross. (b) Will the critical point of ethanol occur at a pressure above or below 1 bar? (c) Will the triple point of ethanol occur at a pressure less than or greater than 1 bar? (d) Ignoring the very slight pressure effect on the liquid curve, determine the pressure and temperature of the triple point of ethanol. Values are for 1 bar pressure and the temperature of equilibrium in each case.

$$C_2H_5OH(s) \longrightarrow C_2H_5OH(l)$$
$$\Delta_r H° = 4.973 \text{ kJ/mol} \quad \Delta_r S° = 31.3 \text{ J/mol·K}$$

$$C_2H_5OH(l) \longrightarrow C_2H_5OH(g)$$
$$\Delta_r H° = 42.3 \text{ kJ/mol} \quad \Delta_r S° = 120.17 \text{ J/mol·K}$$

12.34 Given the following information on the phase changes of propane (C_3H_8): (a) Construct a G–T graph consisting of three curves, one for each of the phases of propane. Be sure to indicate the temperature at any point where two curves cross. (b) Will the critical point of propane occur at a pressure above or below 1 bar? (c) Will the triple point of propane occur at a pressure less than or greater than 1 bar? (d) Ignoring the very slight pressure effect on the liquid curve, determine the pressure and temperature of the triple point of propane. Values are for 1 bar pressure and the temperature of equilibrium in each case.

$$C_3H_8(s) \longrightarrow C_3H_8(l)$$
$$\Delta_r H° = 3.524 \text{ kJ/mol} \quad \Delta_r S° = 41.24 \text{ J/mol·K}$$

$$C_3H_8(l) \longrightarrow C_3H_8(g)$$
$$\Delta_r H° = 16.25 \text{ kJ/mol} \quad \Delta_r S° = 70.35 \text{ J/mol·K}$$

$$C_3H_8(s) \longrightarrow C_3H_8(g)$$
$$\Delta_r H° = 28.5 \text{ kJ/mol} \quad \Delta_r S° = 331.40 \text{ J/mol·K}$$

12.35 Ⓐ Given the following information on the phase changes of 1-propanol (C_3H_7OH): (a) Construct a G–T graph showing three curves, one for each of the phases of 1-propanol. Be sure to indicate the temperature at any point where two curves cross. (b) Will the critical point of 1-propanol occur at a pressure above or below 1 bar? (c) Will the triple point of 1-propanol occur at a pressure less than or greater than 1 bar? (d) Ignoring the very slight pressure effect on the liquid curve, determine the pressure and temperature of the triple point of 1-propanol. Values are for 1 bar pressure and the temperature of equilibrium in each case.

$$C_3H_7OH(s) \longrightarrow C_3H_7OH(l)$$
$$\Delta_r H° = 5.37 \text{ kJ/mol} \quad \Delta_r S° = 36.1 \text{ J/mol·K}$$

$$C_3H_7OH(l) \longrightarrow C_3H_7OH(g)$$
$$\Delta_r H° = 41.44 \text{ kJ/mol} \quad \Delta_r S° = 111.9 \text{ J/mol·K}$$

12.36 Given the following information on the phase changes of butane (C_4H_{10}): (a) Construct a G–T graph showing three curves, one for each of the phases of butane. Be sure to indicate the temperature at any point where two curves cross. (b) Will the critical point of butane occur at a pressure above or below 1 bar? (c) Will the triple point of butane occur at a pressure less than or greater than 1 bar? (d) Ignoring the very slight pressure effect on the liquid curve, determine the pressure and temperature of the triple point of butane. Values are for 1 bar pressure and the temperature of equilibrium in each case.

$$C_4H_{10}(s) \longrightarrow C_4H_{10}(l)$$
$$\Delta_r H° = 4.66 \text{ kJ/mol} \quad \Delta_r S° = 34.56 \text{ J/mol·K}$$

$$C_4H_{10}(l) \longrightarrow C_4H_{10}(g)$$
$$\Delta_r H° = 22.44 \text{ kJ/mol} \quad \Delta_r S° = 82.2 \text{ J/mol·K}$$

Solubility

12.37 For the dissolving of sugar, i.e. sugar$(s) \rightleftharpoons$ sugar(aq), $\Delta_r H°$ and $\Delta_r S°$ are both positive. Using a G–T graph, show why it is that more sugar dissolves in hot water than in cold water. What is important here, $\Delta_r H$ or $\Delta_r S$?

12.38 Ⓐ Data for the solubility of sucrose ($C_{12}H_{22}O_{11}$) in pure water measured at various temperatures are given below.

$T/°C$	$[C_{12}H_{22}O_{11}]_{sat}/(mol/L)$
0	10.5
10	11.2
20	12.0
30	12.7
40	13.9
50	14.8
60	15.5

(a) Make a plot of ln $[C_{12}H_{22}O_{11}]_{sat}$ vs. $1/T$ (kelvins). (b) Use your plot to determine $\Delta_r H°$ and $\Delta_r S°$ for the reaction as written:

$$C_{12}H_{22}O_{11}(s) \longrightarrow C_{12}H_{22}O_{11}(aq)$$

12.39   Data for the solubility of copper sulfate ($CuSO_4$) in pure water measured at various temperatures are given below:

$T/°C$	$[CuSO_4]_{sat}/(mol/L)$
0	0.08
10	0.09
20	0.10
25	0.11
30	0.12
40	0.14
50	0.16
60	0.18
70	0.20
80	0.23
90	0.25

(a) Make a plot of ln $[CuSO_4]_{sat}$ vs. $1/T$ (kelvins). (b) Use your plot to determine $\Delta_r H°$ and $\Delta_r S°$ for the reaction as written:

$$CuSO_4(s) \longrightarrow CuSO_4(aq)$$

12.40   Ⓐ Almost all gases are more soluble in cold water than in hot water. Explain this using a G–T graph.

12.41   You need to make a 250-mL solution of the amino acid serine ($C_3H_7NO_3$, MW 105) in ethanol that will be saturated at 65.0°C. Looking in the *CRC Handbook of Chemistry and Physics*, you find the following information regarding serine/ethanol solutions:

$T/°C$	solubility /(g/100 mL)
25.1	0.0840
45.1	0.1850

Based on this information determine (a) the solubility of serine in ethanol at these two temperatures in mol/L; (b) $\Delta_r H°$ and $\Delta_r S°$ for the dissolving of serine in ethanol; and, from that, (c) how much serine you will need *for this 250-mL sample*. [*Hint:* In part (b) Equations 11.3 and 11.4 apply.]

Impure Liquids

12.42   Ⓐ Determine the boiling point of a solution created by dissolving in 2000 g of water (a) 50.0 g of glucose, $C_6H_{12}O_6$, a nonelectrolyte; (b) 50.0 g of KBr; (c) 50.0 g of $MgCl_2$.

12.43   A 250.0-g sample of the automotive antifreeze ethylene glycol, a nonelectrolyte, is dissolved in 750.0 g of water. The freezing point of water is lowered by 8.75°C. (a) Estimate the molecular mass of ethylene glycol from this information. (b) The empirical formula of ethylene glycol is CH_3O. What is its molecular formula?

12.44   The equilibrium vapor pressure of water is 17.54 torr at 20°C and 355.10 torr at 80°C. Determine the vapor pressure

of water above a solution made by adding 50.0 mL of glycerin, $C_3H_8O_3$, a nonelectrolyte with a density of 1.26 g/mL, to 500.0 mL of pure water at (a) 20°C, (b) 80°C.

Brain Teasers

12.45   Explain why there are no substances in Appendix D that show values for both solid and liquid phases, while there are several with both liquid and gas phases listed.

12.46   Do some research and explain the current ideas relating to the *real* reason there is so little friction between ice skates and ice. What are the best temperature conditions for ice skating? (See James White, *The Physics Teacher, 30*, 495, **1992**.)

12.47   Explain how a candle works.

12.48   Explain why, if you pour a volatile liquid such as gasoline out of a container and then close the container, when you come back later you should be careful, because when you open the container, it will probably be at higher than atmospheric pressure.

12.49   Use the table at the web site *http://vortex.atmos.uah .edu/atmos/science/StdAtm.html* relating pressure to altitude to estimate the boiling point of water on the top of Mt. Everest. Note that 100 kPa (kilopascals) = 1 bar.

12.50   The following weather data was gathered at 3:00 PM local time on March 29, 1999.

Location	Temp.	% Humidity	Low Temp.
St. Paul, MN, USA	54°F	26	30°F
Darwin, Australia	31°C	70	24°C
Alert, NT, Canada	−34°C	66	−37°C
Cape Town, South Africa	28°C	82	18°C

(a) Determine the equilibrium vapor pressure of water at each location at 3:00 PM. [For Alert, use $\Delta_r H° = 50$ kJ/mol and $\Delta_r S° = 143.2$ J/mol·K for $H_2O(s) \longrightarrow H_2O(g)$.] (b) Determine the actual vapor pressure in each case at 3:00 PM. (c) Determine the dew point, in the units in which the temperature is given, in each case. (d) In which locations was the formation of frost or dew likely overnight?

12.51   In this problem we will be concerned with the equilibrium between gaseous nitrogen and dissolved nitrogen in body tissue, such as fat and muscle, as it relates to decompression sickness.[6]

---

[6] Decompression sickness or the *bends* is a condition that can affect scuba divers, construction workers, high altitude pilots, and astronauts. During a dive, a scuba diver will breathe compressed air, the major components of which are nitrogen (78 mol%) and oxygen (21 mol%). At the high pressure of the deep sea, these gases are absorbed into blood and body tissue. Decompression sickness occurs when a person moves quickly from a high-pressure environment, such as a deep sea dive, to a low-pressure environment, such as the water surface. If the transition between the two pressure regimes is too rapid, bubbles of nitrogen gas, called *emboli*, can form in various body fluids including blood, joint fluid, and spinal fluid. This condition is potentially fatal.

$$N_2(g) \rightleftharpoons N_2(\text{tissue}) \qquad K \approx 6.5 \times 10^{-4} \text{ at } 25\,^\circ\text{C}$$

(a) Draw a G–T graph for the situation where equilibrium is reached at 25°C and the normal N_2 partial pressure of 0.78 bar, assuming that the forward reaction is exothermic. (b) What is the normal concentration of N_2 in body tissue? (c) Show the effect on your curves and $[N_2]_{\text{tissue}}$ of a descent to 50 meters below sea level (4 bar total air pressure). Is the diver in danger of forming emboli during the descent? (c) After spending enough time at 50 meters for the dissolved N_2 and gaseous N_2 to equilibrate, provided the total pressure is still 4 bar, what is $[N_2]_{\text{tissue}}$? (d) The diver then ascends to the surface. Sketch a G–T graph for the situation just before the ascent is started. What is the favored reaction upon ascent? (e) When might astronauts have to worry about decompression sickness?

12.52   One concern for pilots who fly at very high altitude is the effect of low pressure on their bodies. (a) Assuming that all body fluids are pure water, at what atmospheric pressure will those fluids boil at body temperature, 37°C? (b) Use the table at the web site *http://vortex.atmos.uah.edu/atmos/science/StdAtm.html* to determine the altitude (in feet) at which body fluids will spontaneously boil, causing bubbles of water vapor to form in the tissues and blood. The conversion factor between kPa (kilopascal) and bar is 100 kPa = 1 bar. (c) We know that our body fluids are not really pure water. For example, the concentration of ions in blood is approximately 150 millimolar. What precise effect will this concentration of salts in body fluids have on the maximum safe altitude for a pilot not wearing a pressurized suit? (d) Compare your answers for (b) and (c) to *Armstrong's Line* (63,000 ft), which is the experimentally determined value for the maximum altitude at which a pilot not wearing a pressurized suit may fly without fear of death by boiling.

12.53   Derive the relationship shown below from Equation 12.5. [*Hint:* You will have to play some pretty fancy tricks with natural logs to accomplish this!]

$$k_f \approx \frac{RT_{\text{mp}}}{\Delta_{\text{fus}}S^\circ(\text{mol}_{\text{liq}}/\text{kg}_{\text{liq}})}$$

12.54   *In order to complete this problem, you will need to do a few experiments using a small bottle containing an essential oil, such as vanilla or a perfume with a distinct aroma.* Open the bottle containing your oil and, using your built-in-essential-oil-pressure-meter (i.e., your nose), record in your mind what the aroma is and how strong it feels at room temperature. The strength of the aroma is a direct measure of the vapor pressure of the oil. Put the cap back on the bottle. Carefully draw a G–T graph which describes the equilibrium situation in the bottle. Now place the capped bottle in a refrigerator freezer compartment for at least two hours. Quickly remove the bottle, uncap it, and record in your mind the vapor pressure relative to the room temperature bottle. (Is it stronger or weaker—higher vapor pressure or lower?) Draw another G–T graph using a solid line to describe the equilibrium situation in the cold bottle and a dotted line to indicate the situation at room temperature. Now place the capped bottle on a warm stove (*not on the burner, mind you!*) or in a warm, sunny spot, or in a pan of warm water for a few minutes. Quickly remove the bottle, *carefully* uncap it, and record in your mind the vapor pressure relative to the room temperature bottle. (Is it stronger or weaker—higher vapor pressure or lower?) Draw another G–T graph, this time using a solid line to describe the equilibrium situation in the warm bottle and a dotted line to indicate the situation in it at room temperature. With a small piece of tissue paper, remove some of the oil from the bottle. Recap the bottle and discard the tissue (in some other room!). After about an hour or more, open the bottle and compare the vapor pressure with your initial reading. Explain why the G–T graph for this situation is identical to the G–T graph for the original bottle. Now go to some place that is quiet. Turn off the lights. Sitting in a comfortable chair, uncap the bottle and simply enjoy its presence. See if you can detect the increasing pressure of the oil's vapor in the room. Draw a G–T graph depicting the changing situation in the room in relation to the oil itself. What would happen if you left the cap off indefinitely? Would the vapor and liquid ever reach equilibrium?

12.55   You come home one evening in late February to find your driveway covered in water to a depth of 1/8 inch after a brief rain shower that afternoon. The temperature is supposed to drop down to 28°F overnight, and you don't want your driveway to be a skating rink in the morning. The driveway measures 10 ft by 30 ft and the density of water is 1.00 g/mL. Assuming any salt you add will be evenly dispersed over the driveway and will completely dissolve, how many pounds of salt do you need to keep the driveway from freezing if you use (a) NaCl? (b) CaCl$_2$?

12.56   Explain using a G–T graph why it might be that an impure *solid* melting to give a liquid mixture might melt over a broad range of temperature instead of having one sharp melting point. Where is the entropy effect coming into play here?

CHAPTER 13

Applications of Gibbs Energy: Electrochemistry

Note: This chapter is not intended to be a full chapter on electrochemistry as might be found in a first-year chemistry textbook. Rather, it focuses on how Gibbs energy and entropy relate to electrochemistry.

13.1 Introduction

How does electricity fit into our picture of energy, work, heat, and equilibrium? Batteries clearly involve chemical reactions. We say that batteries "store" energy. What kind of energy? Internal energy, U? Enthalpy, H? Gibbs energy, G? What is the connection between this energy and a battery's voltage? What do you suppose it means when a battery is *dead*?[1] In this chapter you will see that electricity—basically the movement of electrons—is considered to be a new kind of work, w_{elec}.

The presence of this new form of work changes the relationship between ΔH and q, and that results in a slightly different expression for Gibbs energy. In fact, "$-\Delta_r G$" will turn out to be the maximum amount of electrical work that can be obtained from a chemical system per mole of reaction.

An interesting and very important corollary of the fact that $\Delta_r G$ and electrical work are connected will be that *any* system that is not at equilibrium could potentially be used to generate electricity. We are currently seeing a revolution in battery technology driven by environmental concerns. It is likely that batteries in use ten years from now will be completely different from the ones in use today. In fact, twenty years from now you may find yourself filling up your car *fuel cell* with clean-burning gasoline, and the internal combustion engine may be a thing of the past. You can bet that any chemist working to develop new battery technology knows a lot about Gibbs energy. Let's see why . . . .

13.2 Review: Gibbs Energy and Entropy

To start, consider again the combustion of hydrogen to give liquid water.[2] We have

$$2\,H_2(g) + O_2(g) \longrightarrow 2\,H_2O(l) \qquad Q = \frac{1}{P_{H_2}^2 P_{O_2}}$$

---

[1] If you said, "When a battery is dead it must be at equilibrium," you were right.

[2] This reaction was used by the Apollo astronauts to generate electricity in space. As a bonus, they got to drink the water! The fuel cell they used weighs about 500 pounds and is on display at the Smithsonian Institute, in Washington, DC.

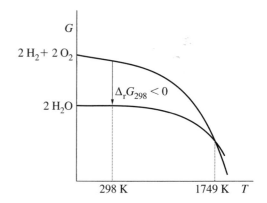

Figure 13.1
G–T graph for the combustion of H_2 to produce water. At 298 K, the forward reaction is favored since $\Delta_r G < 0$.

for which $\Delta_r H^\circ = -572$ kJ/mol and $\Delta_r S^\circ = -327$ J/mol·K. We won't worry about concentrations or pressures at first. Instead, we'll focus on the standard state, imagining that we can keep the pressure of both H_2 and O_2 constant at 1 bar. Thus, at 25°C, we have, per mole of reaction (i.e., for each mole of O_2 consumed):

$$\Delta_r G = \Delta_r G^\circ = \Delta_r H^\circ - T \Delta_r S^\circ$$
$$= -572{,}000 \text{ J/mol} - (298 \text{ K})(-327 \text{ J/mol·K})$$
$$= -475 \text{ kJ/mol}$$

In terms of a G–T graph, we have Figure 13.1. The products, $2\,H_2O$, with more bonding and lower enthalpy, start lower on the graph. The reactants, $H_2 + 2\,O_2$, being gases, have more closely spaced energy levels, more entropy in the standard state, and a steeper curve. Clearly the value of $\Delta_r G$ is going to be dependent upon temperature, and around room temperature it is quite negative. This reaction should have no problem going to completion. (All it needs is a spark!)

Let's focus now on just one specific temperature where $\Delta_r G$ is negative—298 K. Let's review why $\Delta_r G$ is negative in this case, and where $\Delta_r G$ is coming from in the first place.

Remember that $\Delta_r G$ doesn't really represent any *actual* change in Gibbs energy (because it presumes a constant reaction quotient). Nonetheless, for the purposes here, let's imagine that we really did have a reaction that involved two moles of H_2 reacting with a mole of O_2 in such a huge reaction system that a loss of a mere mole or two wouldn't change the situation significantly. In that case we can drop the little "r" and just write for our imaginary reaction

$$\Delta G = \Delta H - T \Delta S \qquad \Delta H = -572 \text{ kJ} \qquad \Delta S = -327 \text{ J/K} \qquad \Delta G = -475 \text{ kJ}$$

Consider these quantities; notice that all are negative. If we think of them as arrows, we can perhaps picture what is going on (Figure 13.2). An arrow pointing up on this graph would represent a positive change. In this case *all* the arrows are pointing down, because ΔG, ΔH, and $T \Delta S$ are all negative changes. Notice also that the arrows for ΔG and $T \Delta S$ (a) add up to the arrow for ΔH (b). This is because the expression for ΔG can be rearranged to give

$$\Delta G + T \Delta S = \Delta H$$

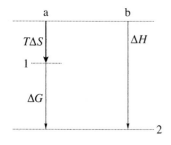

Figure 13.2
Vector graph relating ΔG to ΔH and $T \Delta S$ for the combustion of H_2 at 298 kelvins. Under these conditions, ΔS, ΔH, and ΔG are all negative. The vertical distance from Line 1 to Line 2 is ΔG. Note that $\Delta G + T \Delta S = \Delta H$.

The reason you just might want to think of it this way is that ΔH is always pretty much its standard value, ΔH°. *In this discussion we will assume that ΔH is relatively constant.* Remember, ΔH serves us two ways. First, it represents the amount of energy required to break bonds and move around electrons in a chemical reaction. Second, ΔH is our connection to the entropy of the surroundings. The entropy increase of the surroundings, ΔS_{sur}, is related to the heat going *into* the surroundings ($-q$) and the temperature by

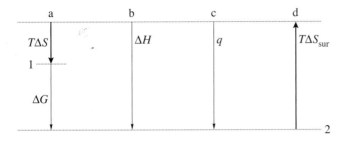

Figure 13.3
Vector graph for the combustion of H_2 at 298 K, now including heat and entropy going into the surroundings. The length of vectors q (c) and $T\Delta S_{sur}$ (d) are fixed by the vector for ΔH (b).

$$\Delta S_{sur} = \frac{q_{sur}}{T} = \frac{-q}{T}$$

and under conditions of constant pressure, since $\Delta H = q$, we have that

$$\Delta H = q = -T\Delta S_{sur}$$

In terms of our arrows we have Figure 13.3. Vector (b) is still ΔH. You can think of it as a rock. It's not going to move at all. When two moles of H_2 combine with one mole of O_2 to make two moles of H_2O, ΔH is the energy change that occurs regardless of the pressure of anything in the reaction, as long as the overall pressure and temperature are maintained constant, or at least returned to their original values. Vector (a) is again the sum of ΔG and $T\Delta S$, and its overall length is fixed at the value of ΔH. Similarly, the lengths of vector (c) and vector (d) are both also fixed by ΔH.

Although the overall lengths of all these vectors are thus fixed by ΔH, the exact position of Line 1 relative to Line 2 is adjustable, depending upon our conditions. Those differences in concentration and pressure show up as adjustments to the change in reaction entropy, ΔS. Remember, the reaction quotient, Q, includes partial pressures of both H_2 and O_2. Depending upon what these are, Q will have different values. Now the effect of all this on entropy change when the H_2 and O_2 react is

$$\Delta S = (1\,\text{mol})[\Delta_r S^\circ - R \ln Q]$$

or

$$T\Delta S = (1\,\text{mol})[T\Delta_r S^\circ - RT \ln Q]$$

Notice that we've added "(1 mol)" here because we are really saying that a mole-scale reaction is occurring *at* this particular reaction position, at this particular value of Q. Depending upon the value of Q, we can imagine the $T\Delta S$ arrow having different lengths, but, interestingly, the arrow for ΔH doesn't get longer. No, the arrow for ΔG simply gets shorter (Figure 13.4). If that $T\Delta S$ arrow grows too long, it isn't hard to see what is going to happen to ΔG. It's going to go to 0. The reaction would be at equilibrium. So the length of the $T\Delta S$ arrow is adjustable, but only within limits. That limit is the length of the ΔH arrow.

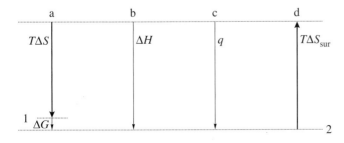

Figure 13.4
Vector graph now showing a more negative $T\Delta S$ arrow, due to the fact that Q is larger, and $\Delta S = (1\,\text{mol})(\Delta_r S^\circ - R \ln Q)$. Too large a value for Q and the entropy loss of the system will grow to match the entropy gain of the surroundings. Then ΔG would be zero, and the reaction would be at equilibrium.

13.3 Including Internal Energy and Electrical Work in the Big Picture

Our objective is to include *work* in this picture, because what we think of as electrons going through a wire is classified as electrical work. For that, we need internal energy, U. If we were to add ΔU to this diagram, what would it look like? Recall that

$$\Delta U = q + w$$

We defined work as $w = -P\Delta V$ before, but now we want to consider more than that. We want to consider the possibility of doing electrical work (Figure 13.5).

The only thing new here is that there is a new way to get energy out of our chemical system—an electrical way, w_{elec}. Electrons passing through a coil of wire set up a magnetic field. This field can move metal parts, such as motors and speakers, thus doing work. (In fact, the movement of electrons from one end of a wire to another is itself considered work.) This work is simply one more way our system can change its internal energy. The old way of doing work, "expansion" work, is still there and still a possibility. We relabel it w_{PV} to distinguish it from the electrical kind, w_{elec}. Thus, we have a slightly new definition of the change in internal energy of the system:

$$\Delta U = q + w = q + w_{elec} + w_{PV} \tag{13.1}$$

But how are we to add this idea to our picture? The key is to realize that ΔH has the peculiar characteristic of removing the pressure–volume work ($-P\Delta V$) from ΔU:

$$\Delta H = \Delta U + P\Delta V$$
$$= \Delta U - w_{PV}$$
$$= (q + w_{PV} + w_{elec}) - w_{PV}$$
$$= q + w_{elec}$$

That probably looks like a sleight of hand. But it's how it works. When electrical work (or for that matter, any other form of work) is involved, ΔH does *not* just equal q. It equals q plus that other non–pressure–volume work. This changes our picture (Figure 13.6). The key is to realize that ΔH is now no longer q, but rather $q + w_{elec}$.

Notice that we have still kept the ΔH arrow its original length. The inclusion of w_{elec} requires that the arrow for q be shorter. The consequence is that the arrow for $T\Delta S_{sur}$ must also be shorter. Take a close look at Figure 13.6. Try to reproduce it yourself. It is generated

Figure 13.5
A chemical system (electrochemical cell) that produces electricity. If electrical work is actually produced *by* the cell, the sign of w_{elec} will be negative, because it will be work going *out* of the system.

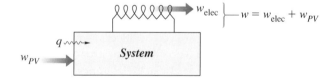

Figure 13.6
Vector graph now showing the full picture, which includes non–pressure–volume work, w_{elec}. From left to right, this graph is constructed based on three connections: (a) the definition of Gibbs energy, (b) the inclusion of w_{elec} in ΔH, and (c) the relationship between the entropy change of the surroundings and heat, $T\Delta S_{sur} = -q$. Due to the presence of w_{elec}, the lengths of the vectors for q and $T\Delta S_{sur}$ are both reduced.

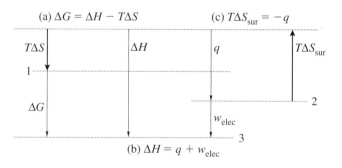

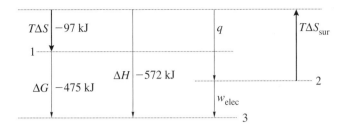

Figure 13.7
The vector graph summarizing the state of affairs for the hydrogen combustion reaction at 298 K. This reaction will proceed only if the arrow for $T \Delta S_{sur}$, *which is shorter than when no electrical work is done,* is still longer than the arrow for $T \Delta S$.

using only three ideas: (a) the definition of Gibbs energy, (b) the idea that now $\Delta H = q + w_{elec}$, and (c) the fact that $T \Delta S_{sur} = -q$. Notice that the overall length of the vectors has remained the same, ΔH. (It's still the same chemical reaction, with the same bonds breaking), *but now less heat is generated*. The idea is that we can do this reaction more than one way—along more than one path. We can get more or less electrical work out of the system, less or more heat. Basically, the bottom line (Line 3) is fixed, but Lines 1 and 2 are *both* adjustable. Isn't that interesting?

Let's use some real numbers based on the hydrogen combustion reaction (Figure 13.7) for standard conditions of 1 bar partial pressure of both H_2 and O_2:

- We have that ΔH is -572 kJ in this case.
- Since $\Delta S = -327$ J/K and we are at 298 K, we have that $T \Delta S = (298 \text{ K}) \times (-327 \text{ J/K}) = -97,000 \text{ J} = -97 \text{ kJ}$. That makes $\Delta G = -572 \text{ kJ} - (-97 \text{ kJ}) = -475$ kJ.
- The vector for heat, q, is somewhat shorter than the others, because now we presume there to be some possibility of electrical work, w_{elec}. The electrical work arrow is negative simply because we want to get some electrical work *out* of this system.
- Finally, the vector representing $T \Delta S_{sur}$ is still just the opposite of the arrow for q. It is fixed by the amount of electrical work we decide we want to have involved, *and it is shorter than it was when no electrical work was done*.

13.4 Electrical Work Is Limited by the Gibbs Energy

We now have everything we need to see how electrical work and Gibbs energy are related. The question is, how can this system be adjusted? *Note that in Figures 13.6 and 13.7 the maximum length of the arrow representing electrical work is the length of the vector for ΔG.* Otherwise Line 2 will rise above Line 1, the arrow for $T \Delta S_{sur}$ will be shorter than the arrow for $T \Delta S$, and the Second Law will be violated! Figure 13.8 shows this limit being approached, then reached.

We can adjust line 1 by playing with the pressures of H_2 and O_2 in the system, just as before. That has the effect of changing the magnitude of $T \Delta S$. In addition, though, we can adjust the position of Line 2. That has the effect of changing the magnitude of $T \Delta S_{sur}$. But there are limits. We still must abide by the Second Law of Thermodynamics!

No matter how the system is adjusted, the sum of the two entropy arrows—one for the system and one for the surroundings—must never go negative. To put it another way, Line 2 may be raised in order to get more electrical work out of the system, but its upper limit is Line 1. That means that the arrow representing w_{elec} may be no longer than the arrow representing ΔG. How would you say this in words? Sometimes it is said, "ΔG is the maximum amount of electrical work," but this isn't quite right, because both values are negative. Maybe the following would do:

When ΔG is negative for a chemical reaction, the maximum amount of electrical work that can be accomplished ($-w_{elec}$) is $-\Delta G$.

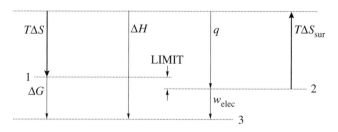

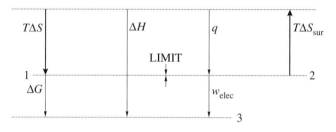

Figure 13.8
Vector graph now showing just the details necessary for understanding the relation between Gibbs energy and electrical work. Both Lines 1 and 2 are adjustable, but only within limits. They must not cross, or the Second Law will be violated. Thus, the arrow for electrical work, w_{elec}, can be no longer than the arrow for Gibbs energy, ΔG.

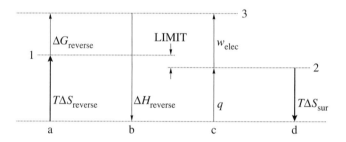

Figure 13.9
Vector graph showing the limitations on electrical work for the reverse reaction in our fuel cell. The values of $\Delta H_{reverse}$, $\Delta S_{reverse}$, and $\Delta G_{reverse}$ are all positive. Nonetheless, Line 2 still must not be above Line 1 or the Second Law will be violated! Thus, now $\Delta G_{reverse}$ is the minimum amount of electrical work needed to run this "reverse" reaction.

13.5 The Gibbs Energy Change *Can* Be Positive

Up until this point we have said that ΔG must be negative for a reaction to occur. But look at Figure 13.9. Here we are looking at the *reverse* reaction:

$$2\,H_2O(l) \longrightarrow 2\,H_2(g) + O_2(g)$$

Everything is reversed. Or is it? Although the entire diagram is flipped in relation to Figure 13.8, the Second Law is not violated as long as, once again, Line 2 is kept below Line 1! Clearly now $\Delta G_{reverse}$ is positive and is now a *minimum* for w_{elec}. But how can $\Delta G_{reverse}$ be positive and the Second Law not be violated? The answer is that w_{elec} is positive—we are putting electrical energy *into* the system. This electrical energy takes the place of some of the heat (q) that would normally have to go into making the reverse reaction happen, and since less heat is required, there's less of a demand on the entropy of the surroundings. The increase in entropy of our system outweighs the smaller decrease in entropy of the surroundings, and by putting *in* electrical work, the reaction can go "backward." We just recharged the fuel cell! We could assert:

When ΔG is positive for a chemical reaction, the minimum amount of electrical work (w_{elec}) that must be added to get the reaction to go is ΔG.

13.6 The Electrical Connection: $-\Delta G = Q_{elec} \times E_{cell} = I \times t \times E_{cell}$

Getting back to reactions involving electricity that are by themselves spontaneous ($\Delta_r G < 0$), when the electrons pass through the wire, energy in the form of heat or work (depending on whether we short-circuit the cell or run a fan or a motor) will be *dissipated*—transferred to the surroundings. This change in electrical energy, as for any energy, can take only two forms: heat and work. How much of each depends upon what is in the circuit. For example, motors and radios are designed to do work (turning the shaft and pushing the speaker); electric heaters provide heat, but do no work at all. The point is, the "maximum amount of electrical work" would be if *no* heat were dissipated at all. So whether the circuit produces a lot of heat or not much heat, the *maximum* amount of electrical work is just the amount of energy dissipated by a circuit.

Now, the amount of energy dissipated by a circuit attached to an electrochemical cell turns out to be the amount of *charge* passed through the wire, Q_{elec} (in *coulombs*), times the *cell potential* (also called *voltage, emf,* or *electromotive force*), E_{cell} (in units of joules/coulomb, also called *volts*). So we have:

$$-\Delta G = (-w_{elec})_{max} = Q_{elec} \times E_{cell} \tag{13.2}$$

Remember that w_{elec} is negative here because our cell is *producing* electricity. In addition, we can relate the amount of charge passed through the wire to the electrical current and the time as follows:

$$Q_{elec} = I \times t \tag{13.3}$$

where

I	is the current (in *amperes*, where 1 amp = 1 coulomb/second)
t	is the time (in seconds)
Q_{elec}	is the amount of charge moved through the wire (in *coulombs*)

Putting these last two equations together, we have

$$\Delta G = -I_{rxn} \times t_{rxn} \times E_{cell} \tag{13.4}$$

where E_{cell} is the voltage of the electrochemical cell and is positive, while ΔG is negative.

For example, say you have a circuit with a motor hooked up to some battery (Figure 13.10). If a current of 0.20 amps ($I = 0.20$ coulomb/second) goes through the motor for 4 seconds, that amounts to 0.8 coulombs of charge. If the motor is hooked up to a 10-volt battery ($E_{cell} = 10$ V = 10 joules/coulomb), then we have that the amount of energy released from the battery is 8 joules:

$$\Delta G = -0.8 \text{ coulomb} \times \frac{10 \text{ joules}}{\text{coulomb}} = -8 \text{ joules}$$

Probably not all of that energy was converted to work, but the *maximum* amount of work that could have been done is 8 J.

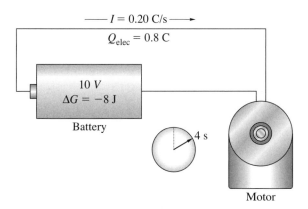

Figure 13.10
A simple electrical circuit involving a battery and a motor. We can calculate the change in Gibbs energy in the battery by knowing its voltage (E_{cell}) and measuring the current and time. Thus, $\Delta G = -I \times t \times E_{cell}$.

13.7 The Chemical Connection: $Q_{rxn} = n \times F$

These units may seem like from another world, because they are! All these ideas of current and charge and potential were all worked out long before anyone had ever even imagined the electron. Scientists imagined charge to be some "corpuscle" of electricity with a "+" associated with it. Now we know that the "charge" going through a wire is really nothing more than electrons. Each unit of charge (1 coulomb) is actually about 6×10^{18} electrons. Or, more precisely in terms of moles of electrons, each mole of electrons amounts to 96,485 coulombs of charge. The number "96,485 coulombs of charge per mole of electrons" is called *Faraday's constant*, F:

$$F = \frac{96,485 \text{ coulombs of charge}}{\text{mole of electrons}}$$

If a chemist says "one mole of electrons went through the wire," then an electrical engineer might say, "OK, for me that's 96,485 coulombs of charge." Mathematically, we can relate what the engineer thinks of as current and charge to what the chemist thinks of as electrons with the aid of Faraday's constant:

$$I \times t = Q_{elec} = n \times F \qquad (13.5)$$

Here:

Q_{elec}	is the amount of charge passed through the wire, in coulombs
I	is the current in amperes (coulombs/second)
t	is the time in seconds
n	is the number of moles of electrons relocated
F	is Faraday's constant, 96,485 coulombs/mole e^-

In chemistry we will always be dealing with the battery reaction, so we will be more specific (Figure 13.11). In all cases we will be able to say that those n moles of electrons came from some sort of *oxidation*, which occurs at the *anode*:

$$A \longrightarrow A^{n+} + ne^- \qquad \text{(A is } oxidized \text{ at the } anode)$$

Realize, those electrons aren't lost. They go through the wire (doing work and generating heat) and end up *reducing* something else at the other end, the *cathode*:

$$B^{n+} + ne^- \longrightarrow B \qquad (B^{n+} \text{ is } reduced \text{ at the } cathode)$$

(Note that the overall charge on B is reduced.) Overall, we have

$$A + B^{n+} \longrightarrow B + A^{n+} \qquad (n \text{ electrons involved})$$

Figure 13.11
An electrochemical cell, or *battery*, is actually two compartments: cathode and anode. Oxidation occurs at the anode ($-$), releasing electrons into the wire. Reduction occurs at the cathode ($+$), where electrons from the circuit are returned to the battery.

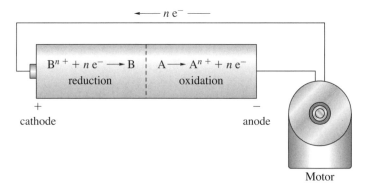

Thus, n comes from figuring out the half-equations for the reaction in the battery. In the case of the hydrogen/oxygen fuel cell, which is run under basic conditions, we could write

oxidation:	$2\,H_2 + 4\,OH^- \longrightarrow 4\,H_2O + 4\,e^-$	(anode)
reduction:	$O_2 + 2\,H_2O + 4\,e^- \longrightarrow 4\,OH^-$	(cathode)

$$2\,H_2 + O_2 \longrightarrow 2\,H_2O \qquad (n = 4)$$

Thus, for the equation as written, $n = 4$. Four moles of electrons will pass through the wire for every two moles of H_2 or one mole of O_2 reacting in the fuel cell.

Note that the magnitude of n depends upon how you write the equation, just like $\Delta_r G$, $\Delta_r S$, and $\Delta_r H$. The following is absolutely fine:

oxidation:	$H_2 + 2\,OH^- \longrightarrow 2\,H_2O + 2\,e^-$	(anode)
reduction:	$^1/_2\,O_2 + H_2O + 2\,e^- \longrightarrow 2\,OH^-$	(cathode)

$$H_2 + ^1/_2\,O_2 \longrightarrow H_2O \qquad (n = 2)$$

Whether we say there will be "2 mol of electrons per mole of H_2" or "4 mol of electrons for every 2 mol of H_2" makes no difference whatsoever in the analysis. These statements amount to the same thing.

13.8 Gibbs Energy and Cell Potential: $\Delta_r G = -nFE_{cell}$

For any overall chemical reaction we can relate the change in Gibbs energy, $\Delta_r G$, to the observed electrical voltage, E_{cell}, as long as we can figure out how many moles of electrons pass through the circuit *per mole of reaction*:

$$-\Delta_r G = Q_{elec} \times E_{cell} = nFE_{cell}$$

or

$$\Delta_r G = -nFE_{cell} \qquad (13.6)$$

or

$$E_{cell} = -\Delta_r G / nF$$

where

$\Delta_r G$	is the reaction Gibbs energy (J/mol)
E_{cell}	is the cell potential of the cell (joules/coulomb = volts)
n	is the number of electrons per mole of reaction (no units)
F	is Faraday's constant (96,485 coulombs/mol)

This, then, is the connection between Gibbs energy and cell potential. However, this equation can be read a different and much more important way. It *also* says that $\Delta_r G$ calculated for a specific equation using tables of $\Delta_f H^\circ$ and S° is related to the "n" that comes from the *properly balanced half-equations*. For example, for hydrogen combustion written

$$2\,H_2(g) + O_2(g) \longrightarrow 2\,H_2O(l) \qquad (n = 4)$$

since we have $n = 4$, as long as we are at 298 K and the standard conditions of 1 bar for all gases, we have $\Delta_r G = \Delta_r G^\circ_{298}$, which in this case is -475 kJ. Thus, at 25°C we have for the voltage of the fuel cell:

$$E_{cell} = \frac{-\Delta_r G}{nF} = \frac{475,000\,\text{J/mol}}{(4)96,485\,\text{coulomb/mol}} = 1.23\,\text{J/coulomb} = 1.23\,\text{volt}$$

That's roughly what one would expect to measure with a voltmeter across the terminals of the fuel cell. Note that had we used instead the following perfectly fine equation for our fuel cell:

$$H_2(g) + \tfrac{1}{2} O_2(g) \longrightarrow H_2O(l) \qquad (n = 2; \; \Delta G^\circ_{298} = -238 \, \text{kJ/mol})$$

then we still would have gotten the same expected voltage, since *both* ΔG°_{298} and n would be halved:

$$E_{\text{cell}} = \frac{-\Delta_r G}{nF} = \frac{238{,}000 \, \text{J/mol}}{(2)96{,}485 \, \text{coulomb/mol}} = 1.23 \, \text{J/coulomb} = 1.23 \, \text{volt}$$

It is very important to remember that this relationship between $\Delta_r G$ and E_{cell} gives us two capabilities, once we know what chemical reaction is going on inside the battery:

- If we know the cell potential (voltage) of a battery, E_{cell}, and we know how to write the equation of the battery reaction (including getting the number of moles of electrons transferred, n) then we can determine $\Delta_r G$ for the battery reaction using $\Delta_r G = -nF E_{\text{cell}}$.
- Conversely, if we could calculate $\Delta_r G$ for the equation of the battery reaction, then we could *predict* the cell potential (voltage) of the battery using the same relationship.

One thing is for sure based on this relationship between Gibbs energy and cell potential: If $\Delta_r G = 0$ (the battery is at equilibrium) then $E_{\text{cell}} = 0$ as well, and the battery is "dead."

13.9 Standard State for Cell Potential: $E^\circ_{\text{cell}, T}$

Just as it's useful to refer to standard states in relation to ΔH and ΔS, we can refer to values of $\Delta_r G$ and E_{cell} for the *standard* conditions of 1 M concentration of all solutes and 1 bar pressure of all gases. Thus, we have

$$\Delta_r H^\circ - T \Delta_r S^\circ = \Delta_r G^\circ_T = -nF E^\circ_{\text{cell}, T}$$

Note that the standard cell potential, $E^\circ_{\text{cell}, T}$, is temperature-dependent because $\Delta_r G^\circ_T$ itself depends upon temperature. The effect can be significant, as a little spreadsheet calculation shows, using the hydrogen combustion reaction as an example (Table 13.1). Remember, this temperature effect is because of that $T \Delta_r S^\circ$ term in $\Delta_r G^\circ$. This temperature effect should not be ignored![3]

Thus, one way to get a standard cell potential is to derive it from $\Delta_r G^\circ_T$. Another method is to use tables of *Standard Reduction Potentials* such as given in Appendix C. These tables are generally valid only at 25°C, and, despite what was just said about temperature effects, if a rough estimate of cell voltage is all that one needs, these potentials can probably be used for any reaction in the 20°C to 40°C range.

All of the entries in a table of standard reduction potentials have electrons as reactants, on the left side of the equation. That is, they are standard *reductions*. Since all electrochemical reactions involve two half-equations, one oxidation and one reduction, only one half-equation will be listed on the table. The other, for the oxidation part of the reaction,

---

[3] Perhaps the temperature effect on E°_{cell} can be ignored for simple sorts of oxidations of one metal by another metal's ion, such as the following, where no bonds are made, no phases are changed, no gases are produced or reacted, and $\Delta_r S^\circ$ is very small:

$$Cu^{2+} + Zn \longrightarrow Cu + Zn^{2+}$$

Table 13.1
Gibbs energy and cell potential for hydrogen combustion

$T/°C$	T/K	$\Delta_r G^\circ_T/(kJ/mol)$	$E^\circ_{cell,T}/V$
0	273	−483	1.25
20	293	−476	1.23
25	298	−475	1.23
100	373	−450	1.17
200	473	−417	1.08
500	773	−319	0.83
$\vdots$			
1476	1749	0	0.00

will be listed backward, with products and reactants switched. For example, with our fuel cell, we would find:

$$O_2 + 2\,H_2O + 4\,e^- \longrightarrow 4\,OH^- \qquad E^\circ_{red} = +0.40\ V$$
$$4\,H_2O + 4\,e^- \longrightarrow 2\,H_2 + 4\,OH^- \qquad E^\circ_{red} = -0.83\ V$$

In order to use this information, we must switch the second equation and reverse its potential. In doing so, we call the voltage a standard *oxidation potential*:

$$O_2 + 2\,H_2O + 4\,e^- \longrightarrow 4\,OH^- \qquad E^\circ_{red} = +0.40\ V$$
$$2\,H_2 + 4\,OH^- \longrightarrow 4\,H_2O + 4\,e^- \qquad E^\circ_{ox} = +0.83\ V$$
$$\overline{}$$
$$2\,H_2 + O_2 \longrightarrow 2\,H_2O \qquad E^\circ_{cell} = +1.23\ V$$

If required, we could then use this value of $E^\circ_{cell,298}$, the fact that $n = 4$, and the definition that 1 volt = 1 joule/coulomb to determine $\Delta_r G^\circ_{298}$ for the reaction:

$$\Delta_r G^\circ_{298} = -nFE^\circ_{cell,298}$$

$$= -(4) \times \frac{96{,}485\ \text{coulomb}}{\text{mol}} \times \frac{1.23\ \text{J}}{\text{coulomb}} = -475{,}000\ \text{J/mol}$$

It's very important to remember never to multiply the table's value for a reduction potential by n. Thus, the value of E°_{cell} does not depend upon how we have written our chemical equation. We could equally well have written:

$$\tfrac{1}{2}\,O_2 + H_2O + 2\,e^- \longrightarrow 2\,OH^- \qquad E^\circ_{red} = +0.40\ V$$
$$H_2 + 2\,OH^- \longrightarrow 2\,H_2O + 2\,e^- \qquad E^\circ_{ox} = +0.83\ V$$
$$\overline{}$$
$$H_2 + \tfrac{1}{2}\,O_2 \longrightarrow H_2O \qquad E^\circ_{cell} = +1.23\ V$$

In this case, though, $n = 2$, and we would correctly calculate that *for the equation as written*:

$$\Delta_r G^\circ_{298} = -nFE^\circ_{cell,298}$$

$$= -(2)\frac{96{,}485\ \text{coulomb}}{\text{mol}} \times \frac{1.23\ \text{J}}{\text{coulomb}} = -238{,}000\ \text{J/K}$$

Notice that we do not multiply the values given in Appendix C by anything, especially not the coefficients in the half-equations. *The voltage of a cell is independent of the stoichiometry we write for the chemical equation.* At 25°C we simply have:[4]

$$E^\circ_{cell} = E^\circ_{red} + E^\circ_{ox}$$

If you really need the best value of a standard cell potential at another temperature, you have to go back to tables of $\Delta_f H^\circ$ and S°, calculate both $\Delta_r H^\circ$ and $\Delta_r S^\circ$ (taking heat capacity into account if you really want to do it right), recalculate $\Delta_r G^\circ_T$ from $\Delta_r H^\circ - T\Delta_r S^\circ$, determine n, and solve $\Delta_r G^\circ_T = -nFE^\circ_{cell,T}$ for $E^\circ_{cell,T}$. Whew!

13.10 Using Standard Reduction Potentials to Predict Reactivity

The fact that standard Gibbs energies and standard cell potentials are related so directly:

$$\Delta_r G^\circ_T = -nFE^\circ_{cell,T}$$

suggests a quick way to determine whether a reaction will occur or not under standard conditions and at 298 K even if we aren't dealing with electrochemical cells.

The question, for example, might be whether metallic iron (Fe) will react with a solution containing silver(I) ions (Ag^+). A quick look at a table of standard reduction potentials gives us the answer, especially if the table is ordered by reduction potential as in Appendix C. The key is in realizing that the reaction will occur under standard conditions only when $E^\circ_{cell} > 0$. Thus, we find for Fe and Ag^+:

$$
\begin{aligned}
Ag^+ + e^- &\longrightarrow Ag \qquad & E^\circ_{red} &= +0.7994\text{ V} \\
Fe^{2+} + 2\,e^- &\longrightarrow Fe \qquad & E^\circ_{red} &= \quad -0.44\text{ V}
\end{aligned}
$$

The second reaction is the one we need to switch in order to make Fe a reactant. We have

$$E^\circ_{cell} = 0.7994 + 0.44 = 1.24\text{ V}$$

The answer to our question is yes, Fe should react spontaneously with Ag^+.

In fact, if the table is ordered as in Appendix C, with the highest reduction potential at the top and the lowest at the bottom, we don't even need to do a calculation to answer our question. Shown in Table 13.2 are several half-equations excerpted from Appendix C. Note that given two of these equations, switching the one lower on the table (with the lower reduction potential) always results in a positive overall cell potential. That's because a larger number minus a smaller number is always positive. The net effect of this is that **anything on the lower right will react with anything above it on the left**. This is very handy information! Zn, on the lower right, is below Cu^{2+}, on the upper left, thus a solution of Cu^{2+} will react with metallic zinc. In addition, we predict that a Zn/Cu^{2+} battery is feasible. Zinc metal does not react in an Na^+ solution, however, and a Zn/Na^+ battery would not work.

You will find it harder to look up standard reduction potentials using Appendix C than using a table that is in alphabetical order, but there are many interesting things that can be learned just scanning Appendix C that are simply not obvious at all from a table that is in alphabetical order. Examples of what you can learn from Appendix C include:

---

[4] To get E°_{ox}, the standard *oxidation* potential, reverse the sign of the number found on the table of standard reduction potentials (just as you reverse the chemical equation found on the table). Thus, if the electrons in the half-equation are on the *right*, *reverse* the sign, if they are on the *left*, *leave* it alone. "Leave it on the left; reverse it on the right." Never, ever, multiply standard potentials by anything. They must be added directly.

Table 13.2
Representative standard reduction potentials

	$E°_{red}/V$
$Au^+ + e^- \longrightarrow Au$	+1.68
$Au^{3+} + 3\,e^- \longrightarrow Au$	+1.50
$O_2(g) + 4\,H^+ + 4\,e^- \longrightarrow 2\,H_2O$	+1.229
$Pt^{2+} + 2\,e^- \longrightarrow Pt$	+1.2
$Ag^+ + e^- \longrightarrow Ag$	+0.7994
$Cu^{2+} + 2\,e^- \longrightarrow Cu$	+0.337
$Zn^{2+} + 2\,e^- \longrightarrow Zn$	−0.763
$Na^+ + e^- \longrightarrow Na$	−2.714

- Elemental fluorine, F_2, heads the table and is one of the most reactive substances known. It reacts spontaneously with *everything* on the right on the table below it.
- Similarly, elemental lithium and all other alkali and alkaline earth metals including potassium, rubidium, barium, strontium, calcium, sodium, and magnesium are at the end of the list, on the right. They are all very reactive elements, reacting with *everything* above them on the left, including the small amount of H^+ in water, which has a reduction potential of 0.00 V.
- Gold (Au) is the only metal appearing on the right above O_2. Thus, gold is the only metal that does not "rust" or *tarnish*. All other metals react with O_2 to make an oxide coating. This is what made gold especially enticing to the ancients.
- Some substances and ions, such as H_2O_2 and Cu^+, appear both on the left and on the right. This means that they are inherently unstable and, with time, will react with themselves. For example, we have

$$Cu^+ + e^- \longrightarrow Cu(s) \qquad E°_{red} = +0.521\ V$$
$$Cu^{2+} + e^- \longrightarrow Cu^+ \qquad E°_{red} = +0.153\ V$$

This leads to

$$2\,Cu^+ \longrightarrow Cu(s) + Cu^{2+} \qquad E°_{cell} = 0.368\ V$$

and the conclusion that you may have considerable trouble making a solution of copper(I) chloride.
- Many of the substances we learn to respect as potentially dangerous appear in the upper left side of this table. Besides F_2 and H_2O_2, we have HClO, MnO_4^- (from $KMnO_4$, usually), ClO_3^- (from $HClO_3$), Cl_2, $Cr_2O_7^{2-}$ (from K_2CrO_7), O_2, ClO_4^- (from $HClO_4$), Br_2, and NO_3^- (from HNO_3). All of these substances are easily reduced and are classified as *oxidizing agents* because there are so many substances below them on the right on the list that will be spontaneously oxidized upon contact with them. Handle with care!

Thus, Appendix C is a gold mine of information, and is more useful for *learning chemistry* than for just looking up numbers. The *CRC Handbook of Chemistry and Physics* has two complete sets of tables, one ordered alphabetically and one ordered by reduction potential. Both are useful in their own way.

13.11 Equilibrium Constants from Cell Potentials:
$$0 = -nFE^\circ_{\text{cell},T} + RT \ln K$$

Since we can relate $\Delta_r G^\circ_T$ to K, the equilibrium constant for a reaction (Section 11.9), we can also relate $E^\circ_{\text{cell},T}$ and K based on the fact that $\Delta_r G^\circ_T = -nFE^\circ_{\text{cell},T}$:

$$K = e^{-\Delta_r G^\circ_T/RT} = e^{nFE^\circ_{\text{cell},T}/RT} \tag{13.7}$$

That is, we can determine equilibrium constants not just from $\Delta_r H^\circ$ and $\Delta_r S^\circ$, but also from $E^\circ_{\text{cell},T}$. (But be careful! For these purposes do not assume $E^\circ_{\text{cell},T}$ is independent of temperature.) For example, given

$$2\,\text{Cu}^+ \longrightarrow \text{Cu}(s) + \text{Cu}^{2+} \qquad E^\circ_{\text{cell}} = 0.368 \text{ V}$$

we can calculate

$$K = e^{nFE^\circ_{\text{cell},T}/RT}$$

$$= e^{\dfrac{(1)(96,485\,\text{coulomb/mol})(0.368\,\text{J/coulomb})}{\frac{8.31451\,\text{J}}{\text{mol·K}}\,298\,\text{K}}} = e^{14.33} = 1.7 \times 10^6$$

We could write for the equilibrium expression then, *even in situations when no electrochemistry is being done,*

$$K = \frac{[\text{Cu}^{2+}]_{\text{eq}}}{[\text{Cu}^+]^2_{\text{eq}}} = 1.7 \times 10^6$$

Thus, a table of standard reduction potentials can be very helpful even if we aren't doing any electrochemistry. This table can be a source of equilibrium constants (at least at 25°C).

Note that the relationship between K and E°_{cell} in Equation 13.7 can be rearranged to be in the form of our "standard model" (see Section 11.12):

$$0 = -nFE^\circ_{\text{cell},T} + RT \ln K$$

Though this may not seem particularly useful in comparison to Equation 13.7, we will see in a moment that it is just a "special case" of a more general expression.

13.12 Actual Cell Voltages and the Nernst Equation:
$$-nFE_{\text{cell}} = -nFE^\circ_{\text{cell},T} + RT \ln Q$$

Remember, in general, whether the system is at equilibrium or not, we have the key relationship relating $\Delta_r G$ to Q:

$$\Delta_r G = \Delta_r H^\circ - T(\Delta_r S^\circ - R \ln Q)$$

where Q is the *reaction quotient*, involving the *actual* pressures of gases (in bar) and concentrations of solutes (in mol/L). Recall that $Q = K$ only at equilibrium, when $\Delta_r G = 0$. In terms of *standard* Gibbs energy change, $\Delta_r G^\circ$, this becomes

$$\Delta_r G = \Delta_r H^\circ - T(\Delta_r S^\circ - R \ln Q)$$

$$\Delta_r G = (\Delta_r H^\circ - T\Delta_r S^\circ) + RT \ln Q$$

$$\Delta_r G = \Delta_r G^\circ_T + RT \ln Q$$

Now combining this with the relationship between $\Delta_r G$ and E_{cell}, we get

$$-nFE_{\text{cell}} = -nFE^{\circ}_{\text{cell},T} + RT \ln Q$$

or

$$E_{\text{cell}} = E^{\circ}_{\text{cell},T} - \frac{RT}{nF} \ln Q \tag{13.8}$$

This is a very famous equation, known as the **Nernst equation**. When pressures of all gases are 1 bar and concentrations of all solutes are 1 M, then $Q = 1$, $\ln Q = 0$, and $E_{\text{cell}} = E^{\circ}_{\text{cell},T}$, the "standard" voltage at the specified temperature. But when the pressures of all gases are *not* 1 bar or the concentrations of all solutes are *not* 1 M (which is usually the case), if you want to know what a battery's voltage will *really* be, then you need to know the following four things:

1. above all, T, the temperature, in kelvins;
2. n, the number of electrons involved per mole of reaction;
3. Q, the reaction quotient, involving pressures of all gases in bar and concentrations of all solutes in mol/L;
4. $E^{\circ}_{\text{cell},T}$, the standard cell voltage at this temperature, either:
    (a) by using Appendix C and $E^{\circ}_{\text{cell}} = E^{\circ}_{\text{red}} + E^{\circ}_{\text{ox}}$, as long as $T \approx 298$ K, or
    (b) by using Appendix D, determining $\Delta_r H^{\circ}$ and $\Delta_r S^{\circ}$ for the reaction, and then using $\Delta_r G^{\circ}_T = \Delta_r H^{\circ} - T\Delta_r S^{\circ}$ to get $\Delta_r G^{\circ}_T$, which can then be used to get $E^{\circ}_{\text{cell},T}$.

Well, that may seem like a lot to have to know, but that's what you have to know. Once you have all that information, you can determine what the actual voltage will be using the Nernst equation.

Sometimes the Nernst equation is written specifically for 25°C and expressed in terms of $2.303 \log_{10}$ instead of $\ln$. Then the constants "$2.303\, RT/F$" are put together as "0.0592":

$$E_{\text{cell}} = E^{\circ}_{\text{cell},T} - \frac{RT}{nF} \ln Q$$

$$= E^{\circ}_{\text{cell},298} - \frac{0.0592}{n} \log Q \quad \textbf{25°C ONLY} \tag{13.9}$$

But beware! This is *only* for 25°C! Hiding in there are *two* terms involving T.

Note that at equilibrium, that is, when $\Delta_r G = 0$, $Q = K$, and $E_{\text{cell}} = 0$, then the Nernst equation reduces to

$$0 = E^{\circ}_{\text{cell},T} - \frac{RT}{nF} \ln K$$

which, when rearranged, gives Equation 13.7. The Nernst equation simply relates to the more general situation when the system may or may not be at equilibrium.

13.13 Detailed Examples

For example, let's determine the equilibrium constant for the following reaction at 25°C using only electrochemical data from Appendix C:

$$2\,\text{Br}^-(aq) + \text{F}_2(g) \longrightarrow \text{Br}_2(l) + 2\text{F}^-(aq)$$

Writing half-equations we find that $n = 2$ and $E^{\circ}_{\text{cell}} = 1.80$ V:

$$
\begin{aligned}
2\,\text{Br}^- &\longrightarrow \text{Br}_2 + 2\,\text{e}^- & E^{\circ}_{\text{ox}} &= -1.07\text{ V} \\
\text{F}_2 + 2\,\text{e}^- &\longrightarrow 2\,\text{F}^- & E^{\circ}_{\text{red}} &= +2.87\text{ V} \\
\hline
2\,\text{Br}^- + \text{F}_2 &\longrightarrow \text{Br}_2 + 2\,\text{F}^- & E^{\circ}_{\text{cell}} &= +1.80\text{ V}
\end{aligned}
$$

From Equation 13.7 we have

$$K = e^{nFE^{\circ}_{\text{cell},T}/RT}$$

Substituting in $n = 2$, $F = 96{,}485$ coulomb/mol, $E^{\circ}_{\text{cell}} = 1.80$ joules/coulomb, $R = 8.3145$ J/mol·K, and $T = 298$ K, we get $K = 7.63 \times 10^{60}$.

A somewhat more complicated example might be the following: An electrochemical cell involves the following overall reaction:

$$\text{Zn}(s) + 2\,\text{H}^+ \longrightarrow \text{Zn}^{2+} + \text{H}_2(g)$$

We wish to use this cell as the basis of a pH meter. The idea is that if we know the pressure of H_2, the concentration of Zn^{2+}, and the voltage, we should be able to determine $[\text{H}^+]$. From that we can get pH, which is $-\log[\text{H}^+]$.

Here's the question: What is the pH in the cathode compartment (where H^+ is being reduced) if the pressure of H_2 is 0.2 bar, $[\text{Zn}^{2+}] = 0.1$ M, the temperature is 25°C, and the meter reads 0.45 V? The point is, we know all except one piece of information ($[\text{H}^+]$ in this case). We should be able to get that if we think about it. First we calculate $E^{\circ}_{\text{cell},298}$ from a table of standard reduction potentials, getting 0.763 volts. Second, using the Nernst equation:

$$E_{\text{cell}} = E^{\circ}_{\text{cell},T} - \frac{RT}{nF} \ln Q$$

and substituting in $E^{\circ}_{\text{cell},298} = 0.763$ joules/coulomb, $R = 8.3145$ J/mol·K, and $T = 298$ K, $n = 2$, and $F = 96{,}485$ coulomb/mol, we get Q = 3.86×10^{10}. Now, the definition of Q in this case is

$$Q = \frac{[\text{Zn}^{2+}]P_{\text{H}_2}}{[\text{H}^+]^2}$$

We can rearrange this equation for $[\text{H}^+]$. Using Q = 3.86×10^{10}, $[\text{Zn}^{2+}] = 0.1$ M, and $P(\text{H}_2)$ = 0.2 bar, we get $[\text{H}^+] = 7.2 \times 10^{-7}M$, and the pH is $-\log(7.2 \times 10^{-7}) = 6.14$.

There are many, many variations on these themes involving determining concentrations, pressures, voltages, and Gibbs energies. Almost always, they boil down to solving for one or another of the variables in one of the following five equations:

$$\Delta_{\text{r}}G = \Delta_{\text{r}}H^{\circ} - T(\Delta_{\text{r}}S^{\circ} - R \ln Q)$$

$$\Delta_{\text{r}}G = \Delta_{\text{r}}G^{\circ}_{T} + RT \ln Q$$

$$\Delta_{\text{r}}G = -nFE_{\text{cell}}$$

$$E_{\text{cell}} = E^{\circ}_{\text{cell},T} - \frac{RT}{nF} \ln Q$$

$$0 = E^{\circ}_{\text{cell},T} - \frac{RT}{nF} \ln K$$

Can you see how one leads to the next and how $\Delta_{\text{r}}G = -nFE_{\text{cell}}$ is the connecting link between the chemistry and the electricity?

13.14 Summary

At least when no electrochemistry is involved, $\Delta_{\text{r}}G$ is a handy quantity that lets us know when the entropy of the universe is increasing ($\Delta_{\text{r}}G < 0$). We have from Section 11.2 that

$$\Delta_{\text{r}}G = \Delta_{\text{r}}H^{\circ} - T(\Delta_{\text{r}}S^{\circ} - R \ln Q)$$

where Q is the reaction quotient, which equals K only when $\Delta_r G = 0$.

Now we have learned that when $\Delta_r G < 0$ the quantity $-\Delta_r G$ also represents the maximum amount of electrical work that can be done by our system per mole of reaction *as long as the conditions do not change*. (For example, you have to keep replenishing reactants and removing products to keep the concentrations the same.) And when $\Delta_r G > 0$, it tells us how much electrical energy is going to be required per mole of reaction to get the reaction to go in the direction we want it to go instead of in reverse.

Most significantly, $\Delta_r G$ and cell potential are related by the number of moles of electrons per mole of reaction (based on the half-equations) and Faraday's constant, Equation 13.6:

$$\Delta_r G = -nFE_{cell}$$

This relationship comes from the fact that there are two distinct ways of talking about the charge, Q_{elec}, going through an electrical circuit: the electrical engineer's way ($I \times t$), and the chemist's way ($n \times F$), Equation 13.5:

$$I \times t = Q_{elec} = n \times F$$

Without too much trouble, this all leads, ultimately, to the Nernst equation, which relates the actual voltage of a battery to the actual concentrations of reactants and products present, Equation 13.7:

$$E_{cell} = E^\circ_{cell,T} - \frac{RT}{nF} \ln Q$$

Note that typically as the reaction proceeds and Q gets larger, the voltage will drop until $E_{cell} = 0$ and $Q = K$.

$$0 = E^\circ_{cell,T} - \frac{RT}{nF} \ln K$$

or

$$K = e^{nFE^\circ_{cell,T}/RT}$$

Thus, the equilibrium state ($\Delta_r G = 0$, $E_{cell} = 0$, and $Q = K$) is simply the special case of a "dead battery."

When the battery is not dead, then we can use the four following key relationships to predict actual values for a reaction (E_{cell}, $\Delta_r G$, or Q) based on knowing standard values (E°_{cell} or $\Delta_r G^\circ$):

$$\Delta_r G = \Delta_r H^\circ - T(\Delta_r S^\circ - R \ln Q)$$
$$\Delta_r G = \Delta_r G^\circ_T + RT \ln Q$$
$$\Delta_r G = -nFE_{cell}$$
$$E_{cell} = E^\circ_{cell,T} - \frac{RT}{nF} \ln Q$$

In addition, tables such as Appendix C can give us much insight into the "why" of many chemical reactions. Substances with large reduction potentials will react as oxidizing agents toward any substance below them and on the right on this table. All of these potentials are just that: *potentials*. They indicate the tendency or potential for a chemical reaction to occur in precisely the way reaction Gibbs energy, $\Delta_r G$, does. A positive cell potential is equivalent to a negative $\Delta_r G$; the reaction will be spontaneous. Once again we find that it is this predictive nature of thermodynamics that makes it so powerful a scientific tool.

Problems

The symbol Ⓐ indicates that the answer to the problem can be found in the "Answers to Selected Exercises" section at the back of the book.

Current and Charge

13.1 Ⓐ A battery delivers 150 mA of current for 24 hr. (a) How much charge is involved? (b) How many electrons are involved?

13.2 A battery delivers 175 mA of current for 48 hr. (a) How much charge is involved? (b) How many electrons are involved?

13.3 Ⓐ A battery delivers 200 mA of current for 12 hr. (a) How much charge is involved? (b) How many electrons are involved?

13.4 A battery delivers 300 mA of current for 24 hr. (a) How much charge is involved? (b) How many electrons are involved?

13.5 Ⓐ How long will it take to *plate out* (that is, produce by reduction at a cathode) each of the following at a current of 2.00 A?

(a) 25.0 g of $Cu(s)$ from $Cu^{2+}(aq)$
(b) 0.25 mol of $Ag(s)$ from $Ag^{+}(aq)$
(c) 1.25 kg of $Al(s)$ from $Al^{3+}(aq)$

13.6 How long will it take to *plate out* (that is, produce by reduction at a cathode) each of the following at a current of 1.00 A?

(a) 25.0 g of $Mg(s)$ from $Mg^{2+}(aq)$
(b) 0.50 mol of $Cr(s)$ from $Cr^{3+}(aq)$
(c) 1.25 kg of $V(s)$ from $V^{2+}(aq)$

13.7 How long will it take to *plate out* (that is, produce by reduction at a cathode) each of the following at a current of 2.00 A?

(a) 25.0 g of $Zr(s)$ from $Zr^{4+}(aq)$
(b) 0.50 mol of $Zn(s)$ from $Zn^{2+}(aq)$
(c) 1.25 kg of $Cd(s)$ from $Cd^{2+}(aq)$

Gibbs Energy and Cell Potential

13.8 Ⓐ If the battery in Problem 13.1 holds a voltage of 1.5 V, what is the Gibbs energy change due to the reaction?

13.9 The Hall–Heroult process was invented in the late 1800s to produce aluminum metal from aluminum oxide via reduction with graphite at high temperature. The overall reaction is

$$2\, Al_2O_3(\text{in } Na_3AlF_6) + 3\, C(\text{graphite}) \longrightarrow 4\, Al(l) + 3\, CO_2(g)$$

for which $E^\circ_{\text{cell}} = -1.26$ V at 1000°C. (a) Determine the minimum amount of electrical work necessary to produce 1.0 kg of $Al(l)$ at 1000°C. (b) Compare this amount of energy to that necessary to melt 1.0 kg of $Al(s)$, for which $\Delta_{\text{fus}}H^\circ = 10.7$ kJ/mol. (c) In light of your answers to (a) and (b), is recycling of aluminum energetically favorable? If so, what is the *minimum* energy saving per kg of aluminum recycled?

Half-Equations and Standard Cell Potentials

13.10 Ⓐ If the battery from Problem 13.1 is a NiCd recharge-able battery for which the chemical reaction is

$$NiO_2(s) + Cd(s) \longrightarrow Cd(OH)_2(s) + Ni(OH)_2(s)(\text{unbalanced})$$

(a) balance this equation in base using the half-equation method. What is *n* for this equation?
(b) How many grams of NiO_2 are required for this discharge?
(c) How many grams of $Cd(OH)_2$ are produced?

13.11 Ⓐ Using Appendix C, determine which of the following reactions will be spontaneous as they are written at 25°C under standard conditions. For those reactions that are spontaneous, determine the maximum amount of electrical work that can be done by that reaction. For those reactions that are not spontaneous, determine the minimum amount of electrical work that will be necessary to drive the reaction.

(a) $Pb^{2+}(aq) + Sn(s) \longrightarrow Pb(s) + Sn^{2+}(aq)$
(b) $2\, V^{2+}(aq) + Zr(s) \longrightarrow Zr^{4+}(aq) + 2\, V(s)$
(c) $2\, Fe^{3+}(aq) + Fe(s) \longrightarrow 3\, Fe^{2+}(aq)$
(d) $Fe(s) + Cu^{2+}(aq) \longrightarrow Fe^{2+}(aq) + Cu(s)$

13.12 Consider the combustion of methane with oxygen, as written:

$$CH_4(g) + 2\, O_2(g) \longrightarrow CO_2(g) + 2\, H_2O(g)$$

(a) Write the two half-equations and determine *n* for the equation as written.
(b) What are $\Delta_r H^\circ$, $\Delta_r S^\circ$, and $\Delta_r G^\circ$ for this reaction at 25°C?
(c) If a fuel cell were designed using this reaction under standard conditions, what would be the voltage of this fuel cell at 25°C?

13.13 Estimate the voltage of the fuel cell in Problem 13.12 at 1000°C.

Using Reduction Potentials to Predict Reactivity

13.14 List three substances that under standard conditions can

(a) reduce $H^+(aq)$ to $H_2(g)$
(b) oxidize $H_2(g)$ to $H^+(aq)$
(c) oxidize $Pt(s)$ to $Pt^{2+}(aq)$
(d) reduce $Mg^{2+}(aq)$ to $Mg(s)$

13.15 Ⓐ Determine which of the following reactions will be spontaneous, as they are written, under standard conditions. Identify the oxidizing and reducing agent in each case.

(a) $2\, Ce^{4+}(aq) + 2\, Tl(s) \longrightarrow 2\, Ce^{3+}(aq) + 2\, Tl^+(aq)$
(b) $Fe^{2+}(aq) + Cu(s) \longrightarrow Fe(s) + Cu^{2+}(aq)$
(c) $Zn(s) + 2\, H^+(aq) \longrightarrow Zn^{2+}(aq) + H_2(g)$
(d) $Sn^{2+}(aq) + 2\, Ag(s) \longrightarrow Sn(s) + 2\, Ag^+(aq)$

13.16 Gold (Au) is very difficult to oxidize, hence its resistance to tarnishing. Explain why gold will react with *aqua regia* (a mixture of concentrated hydrochloric acid and nitric acid named "royal water" by alchemists specifically because of its ability to dissolve gold) but not with either hydrochloric acid or nitric acid alone. [*Hint:* Examine carefully the half-equations below along with others from Appendix C.]

	E°_{red}/V
$Au^+ + e^- \longrightarrow Au$	+1.68
$Au^{3+}(aq) + 3\,e^- \longrightarrow Au(s)$	+1.50
$Cl_2(g) + 2\,e^- \longrightarrow 2\,Cl^-(aq)$	+1.358
$AuCl_4^-(aq) + 3\,e^- \longrightarrow Au(s) + 4\,Cl^-$	+1.00
$NO_3^-(aq) + 4\,H^+(aq) + 3\,e^- \longrightarrow NO(g) + 2\,H_2O(l)$	+0.96

13.17 Ⓐ You know that the poisonous gas NO results from the reaction of nitric acid (HNO_3) with copper, but you are wondering about iron. (a) Write a balanced equation for the reaction that would occur upon dropping a iron nail into a solution of nitric acid. (b) Determine E°_{cell} for this reaction. Is this reaction spontaneous? (c) Determine the value of the equilibrium constant for this reaction at 298 K.

Equilibrium Constants from Cell Potentials

13.18 Ⓐ Determine E°, $\Delta_r G^\circ$, and K for the following reactions at 25°C as written. [*Hint:* For (a) get n and E° first; for (b) get n and $\Delta_r G^\circ$ first.]

(a) $Cu^{2+} + Zn \longrightarrow Zn^{2+} + Cu$

(b) $CH_4(g) + 2\,O_2(g) \longrightarrow CO_2(g) + 2\,H_2O(g)$

(c) $\frac{1}{2}\,CH_4(g) + O_2(g) \longrightarrow \frac{1}{2}\,CO_2(g) + H_2O(g)$

(d) $2\,Zn^{2+} + 2\,Cu \longrightarrow 2\,Cu^{2+} + 2\,Zn$

13.19 Determine E°, $\Delta_r G^\circ$, and K for the following reactions at 25°C as written:

(a) $3\,Sn(s) + Cr_2O_7^{2-}(aq) + 14\,H^+(aq) \longrightarrow$
        $3\,Sn^{2+}(aq) + 2\,Cr^{3+}(aq) + 7\,H_2O(l)$

(b) $N_2(g) + 2\,Mn^{+2}(aq) + 4\,H_2O(l) \longrightarrow$
        $N_2H_5^+(aq) + 2\,MnO_2(s) + 3\,H^+(aq)$

(c) $6\,Sn(s) + 2\,Cr_2O_7^{2-}(aq) + 28\,H^+(aq) \longrightarrow$
        $6\,Sn^{2+}(aq) + 4\,Cr^{3+}(aq) + 14\,H_2O(l)$

(d) $\frac{1}{2}\,N_2(g) + Mn^{+2}(aq) + 2\,H_2O(l) \longrightarrow$
        $\frac{1}{2}\,N_2H_5^+(aq) + MnO_2(s) + \frac{3}{2}\,H^+(aq)$

Actual Cell Voltages: The Nernst Equation

Note: If you have a graphing calculator, you might want to take a look at *http://www.stolaf.edu/depts/chemistry/imt/js/tinycalc/ti-calc.htm* before approaching these problems.

13.20 Ⓐ What is the voltage E for a cell undergoing the reaction

$$Cu^{2+} + Zn \longrightarrow Zn^{2+} + Cu$$

under the following conditions, all at 25°C? In (b)–(d) discuss your answer relative to your answer to (a).

(a) $[Cu^{2+}] = 1.0\ M$, $[Zn^{2+}] = 1.0\ M$

(b) $[Cu^{2+}] = 1.0\ M$, $[Zn^{2+}] = 1 \times 10^{-4} M$

(c) $[Cu^{2+}] = 1.0 \times 10^{-6} M$, $[Zn^{2+}] = 1.0\ M$

(d) $[Cu^{2+}] = 1.0 \times 10^{-6} M$, $[Zn^{2+}] = 1.0 \times 10^{-4} M$

13.21 Ⓐ How low would the concentration of Cu^{2+} have to be make the cell in Problem 13.20 work *backward* if $[Zn^{2+}] = 1.0$ M? [*Hint:* Find $[Cu^{2+}]$ for the condition when $E = 0$.]

13.22 Ⓐ If the concentration of Cu^{2+} were not known for the cell in Problem13.20, but $[Zn^{2+}]$ were 1.0 M and the voltage were found to be 0.40 V, what must be $[Cu^{2+}]$?

13.23 What is the voltage E for an electrochemical cell undergoing the reaction

$$2\,Ag^+(aq) + Ni(s) \longrightarrow 2\,Ag(s) + Ni^{2+}(aq)$$

under the following conditions, all at 25°C? In (b)–(d) discuss your answer relative to your answer to (a).

(a) $[Ag^+] = 1.0\ M$, $[Ni^{2+}] = 1.0\ M$

(b) $[Ag^+] = 1.0\ M$, $[Ni^{2+}] = 1 \times 10^{-4} M$

(c) $[Ag^+] = 1.0 \times 10^{-6} M$, $[Ni^{2+}] = 1.0\ M$

(d) $[Ag^+] = 1.0 \times 10^{-6} M$, $[Ni^{2+}] = 1.0 \times 10^{-4} M$

13.24 How low would the concentration of Ag^+ have to be to make the cell in Problem 13.23 work *backward* if $[Ni^{2+}] = 1.0 M$?

13.25 If the concentration of Ag^+ were not known for the cell in Problem 13.23, but $[Ni^{2+}]$ were 1.0 M and the voltage were found to be 0.40 V, what must be $[Ag^+]$?

13.26 A *concentration cell* is an electrochemical cell involving two solutions (A and B) that differ only in concentration, connected by a salt bridge or porous glass membrane. This particular concentration cell has a porous glass membrane. Compartment A contains a 1.202 M solution of silver nitrate. A solution of silver nitrate of unknown concentration is in compartment B. The electrode in each solution is a silver wire, and the electrodes are connected to one another through a voltmeter, which reads 0.121 V. (a) What must be the concentration of the unknown silver nitrate solution in compartment B? (b) Sketch this cell, with compartment A on the left and B on the right. Indicate the anode, cathode, and the direction of electron, silver ion, and nitrate ion flow. (c) What would the concentration be if the meter reads −0.121 V? [Careful!]

Brain Teasers

13.27 Imagine a coffee heater, operating at 120 V. Its heating action occurs by simply creating a short circuit, converting all of the electrical work into heat. If this heater were placed in a cup of coffee (250 g, mostly water), and it delivered a current of 15 amps: (a) How many degrees would the coffee be heated in 10 seconds? (b) How long would it take to heat the coffee from room temperature (23°C) to a drinkable temperature (75°C)?

13.28 Do a little research and determine, based on the cost of aluminum and the cost of electrical energy, whether recycling of aluminum using the Hall–Heroult process (Problem 13.9) is *economically* feasible.

Symbols and Constants

Note: Some symbols have more than one meaning. In that case they are listed twice. Names, values, and standard units are shown in parentheses. Principal sections where the symbol is introduced are indicated in brackets.

- ° (standard state, pronounced "zero") [7.7, 9.3, 11.3, 11.9, 13.9]

The standard state is necessary as a reference when calculating changes or difference in enthalpy, entropy, and free energy, and cell potential between any two non-standard states. It means:

    for 1 mole
    at 1 bar pressure (solids, liquids, or gases)
    or 1 mol/L concentration (solutes)

- Δ (delta, "change in") [2.8, 4.5, 4.6, 4.7, 5.2]

ΔX describes the *change* in X under certain specified or implied conditions. For example, in the definition of work,

$$\text{work} = \frac{\text{force}}{\text{area}} \times (\text{area} \times \text{distance})$$

$$w = -P\Delta V$$

Eq. 4.4, p. 82

by ΔV we mean the *change* in volume ("after" minus "before").

- Δ_r (delta-r, reaction "difference") [4.10, 5.2, 7.10, 9.4, 10.2]

Always the "difference of products minus reactants for a given writing of the balanced chemical equation." The subscript "r" in $\Delta_r X$ indicates that we are describing the difference in X between products and reactants for a given balanced chemical equation under specific ("constant") temperature, pressure, volume, and set of reactant/product concentrations. For example, when we say for the chemical reaction as written

$$2\,A \longrightarrow B$$

that "$\Delta_r H^o_{298} = 5.4$ kJ/mol," we mean the *difference* in standard molar enthalpy between a mole of B and 2 moles of A is 5.4 kJ at 298 K in a volume that would contain both A and B at 1 bar partial pressure each (2 bar total pressure). (The "/mol" here refers to "per mole of reaction, as written.")

    Under certain conditions of reaction, $\Delta_r X$ can be interpreted also as the *change* in X (ΔX) for the number of moles of reactant specified in the chemical equation reacting to form the specified number of moles of product. So, in the above example, under conditions of constant pressure, temperature, and reactant/product concentrations or partial pressures, 5.4 kJ of energy will be drawn from the surroundings for every mole of B formed.

- ε_i (energy of level i, joules) [2.8, 3.3–3.8]

Energy, being quantized, takes only specific discrete values, depending upon the *quantum number i* (or *n*, sometimes). The exact relationship between the energy and quantum number depends upon the energy type—electronic, vibrational, rotational, or translational. All four of these are related to de Broglie's "particle in a box" idea:

Eq. 3.7, p. 49

$$\varepsilon_n = \frac{h^2}{2m_e\lambda^2} = \frac{h^2}{2m_e(2d/n)^2}$$

$$= n^2\frac{h^2}{8}\left(\frac{1}{m_e d^2}\right) \quad \text{where } n = 1, 2, 3, \dots$$

Specifically, we have

Eq. 3.8, p. 51

$$\varepsilon_n = -\left(\frac{1}{n^2}\right)\frac{h^2}{8\pi^2}\left(\frac{1}{m_e a_0^2}\right) \quad \text{where } n = 1, 2, 3, \dots \quad \text{(electronic; H atom only)}$$

Eq. 3.13, p. 57

$$\varepsilon_i = \left(i + \frac{1}{2}\right)h\nu \quad \text{where } \nu = \frac{1}{2\pi}\sqrt{\frac{k_f}{\mu}} \text{ and } i = 0, 1, 2, \dots \quad \text{(vibration)}$$

Eq. 3.14, p. 59

$$\varepsilon_i = i(i+1)\frac{h^2}{8\pi^2}\left(\frac{1}{\mu R^2}\right) \quad \text{where } i = 0, 1, 2, \dots \quad \text{(rotation)}$$

Eq. 3.15, p. 60

$$\varepsilon_{n_x,n_y,n_z} = (n_x^2 + n_y^2 + n_z^2)\frac{h^2}{8}\left(\frac{1}{md^2}\right) \quad \text{where } n = 1, 2, 3, \dots \quad \text{(translation)}$$

The key to understanding these equations is that each involves a set of *constraints*. In each case (even vibration, though not so evident by this particular equation), mass-related and distance-related factors appear in the denominator. Of two related systems, the one with the bigger box and more mass (or reduced mass, μ, in the case of vibration and rotation) will have the more closely spaced energy levels.

- λ (lambda, wavelength) [3.3, 3.12]

A quantity related inversely to frequency of light, ν ("nu"), as

Eq. 3.2, p. 48

$$\lambda\nu = c \quad \text{or} \quad \nu = c/\lambda$$

De Broglie found that for an electron with mass m_e, $9.1093897 \times 10^{-31}$ kg, and velocity v_e (vee):

Eq. 3.4, p. 48

$$\lambda = \frac{h}{m_e v_e} \quad \text{or} \quad v_e = \frac{h}{m_e\lambda} \quad \text{(de Broglie, 1924)}$$

where h is Planck's constant, $6.6260755 \times 10^{-34}$ joule-seconds.

- μ (mu, reduced mass, kg) [3.6, 3.7]

For a two-mass system, we have

by definition: $\quad \dfrac{1}{\mu} = \dfrac{1}{m_1} + \dfrac{1}{m_2} \quad$ or $\quad \mu = \dfrac{m_1 m_2}{m_1 + m_2}$

Eq. 3.10, p. 55

When two masses are the same, then $\mu = m/2$, and when they are different, then $\mu < m_1 < m_2$. As m_2 gets larger and larger, μ approaches the value of m_1, the less massive of the two. This accounts for the fact that in relative motion of two masses, it is the lighter mass doing most of the moving.

- ν (nu, frequency, s^{-1}) [3.3–3.8]

The frequency of light associated with its energy and equivalent mass:

$$E = h\nu \text{ (Planck, 1900)} = mc^2 \text{ (Einstein, 1905)}$$

Eq. 3.1, p. 48

where h is Planck's constant, $6.6260755 \times 10^{-34}$ joule-seconds. Also related to wavelength, λ, as

$$\lambda\nu = c \quad \text{or} \quad \nu = c/\lambda$$

Eq. 3.2, p. 48

Frequency shows up also in relation to vibrational energy, where it is related to the spring force constant of the bond, k_f, and the reduced mass, μ:

$$\varepsilon_i = \left(i + \dfrac{1}{2}\right) h\nu \quad \text{where } \nu = \dfrac{1}{2\pi}\sqrt{\dfrac{k_f}{\mu}} \text{ and } i = 0, 1, 2, \dots \quad \text{(vibration)}$$

Eq. 3.13, p. 57

- a_0 (Bohr radius, $5.29177249 \times 10^{-11}$ m) [3.5]

The average distance of the electron to the proton in the ground state hydrogen atom. For other levels, the average radius is $n^2 a_0$. Shows up in the electronic energy equation for hydrogen:

$$\varepsilon_n = -\left(\dfrac{1}{n^2}\right) \dfrac{h^2}{8\pi^2}\left(\dfrac{1}{m_e a_0{}^2}\right)$$

Eq. 3.8, p. 51

where $n = 1, 2, 3, \dots \quad$ (electronic; H atom only)

- BDE (mean bond dissociation energy, joules/mol) [5.3–5.6]

Mean bond dissociation energies (Chapter 5) allow a quick estimate of differences in internal energy $\Delta_r U$ for a chemical reaction as written in a specific equation using Hess's Law:

$$\Delta_r U = \Sigma BDE_{\text{reactants}} + (-\Sigma BDE_{\text{products}})$$

Eq. 5.1, p. 98

In order to use bond dissociation energies, we have to be able to write the Lewis structures for all of the substances in the chemical equation, and we have to ignore the phase (solid, liquid, or gas) completely.

- c (speed of light in a vacuum, 2.99792458×10^8 m/s) [3.3]

The speed of light plays a role in relating energy to equivalent mass and wavelength to frequency.

Eq. 3.1, p. 48

$$E = h\nu \text{ (Planck, 1900)} = mc^2 \text{ (Einstein, 1905)}$$

Eq. 3.2, p. 48

$$\lambda\nu = c \quad \text{or} \quad \nu = c/\lambda$$

- C (heat capacity, J/K) [4.6, 4.9, 4.10]

For simple heating, as in the surroundings, the heat capacity relates the heat put in to the change in temperature:

Eq. 4.3, p. 80

$$q = C\Delta T$$

The higher the heat capacity, the more heat is required to change the temperature a given amount. If a gas is involved, then depending upon how a process is done, at constant volume or constant pressure, the heat capacity may be different. We write C_v and C_p, respectively. When more than one substance is being heated, the total heat capacity is the sum of the heat capacities of the individual substances being heated:

$$C = C_A + C_B + C_C + \cdots$$

Related also to heat capacity is *specific heat,* which for a pure substance is the amount of heat required *per gram* of substance to raise its temperature 1 K (or 1°C).

- C_p^n (the combination of n things taken p at a time) [2.5]

Eq. 2.1, p. 27

$$C_p^n = \frac{n(n-1)(n-2)\cdots(n-p+1)}{1\cdot 2\cdot 3\cdots\cdots p}$$

- d (distance, meters) [3.3–3.8]

Distance appears in the context of de Broglie's box, where it is in the denominator, indicating that an increase in d results in the energy levels becoming more closely spaced:

Eq. 3.7, p. 49

$$\varepsilon_n = \frac{h^2}{2m_e\lambda^2} = \frac{h^2}{2m_e(2d/n)^2}$$

$$= n^2\frac{h^2}{8}\left(\frac{1}{m_e d^2}\right)$$

where $n = 1, 2, 3, \ldots$

Here h is Planck's constant, $6.6260755 \times 10^{-34}$ joule-seconds, and m is the mass of the particle in the box. In the context of vibrations, d is the *average displacement* during the vibration, which is related to the quantum number, i; Planck's constant, h; the force constant, k_f; and the reduced mass, μ; as

$$d^2 = \left(i + \frac{1}{2}\right) \frac{h}{2\pi(\mu k_f)^{1/2}}$$

Eq. 3.12, p. 56

For a typical diatomic molecule such as HCl, d is on the order of 1/10 the length of the bond, but grows to be larger as more energy is put into the system.

- e (base of the natural logarithm system) [2.8, 7.2, 8.4, 11.2]

A simple definition of e can be given in terms of factorials:

$$e = 1/0! + 1/1! + 1/2! + 1/3! + \cdots = 2.7182818284590\ldots$$

The nunber e is a convenient number for expressing exponential relationships. If $y = e^x$ then $x = \ln y$. In thermodynamics, we see the number e used in two places. First, we see it in the Boltzmann law,

$$\frac{n_j}{n_i} = e^{-(\varepsilon_j - \varepsilon_i)/kT} = e^{-\Delta\varepsilon_{ij}/kT}$$

Eq. 2.3, p. 31

to relate the number of particles in one level to the number of particles in another. Secondly, we see it in relation to natural logs in Boltzmann's definition of entropy:

$$S = k \ln W$$

Eq. 7.3, p. 125

Ultimately, this ends up in the master equation for reaction Gibbs energy:

$$\Delta_r G = \Delta_r H^\circ - T[\Delta_r S^\circ - R \ln Q]$$

Eq. 11.1, p. 194

where it appears as part of the adjustment to standard entropy for nonstandard pressures of gases and concentrations of solutes.

- E (energy of a photon, joules) [3.3]

The energy associated with radiation (light, heat, sound, etc.) is related to frequency, ν, by Planck's constant, $h = 6.6260755 \times 10^{-34}$ joule-seconds, and to mass, m, by the speed of light, $c = 2.99792458 \times 10^8$ m/s:

$$E = h\nu \text{ (Planck, 1900)} = mc^2 \text{ (Einstein, 1905)}$$

Eq. 3.1, p. 48

- E_{cell} (the cell potential or electromotive force, volts or joules/coulomb) [13.6–13.12]

The cell potential is related to the difference in Gibbs energy for a chemical reaction as

$$-\Delta_r G = Q_{elec} \times E_{cell} = nFE_{cell}$$

Eq. 13.6, p. 243

or

$$\Delta_r G = -nFE_{cell}$$

or

$$E_{cell} = -\Delta_r G/nF$$

Here n is the number of electrons appearing in either of the balanced *half-equations* for the reaction and F is Faraday's constant, 96,485 coulombs/mol e$^-$. Including the idea of a standard state, and including the effect of pressures and concentrations on molar entropy of reaction, we get the Nernst equation:

Eq. 13.8, p. 249

$$-nFE_{cell} = -nFE^\circ_{cell,T} + RT \ln Q$$

or

$$E_{cell} = E^\circ_{cell,T} - \frac{RT}{nF} \ln Q$$

At equilibrium, when the cell is dead, we have $E_{cell} = 0$, and $Q = K$:

Eq. 13.7, p. 248

$$K = e^{-\Delta_r G^\circ_T / RT} = e^{nFE^\circ_{cell,T}/RT}$$

Thus, we can relate the *standard* cell potential to equilibrium constants for the reaction going on inside the cell.

It's a good idea to remember that the standard state does *not* include temperature. You can have a standard cell potential at any temperature; if gotten from tables of standard reduction potentials, then it is good only for 25°C. In that case, a common form of the Nernst equation is

Eq. 13.9, p. 249

$$E_{cell} = E^\circ_{cell,T} - \frac{RT}{nF} \ln Q$$

$$= E^\circ_{cell,298} - \frac{0.0592}{n} \log Q \quad \textbf{25 °C } \textit{only}$$

When calculating E_{cell} from standard reduction potentials, write the half-equations first. For each half-equation, if the electrons are on the left, that is reduction—leave the sign of the voltage as it is on the table. If the electrons are on the right, that is oxidation—reverse the sign of the voltage that is on the table and add the two numbers to get the cell potential, E_{cell}.

- E_i (molar energy of energy level i, joules/mole) [2.8]

Sometimes the energies of the levels in a Boltzmann distribution, ε_i, are put in terms of "per mole." In that case, we use E (at least in this book). Then the Boltzmann law takes the following form, where R is the ideal-gas constant:

Eq. 2.6, p. 33

$$\frac{n_j}{n_i} = e^{-(E_j - E_i)/RT} = e^{-\Delta E_{ij}/RT}$$

Here ΔE_{ij} is the difference in energy measured *from* level i *to* level j, $E_j - E_i$.

- F (Faraday's constant, 96,485 coulombs/mole e$^-$) [13.7]

Faraday's constant relates the number of moles of electrons to charge Q_{elec} (current times time) as

Eq. 13.5, p. 242

$$I \times t = Q_{elec} = n \times F$$

See Q_{elec}.

- G (Gibbs energy, joules) [10.2–10.10]

Gibbs energy is defined as

Eq. 10.4, p. 175

$$G = H - TS$$

And changes in Gibbs energy at constant temperature are then

$$\Delta G = \Delta H - \Delta(TS)$$

$$= \Delta H - T\Delta S \quad \text{(constant temperature only)}$$

Eq. 10.5, p. 175

Ultimately, this ends up to be the master equation for Gibbs energy differences between reactants and products in chemical reactions (based on a specific writing of the chemical equation):

$$\Delta_r G = \Delta_r H^{\circ} - T[\Delta_r S^{\circ} - R \ln Q]$$

Eq. 11.1, p. 194

Although the incessant increase in entropy of the universe is all we need to explain all the phenomena discussed in this book, a rearrangement of the Second Law including the surroundings gives, for constant pressure and temperature;

$$\Delta S_{universe} = \Delta S + \Delta S_{sur} = \Delta S - \frac{\Delta H}{T} > 0$$

Eq. 9.4, p. 161

and except in the case of electrochemistry, we aren't interested in the magnitude of the actual *change* in Gibbs energy for a reaction, merely its sign. When $\Delta G < 0$, then the reaction will go as written; when $\Delta G > 0$, then it will go *backward*.

When we consider the possibility of getting electrical work out of a chemical reaction, then we find that (Chapter 13)

$$-\Delta G = (-w_{elec})_{max} = Q_{elec} \times E_{cell}$$

Eq. 13.2, p. 241

where ΔG is the *change* in Gibbs energy for the reaction of a specific mass of reactants, Q_{elec} is the charge going through a wire, and E_{cell} is the cell potential (voltage). This change is specifically under the conditions of constant temperature and pressure.

When we consider a balanced chemical equation for which we can define the number of moles of electrons transferred per mole of reaction, n, this becomes

$$-\Delta_r G = Q_{elec} \times E_{cell} = nFE_{cell}$$

or

$$\Delta_r G = -nFE_{cell}$$

or

$$E_{cell} = -\Delta_r G/nF$$

Eq. 13.6, p. 243

which then leads to the Nernst equation (see E_{cell}).

- h (Planck's constant, $6.6260755 \times 10^{-34}$ joule-seconds) [3.3–3.8]

The constant relating frequency, v, to energy of *light*, in

$$E = hv \text{ (Planck, 1900)} = mc^2 \text{ (Einstein, 1905)}$$

Eq. 3.1, p. 48

The constant h also shows up wherever the energy of an energy level, ε, in a real molecular system needs calculation (see ε, Chapter 3 only).

- H (enthalpy, joules) [9.2, 13.3]

Enthalpy is defined as

$$H = U + PV$$

Eq. 9.1, p. 160

which is particularly meaningful for changes at constant pressure, in which case ΔH is the heat going into the system:

Eq. 9.2, p. 160

$$\Delta H = q$$

In short, at constant pressure, $-\Delta H / T$ gives us the entropy change of the surroundings, so that the Second Law of Thermodynamics relating the total entropy change of a *spontaneous* process, including the surroundings, is given in Section 9.2 as

Eq. 9.4, p. 161

$$\Delta S_{\text{universe}} = \Delta S + \Delta S_{\text{sur}} = \Delta S - \frac{\Delta H}{T} > 0$$

Since enthalpy is associated so closely with energy, we consider it to be related to bonding. Substances with strong bonds have low enthalpy; substances with weak bonds have high enthalpy.

- I (electrical current, amperes = coulombs/second) [13.6]

The rate of flow of charge through an electrical circuit. Multiplied by time gives charge:

Eq. 13.3, p. 241

$$Q_{\text{elec}} = I \times t$$

See Q_{elec}.

- k (Boltzmann's constant, 1.38066×10^{-23} joules/kelvin) [2.8, 7.2]

Boltzmann's constant is the scaling factor applied to temperature in the Boltzmann law:

Eq. 2.3, p. 31

$$\frac{n_j}{n_i} = e^{-(\varepsilon_j - \varepsilon_i)/kT} = e^{-\Delta \varepsilon_{ij}/kT}$$

It also relates the ideal-gas constant R to Avogadro's number, N_{av}:

Eq. 2.5, p. 33

$$k = \frac{R}{N_{\text{av}}} = \frac{8.31451 \, \text{J/mol} \cdot \text{K}}{6.02214 \times 10^{23}/\text{mol}} = 1.38066 \times 10^{-23} \, \text{J/K}$$

In thermodynamics k shows up everywhere we are talking about individual particles instead of moles. Especially, it relates the natural log of the thermodynamic probability of a state to that state's entropy:

Eq. 7.3, p. 125

$$S = k \ln W$$

- K or K_{eq} (the equilibrium constant, *unitless*) [1.13, 11.1]

The equilibrium constant arises from the relationship among $\Delta_r G$, $\Delta_r H^{\circ}$, $\Delta_r S^{\circ}$, and the reaction quotient, Q, for a chemical reaction, *based on a specific writing of the balanced chemical equation*:

Eq. 11.1, p. 194

$$\Delta_r G = \Delta_r H^{\circ} - T[\Delta_r S^{\circ} - R \ln Q]$$

When $\Delta_r G = 0$, conditions are such that the forward reaction is just as probable as the reverse reaction. The system is at equilibrium at some temperature T_{eq}. Rearranging this equation so that Q is on the left by itself gives

Eq. 11.2, p. 194

$$Q = K_{\text{eq}} = e^{\frac{-(\Delta_r H^{\circ} - T_{\text{eq}} \Delta_r S^{\circ})}{R T_{\text{eq}}}}$$

Thus, K is a function of the temperature and the *standard* differences for the reaction as written. More succinctly, we can write

$$\Delta_r S_{\text{universe}} = -\Delta_r G/T = R \ln \frac{K}{Q}$$

Eq. 11.9, p. 205

Clearly from this, only when $Q < K$ will $\Delta_r S_{\text{universe}}$ be positive and a chemical reaction observed be spontaneous.

- k_b (boiling point elevation constant, K·kg solvent/mol solute) [12.7]

A practical constant for the approximate determination of the increase in boiling point of a solution when it contains a nonvolatile solute. Multiplied by the *molality* of the solute (moles of solute per kg of solvent).

$$\Delta T_b \approx k_b\, m_X$$

Eq. 12.9, p. 227

From a theoretical point of view the value of k_b can be derived as

$$k_b \approx \frac{R T_{\text{bp}}}{\Delta_{\text{vap}} S^\circ (\text{mol}_{\text{liq}}/\text{kg}_{\text{liq}})}$$

Eq. 12.10, p. 228

where T_{bp} is the normal boiling point for the solvent. For water, the value of k_b is 0.52 with units "kelvins/(mol of solute per kg of water)."

- k_f (freezing point depression constant, K·kg solvent/mol solute) [12.7]

A practical constant for the approximate determination of the decrease in freezing point of a solution when it contains a nonvolatile solute. Multiplied by the *molality* of the solute (moles of solute per kg of solvent).

$$\Delta T_f \approx k_f\, m_X$$

Eq. 12.6, p. 226

The value of k_f can be derived as

$$k_f \approx \frac{R T_{\text{mp}}}{\Delta_{\text{fus}} S^\circ (\text{mol}_{\text{liq}}/\text{kg}_{\text{liq}})}$$

Eq. 12.7, p. 227

where T_{mp} is the normal melting point for the solvent. For water, the value of k_f is 1.86 in units of "kelvins/(moles of solute per kg of water)."

- k_f (spring force constant for a vibration, kg/s^2) [3.6]

The parameter k_f shows up in calculating the energy of the levels of vibrational energy for a diatomic molecule:

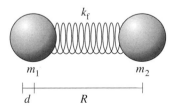

Eq. 3.13, p. 57

$$\varepsilon_i = \left(i + \frac{1}{2}\right) h\nu \quad \text{where } \nu = \frac{1}{2\pi}\sqrt{\frac{k_f}{\mu}} \text{ and } i = 0, 1, 2, \ldots \quad (\text{vibration})$$

where μ is the reduced mass, and h is Planck's constant, $6.6260755 \times 10^{-34}$ joule-seconds. It is related to the average displacement, d, as

Eq. 3.12, p. 56

$$d^2 = \left(i + \frac{1}{2}\right) \frac{h}{2\pi (\mu k_f)^{1/2}}$$

- m (mass, g or kg) [3.3–3.8]

The effect of increased mass on energy levels is to generally make the levels more closely spaced. For vibrations and rotations we need to compare *reduced mass*, μ, but for translation we use atomic mass. Although there is no obvious connection between mass and electronic energy level separations, it is generally observed that electronic energy levels get closer together as you go down the periodic table, so in a sense, even for electronic levels we see the same effect.

- m_e (mass of the electron, $9.1093897 \times 10^{-31}$ kg) [3.3, 3.5]

The mass of the electron shows up in the de Broglie equation:

Eq. 3.4, p. 48

$$\lambda = \frac{h}{m_e v_e} \quad \text{or} \quad v_e = \frac{h}{m_e \lambda} \quad \text{(de Broglie, 1924)}$$

and also in the equation for electronic energy:

Eq. 3.8, p. 51

$$\varepsilon_n = -\left(\frac{1}{n^2}\right) \frac{h^2}{8\pi^2} \left(\frac{1}{m_e a_0^2}\right) \quad \text{where } n = 1, 2, 3, \ldots \quad \text{(electronic; H atom only)}$$

- m_x (molality of substance "X", mol/kg of solvent) [12.7]

Used only in calculating freezing point depression and boiling point elevation. See k_b and k_f.

- n (moles, mol) [1.11, 4.7, 8.4, 8.6, 8.7, 13.7, 13.8, 13.12]

The number of moles of particles, atoms, molecules, or electrons. To convert to actual number, multiply by Avogadro's number, $N_{av} = 6.0221367 \times 10^{23}$/mol. Shows up as moles of gas in the ideal-gas equation:

$$PV = nRT$$

or in its Δ form, which helps for calculating work involved due to changes in the number of moles of gas in a chemical reaction carried out at constant pressure and temperature:

Eq. 4.6, p. 82

$$w = -P\Delta V = -(\Delta n)RT \quad \text{(constant temperature)}$$

The number n also shows up in calculating the change in entropy for a number of particles diffusing from one volume to another:

Eq. 8.4, p. 145

$$\Delta S = nR \ln \frac{V_2}{V_1}$$

or from one pressure to another:

Eq. 8.8, p. 147

$$\Delta S = nR \ln \frac{V_2}{V_1} = nR \ln \frac{P_1}{P_2} = -nR \ln \frac{P_2}{P_1}$$

or from one concentration to another:

Eq. 8.9, p. 147

$$\Delta S = nR \ln \frac{V_2}{V_1} = nR \ln \frac{[X]_1}{[X]_2} = -nR \ln \frac{[X]_2}{[X]_1}$$

It also shows up as the number of moles of electrons per mole of reaction (a unitless number) in the relation between reaction Gibbs energy and cell potential:

$$-\Delta_r G = Q_{elec} \times E_{cell} = nFE_{cell}$$

Eq. 13.6, p. 243

or

$$\Delta_r G = -nFE_{cell}$$

or

$$E_{cell} = -\Delta_r G/nF$$

and the Nernst Equation

$$-nFE_{cell} = -nFE^{\circ}_{cell,T} + RT \ln Q$$

Eq. 13.8, p. 249

or

$$E_{cell} = E^{\circ}_{cell,T} - \frac{RT}{nF} \ln Q$$

- n (quantum number for electronic energy) [3.5]

See ε (epsilon).

- n_i (number of particles in energy level i) [2.8, 2.12, 2.13]

When energy is quantized, we talk of energy "levels." A particle with a certain amount of energy ε_i is said to be "in" level i. The count of the number of particles in this particular energy state is n_i. The relative number of particles in one energy state (n_j) vs. another (n_i) is related to both the difference in energy between the two states, $\varepsilon_j - \varepsilon_i = \Delta\varepsilon_{ij}$, and the absolute temperature in kelvins as

$$\frac{n_j}{n_i} = e^{-(\varepsilon_j-\varepsilon_i)/kT} = e^{-\Delta\varepsilon_{ij}/kT}$$

Eq. 2.3, p. 31

Other valuable equations relating specifically to n particles in an evenly spaced energy system with level separation $\Delta\varepsilon$ include

$$n_{j \text{ or above}} = \frac{n_j}{n_0}n$$

Eq. 2.9, p. 38

$$n_0 = n\left(1 - e^{-\Delta\varepsilon/kT}\right)$$

Eq. 2.10, p. 38

$$f_{reactive} = \frac{n_{j \text{ or above}}}{n} = \frac{n_j}{n_0} = e^{-E_a/RT}$$

Eq. 2.11, p. 39

- n_x, n_y, n_z (quantum numbers for translational energy) [3.8]

See ε (epsilon).

- N_{av} (Avogadro's number, 6.0221367×10^{23}/mol) [2.8]

The number of particles in a standard amount called the *mole*. N_{av} shows up in thermodynamics in relating Boltzmann's constant, k, to the ideal-gas constant, R:

Eq. 2.5, p. 33
$$k = \frac{R}{N_{av}} = \frac{8.31451 \text{ J/mol} \cdot \text{K}}{6.02214 \times 10^{23}/\text{mol}} = 1.38066 \times 10^{-23} \text{ J/K}$$

- P (pressure, bar) [4.7, 8.3]

Pressure for an ideal gas is related to volume, number of moles of gas, and temperature as

$$PV = nRT$$

Pressure comes into play in thermodynamics in two places. First, the work done on a system when the external pressure is constant is calculated from the external pressure and the volume change as

Eq. 4.4, p. 82
$$\text{work} = \frac{\text{force}}{\text{area}} \times (\text{area} \times \text{distance})$$
$$w = -P\Delta V$$

Most importantly, the pressure of a gas has an effect on its entropy:

Eq. 8.8, p. 147
$$\Delta S = nR \ln \frac{V_2}{V_1} = nR \ln \frac{P_1}{P_2} = -nR \ln \frac{P_2}{P_1}$$

The term $R \ln P$ shows up as a downward adjustment to its *standard molar entropy*:

Eq. 8.1, p. 142
$$S_x = S_x^{\circ} - R \ln(P_x/\text{bar})$$

- P (probability) [1.3, 8.4]

The probability of an event is the ways of that happening over the total number of ways possible:

Eq. 1.1, p. 2
$$P_x = \frac{(\text{Ways of getting } x)}{(\text{Ways total})}$$

Probability is behind all physical processes, because all processes naturally proceed to the most probable distribution of energy for the universe. The probability of diffusion is behind the volume effect on the entropy of an ideal gas.

- q (heat, joules or "quanta") [4.3, 4.6, 7.1, 7.5, 9.1, 9.2]

The amount of energy put into a system in lifting its particles to higher energy levels. Heat, work, and internal energy U are related by the First Law of Thermodynamics:

Eq. 4.2, p. 80
$$\Delta U = q + w$$

which states that the energy put into a system is the sum of the heat put in plus the work put in. For simple heating, as in the surroundings, q is related to the temperature change via the heat capacity:

Eq. 4.3, p. 80
$$q = C\Delta T$$

Most importantly, for a reversible process, the entropy change of the surroundings is given by

$$\Delta S_{sur} = \frac{q_{sur}}{T}$$

Eq. 7.2, p. 124

In addition, at constant pressure, we have

$$\Delta H = q$$

Eq. 9.2, p. 160

This leads to the idea that the change in entropy of the surroundings for a chemical reaction when the pressure is constant is:

$$\Delta S_{sur} = \frac{q_{sur}}{T} = \frac{-q}{T} = \frac{-\Delta H}{T}$$

Eq. 9.3, p. 160

and provides the basis for understanding why reactions that release heat (make bonds, $\Delta H < 0$) are generally favorable: at least at low temperature their contribution to the entropy of the surroundings outweighs any entropy loss that may occur in the system due to the reaction itself.

- Q (reaction quotient, *unitless*) [1.13, 8.9, 11.2]

Considering gases and solutions to be ideal, the reaction quotient is written by taking products to their stoichiometric powers over reactants to their stoichiometric powers, using pressures in bar for gases and concentrations in M for solutes. Liquids and solids are ignored. Q derives from the effect of pressure and concentration on entropy, and the way logarithms combine when we calculate the difference in molar entropy between products and reactants in a reaction as written using a specific equation. For example, for the general case

$$a\text{A} + b\text{B} \longrightarrow c\text{C} + d\text{D}$$

we have

$$\Delta_r S = \Delta_r S^\circ - R \ln Q$$

Eq. 8.10, p. 150

where we define the reaction quotient, Q, to be

$$Q = \frac{C^c D^d}{A^a B^b}$$

The numbers for A, B, C, and D here would be pressures in bar for gases and concentrations in mol/L for solutes. Combining this with the definition of Gibbs energy, we get the master equation for reaction Gibbs energy:

$$\Delta_r G = \Delta_r H^\circ - T[\Delta_r S^\circ - R \ln Q]$$

Eq. 11.1, p. 194

When $\Delta_r G = 0$, the system is at equilibrium; in that case $Q = K$. When a reaction is done electrochemically, then this becomes the Nernst equation (see E_{elec}).

- Q_{elec} (electrical charge, coulombs) [13.7]

Charge in chemistry is another way of referring to electrons. Charge can be defined in terms of electrical current and time or in terms of number of electrons:

$$I \times t = Q_{elec} = n \times F$$

Eq. 13.5, p. 242

where Q_{elec} is the amount of charge passed through the wire in coulombs; I is the current in amperes (coulombs/second); t is the time in seconds; n is the number of moles of electrons relocated; and F is Faraday's constant, 96,485 coulombs/mole e$^-$.

- R (ideal-gas constant, 8.31451 J/mol·K = 0.0831451 L·bar/mol·K = 0.0820578 L·atm/mol·K) [2.8, 4.7, 8.3]

The ideal-gas constant, R, is really Avogadro's number times Boltzmann's constant:

$$R = N_{av}k$$

It shows up in the ideal-gas equation,

$$PV = nRT$$

and in the effect of pressure and concentration on molar entropy (Equations 8.1 and 8.2):

Eq. 8.1, p. 142

$$S_x = S_x^\circ - R\ln(P_x/\text{bar})$$

Eq. 8.2, p. 142

$$S_x = S_x^\circ - R\ln([X]/M)$$

The ideal-gas constant generally shows up multiplied by temperature, as in:

Eq. 11.1, p. 194

$$\Delta_r G = \Delta_r H^\circ - T[\Delta_r S^\circ - R\ln Q]$$

Eq. 11.2, p. 194

$$Q = K_{eq} = e^{\frac{-(\Delta_r H^\circ - T_{eq}\Delta_r S^\circ)}{RT_{eq}}}$$

- R (interatomic distance, m or pm) [3.7]

R is used as the variable indicating the interatomic distance in a rigid rotor when calculating the energy levels of rotation:

Eq. 3.14, p. 59

$$\varepsilon_i = i(i+1)\frac{h^2}{8\pi^2}\left(\frac{1}{\mu R^2}\right) \quad \text{where } i = 0, 1, 2, \ldots \quad (\text{rotation})$$

- S (entropy, joules/kelvin) [everywhere beginning in 7.1]

Entropy is clearly the most important concept in this book. Entropy defined in the language of Boltzmann is

Eq. 7.3, p. 125

$$S = k\ln W$$

where k is Boltzmann's constant, 1.38066×10^{-23} joule/kelvin, and W is the thermodynamic probability. Changes in entropy then become

Eq. 7.4, p. 126

$$\Delta S = S_2 - S_1 = k \ln W_2 - k \ln W_1 = k \ln \frac{W_2}{W_1}$$

Due to the volume effect on entropy, the entropy of a gas depends upon its pressure, and the entropy of a solute depends upon its concentration:

$$\Delta S = nR \ln \frac{V_2}{V_1}$$

Eq. 8.4, p. 145

$$\Delta S = nR \ln \frac{V_2}{V_1} = nR \ln \frac{P_1}{P_2} = -nR \ln \frac{P_2}{P_1}$$

Eq. 8.8, p. 147

$$\Delta S = nR \ln \frac{V_2}{V_1} = nR \ln \frac{[X]_1}{[X]_2} = -nR \ln \frac{[X]_2}{[X]_1}$$

Eq. 8.9, p. 147

We define a standard state as 1 bar pressure for gases and 1 mol/L concentration for solutes and write (Equations 8.1 and 8.2):

$$S_x = S_x^\circ - R \ln(P_x/\text{bar})$$

Eq. 8.1, p. 142

$$S_x = S_x^\circ - R \ln([X]/M)$$

Eq. 8.2, p. 142

Entropy rules the universe:

$$\Delta S_{\text{universe}} = \Delta S + \Delta S_{\text{sur}} > 0$$

Eq. 7.1, p. 124

for all spontaneous processes. The problem is, this includes surroundings. To account for the surroundings, we use enthalpy and write, for constant-pressure changes,

$$\Delta S_{\text{sur}} = \frac{q_{\text{sur}}}{T} = \frac{-q}{T} = \frac{-\Delta H}{T}$$

Eq. 9.3, p. 160

This allows us to define Gibbs energy and write, for a specific writing of a balanced chemical equation,

$$\Delta_r G = \Delta_r H^\circ - T[\Delta_r S^\circ - R \ln Q]$$

Eq. 11.1, p. 194

where Q is the reaction quotient. The value for reaction Gibbs energy thus obtained is a general measure of the *potential* for the entropy change in the universe due to a chemical reaction under any conditions. Ultimately, we can write

$$\Delta_r S_{\text{universe}} = -\Delta_r G/T = R \ln \frac{K}{Q}$$

Eq. 11.9, p. 205

- T (absolute temperature, kelvin) [2.8, 2.9]

Absolute 0 is defined as the hypothetical temperature at which all substances exist only in their overall ground state, with all particles in the lowest energy level. This derives from the Boltzmann law:

$$\frac{n_j}{n_i} = e^{-(\varepsilon_j - \varepsilon_i)/kT} = e^{-\Delta\varepsilon_{ij}/kT}$$

Eq. 2.3, p. 31

where n_j/n_i approaches 0 as T approaches absolute zero. $0°C = 273.15$ K. The effect of increasing the temperature of a system is to put more particles in upper energy levels at the expense of particles in lower levels.

• U (internal energy, joules) [4.1]

Internal energy is defined as the sum of all the energies of all the particles in all the levels of a system:

Eq. 4.1, p. 77

$$U = n_0\varepsilon_0 + n_1\varepsilon_1 + n_2\varepsilon_2 + n_3\varepsilon_3 + \cdots = \sum_{i=0}^{\infty} n_i\varepsilon_i$$

It is related to heat and work by the First Law of Thermodynamics:

Eq. 4.2, p. 80

$$\Delta U = q + w$$

That is, the change in internal energy of a system is the sum of the heat put into the system and the work put into the system.

• w (work, joules) [4.3, 4.7, 4.8, 4.12, 13.3]

The two kinds of work we deal with here are pressure–volume work:

Eq. 4.4, p. 82

$$\text{work} = \frac{\text{force}}{\text{area}} \times (\text{area} \times \text{distance})$$

$$w = -P\Delta V$$

and electrical work:

Eq. 13.2, p. 241

$$-\Delta G = (-w_{\text{elec}})_{\text{max}} = Q_{\text{elec}} \times E_{\text{cell}}$$

Work due to volume changes can be largely ignored in relation to chemistry because for most chemical reactions the change in energy in the form of heat due to bond making and breaking is far more important. More importantly in terms of chemistry, electrical work provides the basis for relating Gibbs energy to cell potential, E_{cell}. There are other kinds of work as well, but they aren't discussed in this book.

A handy expression for ideal gases that allows us to determine the work due to changes in numbers of moles of gas in a chemical reaction at constant pressure and temperature is

Eq. 4.6, p. 82

$$w = -P\Delta V = -(\Delta n)RT \quad \text{(constant temperature)}$$

• W (thermodynamic probability) [2.4, 2.11, 7.2, 7.3]

The thermodynamic probability of a state is the count of all the ways to be in that state. For a system of particles with distributed energy such that n_i particles are in energy level i, we have

Eq. 2.2, p. 27

$$W = \frac{n!}{n_0! \, n_1! \, n_2! \cdots}$$

Thermodynamic probability is related to entropy by taking the natural logarithm and multiplying by Boltzmann's constant, k:

Eq. 7.3, p. 125

$$S = k \ln W$$

where k is 1.38066×10^{-23} joules/kelvin. Changes in entropy then become

Eq. 7.4, p. 126

$$\Delta S = S_2 - S_1 = k \ln W_2 - k \ln W_1 = k \ln \frac{W_2}{W_1}$$

and, ultimately, our whole idea of an equilibrium constant is fundamentally tied to the relative thermodynamic probabilities through its relationship to the *standard* change in entropy of the universe:

$$K = e^{\Delta_r S^o_{universe}/R}$$

Eq. 11.10, p. 205

- x (mole fraction) [8.5, 12.7]

The fraction of total moles that are of type A is

$$x_A = \frac{n_A}{n}$$

Eq. 8.6, p. 146

Mole fractions show up two places in this book. First, they show up in the entropy of mixing:

$$\Delta S_{mix} = nR \left[x_A \ln \frac{1}{x_A} + x_B \ln \frac{1}{x_B} \right]$$

Eq. 8.7, p. 146

Then they show up again in Chapter 12, in the reaction quotient, Q, when discussing the increase in entropy of the liquid due to mixing in relation to freezing point depression, boiling point elevation, and the vapor pressures of solutions (Raoult's law):

$$S_{liq} = S^o_{liq} + R \ln \left(\frac{n_{H_2O} + n_A}{n_{H_2O}} \right)$$

Eq. 12.4, p. 225

$$= S^o_{liq} + R \ln \left(\frac{1}{x_{H_2O}} \right)$$

$$= S^o_{liq} - R \ln x_{H_2O}$$

Mathematical Tricks

The following relationships are good to know. They are used in various places throughout this book.

Relationship	Example
$n! = 1 \cdot 2 \cdot 3 \cdot \cdots \cdot n$	$5! = 1 \cdot 2 \cdot 3 \cdot 4 \cdot 5 = 120$
$\ln(xy) = \ln(x) + \ln(y)$	$\ln(6) = \ln(2) + \ln(3)$
$\ln(x^a) = a \ln x$	$\ln(4) = 2\ln(2)$
$\ln(1/x) = -\ln x$	$\ln(1/2) = -\ln(2)$
$e^{(x+y)} = e^x e^y$	$e^5 = (e^2)(e^3)$
$e^{xy} = (e^x)^y$	$e^6 = (e^2)^3$
$e^{-x} = 1/e^x$	$e^{-2} = 1/e^2$
$\ln(n!) \approx n \ln n - n$	$\ln(100!) \approx 100 \ln 100 - 100$
$\ln(1+x) \approx x$ when $x \ll 1$	$\ln(1.001) \approx 0.001$
$\ln(x) \approx 2.303 \log_{10}(x)$	$\ln(100) \approx 2.303 \log_{10}(100) = 2.303 \times 2 = 4.606$

Special Numbers
$\ln(e) = 1$
$\log_{10}(10) = 1$
$e^x = 1 + x/1! + x^2/2! + x^3/3! + \cdots$
$e^0 = 1$
$e^{-x/T}$ approaches $e^{-\infty} = 0$ as T approaches 0
$e^{-x/T}$ approaches $e^0 = 1$ as T approaches infinity

Table of Standard Reduction Potentials

Reactions in Acidic Aqueous Solution at 25°C		Standard Reduction Potential, $E°$/volts
$F_2(g) + 2\,e^-$	$\longrightarrow 2\,F^-(aq)$	2.87
$O_3(g) + 2\,H^+(aq) + 2\,e^-$	$\longrightarrow O_2(g) + H_2O$	2.07
$Co^{3+}(aq) + e^-$	$\longrightarrow Co^{2+}(aq)$	1.842
$Pb^{4+}(aq) + 2\,e^-$	$\longrightarrow Pb^{2+}(aq)$	1.8
$H_2O_2(aq) + 2\,H^+(aq) + 2\,e^-$	$\longrightarrow 2\,H_2O$	1.776
$N_2O(g) + 2\,H^+(aq) + 2\,e^-$	$\longrightarrow N_2(g) + H_2O$	1.77
$NiO_2(s) + 4\,H^+(aq) + 2\,e^-$	$\longrightarrow Ni^{2+}(aq) + 2\,H_2O$	1.7
$PbO_2(aq) + HSO_4^-(aq) + 3\,H^+(aq) + 2\,e^-$	$\longrightarrow PbSO_4(s) + 2\,H_2O$	1.685
$Au^+(aq) + e^-$	$\longrightarrow Au(s)$	1.68
$MnO_4^-(aq) + 4\,H^+(aq) + 3\,e^-$	$\longrightarrow MnO_2(s) + 2\,H_2O$	1.679
$2\,HClO\,(aq) + 2\,H^+(aq) + 2\,e^-$	$\longrightarrow Cl_2(g) + 2\,H_2O$	1.63
$Ce^{4+}(aq) + e^-$	$\longrightarrow Ce^{3+}(aq)$	1.61
$NaBiO_3(s) + 6\,H^+(aq) + 2\,e^-$	$\longrightarrow Bi^{3+}(aq) + Na^+(aq) + 3\,H_2O$	≈ 1.6
$Au^{3+}(aq) + 3\,e^-$	$\longrightarrow Au(s)$	1.50
$MnO_4^-(aq) + 8\,H^+(aq) + 5\,e^-$	$\longrightarrow Mn^{2+}(aq) + 4\,H_2O$	1.491
$2\,ClO_3^-(aq) + 12\,H^+(aq) + 10\,e^-$	$\longrightarrow Cl_2(g) + 6\,H_2O$	1.47
$BrO_3^-(aq) + 6\,H^+(aq) + 6\,e^-$	$\longrightarrow Br^-(aq) + 3\,H_2O$	1.44
$Cl_2(g) + 2\,e^-$	$\longrightarrow 2\,Cl^-(aq)$	1.3583
$Cr_2O_7^{2-}(aq) + 14\,H^+(aq) + 6\,e^-$	$\longrightarrow 2\,Cr^{3+}(aq) + 7\,H_2O$	1.33
$N_2H_5^+(aq) + 3\,H^+(aq) + 2\,e^-$	$\longrightarrow 2\,NH_4^+(aq)$	1.275
$MnO_2(s) + 4\,H^+(aq) + 2\,e^-$	$\longrightarrow Mn^{2+}(aq) + 2\,H_2O$	1.229
$O_2(g) + 4\,H^+(aq) + 4\,e^-$	$\longrightarrow 2\,H_2O$	1.229
$2\,IO_3^-(aq) + 12\,H^+(aq) + 10\,e^-$	$\longrightarrow I_2(aq) + 6\,H_2O$	1.195
$ClO_4^-(aq) + 2\,H^+(aq) + 2\,e^-$	$\longrightarrow ClO_3^-(aq) + H_2O$	1.19
$Pt^{2+}(aq) + 2\,e^-$	$\longrightarrow Pt(s)$	1.188
$Br_2(l) + 2\,e^-$	$\longrightarrow 2\,Br^-(aq)$	1.065
$AuCl_4^-(aq) + 3\,e^-$	$\longrightarrow Au(s) + 4\,Cl^-(aq)$	1.00
$NO_3^-(aq) + 4\,H^+(aq) + 3\,e^-$	$\longrightarrow NO(g) + 2\,H_2O$	0.96
$NO_3^-(aq) + 3\,H^+(aq) + 2\,e^-$	$\longrightarrow HNO_2(aq) + H_2O$	0.94
$Pd^{2+}(aq) + 2\,e^-$	$\longrightarrow Pd(s)$	0.915
$2\,Hg^{2+}(aq) + 2\,e^-$	$\longrightarrow Hg_2^{2+}(aq)$	0.908
$Hg^{2+}(aq) + 2\,e^-$	$\longrightarrow Hg(l)$	0.854
$2\,NO_3^-(aq) + 4\,H^+(aq) + 2\,e^-$	$\longrightarrow N_2O_4(g) + 2\,H_2O$	0.80
$Ag^+(aq) + e^-$	$\longrightarrow Ag(s)$	0.7994
$Hg_2^{2+}(aq) + 2\,e^-$	$\longrightarrow 2\,Hg(l)$	0.792
$Fe^{3+}(aq) + e^-$	$\longrightarrow Fe^{2+}(aq)$	0.770
$SbCl_6^-(aq) + 2\,e^-$	$\longrightarrow SbCl_4^-(aq) + 2\,Cl^-(aq)$	0.75
$PtCl_4^{2+}(aq) + 2\,e^-$	$\longrightarrow Pt(s) + 4\,Cl^-(aq)$	0.73

Reactions in Acidic Aqueous Solution at 25°C		Standard Reduction Potential, $E°$/volts
$O_2(g) + 2\,H^+(aq) + 2\,e^-$	$\longrightarrow H_2O_2(aq)$	0.682
$PtCl_6^{2-}(aq) + 2\,e^-$	$\longrightarrow PtCl_4^{2-}(aq) + 2\,Cl^-(aq)$	0.68
$H_3AsO_4(aq) + 2\,H^+(aq) + 2\,e^-$	$\longrightarrow H_3AsO_3(aq) + H_2O$	0.58
$I_2(s) + 2\,e^-$	$\longrightarrow 2\,I^-(aq)$	0.535
$TeO_2(s) + 4\,H^+(aq) + 4\,e^-$	$\longrightarrow Te(s) + 2\,H_2O$	0.529
$Cu^+(aq) + e^-$	$\longrightarrow Cu(s)$	0.518
$RhCl_6^{3-}(aq) + 3\,e^-$	$\longrightarrow Rh(s) + 6\,Cl^-(aq)$	0.44
$Cu^{2+}(aq) + 2\,e^-$	$\longrightarrow Cu(s)$	0.337
$Hg_2Cl_2(s) + 2\,e^-$	$\longrightarrow 2\,Hg\,(l) + 2\,Cl^-(aq)$	0.268
$AgCl(s) + e^-$	$\longrightarrow Ag(s) + Cl^-(aq)$	0.222
$SO_4^{2-}(aq) + 4\,H^+(aq) + 2\,e^-$	$\longrightarrow SO_2(g) + 2\,H_2O$	0.20
$SO_4^{2-}(aq) + 4\,H^+(aq) + 2\,e^-$	$\longrightarrow H_2SO_3(aq) + H_2O$	0.17
$Cu^{2+}(aq) + e^-$	$\longrightarrow Cu^+(aq)$	0.159
$Sn^{4+}(aq) + 2\,e^-$	$\longrightarrow Sn^{2+}(aq)$	0.154
$S_8(s) + 16\,H^+(aq) + 16\,e^-$	$\longrightarrow 8\,H_2S(aq)$	0.141
$AgBr(s) + e^-$	$\longrightarrow Ag(s) + Br^-(aq)$	0.0713
$2\,H^+(aq) + 2\,e^-$	$\longrightarrow H_2(g)$	0
$N_2O(g) + 6\,H^+(aq) + H_2O + 4\,e^-$	$\longrightarrow 2\,NH_3OH^+(aq)$	−0.05
$Pb^{2+}(aq) + 2\,e^-$	$\longrightarrow Pb(s)$	−0.1263
$Sn^{2+}(aq) + 2\,e^-$	$\longrightarrow Sn(s)$	−0.1364
$AgI(s) + e^-$	$\longrightarrow Ag(s) + I^-(aq)$	−0.152
$Ni^{2+}(aq) + 2\,e^-$	$\longrightarrow Ni(s)$	−0.231
$SnF_6^{2-}(aq) + 4\,e^-$	$\longrightarrow Sn(s) + 6\,F^-(aq)$	−0.25
$Co^{2+}(aq) + 2\,e^-$	$\longrightarrow Co(s)$	−0.277
$Tl^+(aq) + e^-$	$\longrightarrow Tl(s)$	−0.336
$PbSO_4(s) + 2\,e^-$	$\longrightarrow Pb(s) + SO_4^{2-}(aq)$	−0.355
$Se(s) + 2\,H^+(aq) + 2\,e^-$	$\longrightarrow H_2Se(g)$	−0.37
$Cd^{2+}(aq) + 2\,e^-$	$\longrightarrow Cd(s)$	−0.402
$Cr^{3+}(aq) + e^-$	$\longrightarrow Cr^{2+}(aq)$	−0.41
$Fe^{2+}(aq) + 2\,e^-$	$\longrightarrow Fe(s)$	−0.440
$2\,CO_2(g) + 2\,H^+(aq) + 2\,e^-$	$\longrightarrow H_2C_2O_4(aq)$	−0.49
$Ga^{3+}(aq) + 3\,e^-$	$\longrightarrow Ga(s)$	−0.53
$HgS(s) + 2\,H^+(aq) + 2\,e^-$	$\longrightarrow Hg(l) + H_2S(g)$	−0.72
$Cr^{3+}(aq) + 3\,e^-$	$\longrightarrow Cr(s)$	−0.74
$Zn^{2+}(aq) + 2\,e^-$	$\longrightarrow Zn(s)$	−0.7628
$Cr^{2+}(aq) + 2\,e^-$	$\longrightarrow Cr(s)$	−0.91
$FeS(s) + 2\,e^-$	$\longrightarrow Fe(s) + S^{2-}(aq)$	−1.01
$V^{2+}(aq) + 2\,e^-$	$\longrightarrow V(s)$	−1.18
$Mn^{2+}(aq) + 2\,e^-$	$\longrightarrow Mn(s)$	−1.182
$CdS(s) + 2\,e^-$	$\longrightarrow Cd(s) + S^{2-}(aq)$	−1.21
$ZnS(s) + 2\,e^-$	$\longrightarrow Zn(s) + S^{2-}(aq)$	−1.44
$Zr^{4+}(aq) + 4\,e^-$	$\longrightarrow Zr(s)$	−1.539
$Al^{3+}(aq) + 3\,e^-$	$\longrightarrow Al(s)$	−1.66
$Mg^{2+}(aq) + 2\,e^-$	$\longrightarrow Mg(s)$	−2.375
$Na^+(aq) + e^-$	$\longrightarrow Na(s)$	−2.711
$Ca^{2+}(aq) + 2\,e^-$	$\longrightarrow Ca(s)$	−2.868
$Sr^{2+}(aq) + 2\,e^-$	$\longrightarrow Sr(s)$	−2.886
$Ba^{2+}(aq) + 2\,e^-$	$\longrightarrow Ba(s)$	−2.912
$Rb^+(aq) + e^-$	$\longrightarrow Rb(s)$	−2.924
$K^+(aq) + e^-$	$\longrightarrow K(s)$	−2.925
$Li^+(aq) + e^-$	$\longrightarrow Li(s)$	−3.045

Reactions in Basic Aqueous Solution at 25°C	Standard Reduction Potential, $E°$/volts
$ClO^-(aq) + H_2O + 2\,e^- \longrightarrow Cl^-(aq) + 2\,OH^-(aq)$	0.89
$HOO^-(aq) + H_2O + 2\,e^- \longrightarrow 3\,OH^-(aq)$	0.88
$2NH_2OH(aq) + 2\,e^- \longrightarrow N_2H_4(aq) + 2\,OH^-(aq)$	0.74
$ClO_3^-(aq) + 3\,H_2O + 6\,e^- \longrightarrow Cl^-(aq) + 6\,OH^-(aq)$	0.62
$MnO_4^-(aq) + 2\,H_2O + 3\,e^- \longrightarrow MnO_2(s) + 4\,OH^-(aq)$	0.588
$MnO_4^-(aq) + e^- \longrightarrow MnO_4^{2-}(aq)$	0.564
$NiO_2(s) + 2\,H_2O + 2\,e^- \longrightarrow Ni(OH)_2(s) + 2\,OH^-(aq)$	0.49
$Ag_2CrO_4(s) + 2\,e^- \longrightarrow 2\,Ag(s) + CrO_4^{2-}(aq)$	0.446
$O_2(g) + 2\,H_2O + 4\,e^- \longrightarrow 4\,OH^-(aq)$	0.40
$ClO_4^-(aq) + H_2O + 2\,e^- \longrightarrow ClO_3^-(aq) + 2\,OH^-(aq)$	0.36
$Ag_2O(s) + H_2O + 2\,e^- \longrightarrow 2\,Ag(s) + 2\,OH^-(aq)$	0.34
$2\,NO_2^-(aq) + 3\,H_2O + 4\,e^- \longrightarrow N_2O\,(g) + 6\,OH^-(aq)$	0.15
$N_2H_4(aq) + 2\,H_2O + 2\,e^- \longrightarrow 2\,NH_3(aq) + 2\,OH^-(aq)$	0.10
$Co(NH_3)_6^{3+}(aq) + e^- \longrightarrow Co(NH_3)_6^{2+}(aq)$	0.10
$HgO(s) + H_2O + 2\,e^- \longrightarrow Hg\,(l) + 2\,OH^-(aq)$	0.0984
$O_2(aq) + H_2O + 2\,e^- \longrightarrow HOO^-(aq) + OH^-(aq)$	0.076
$NO_3^-(aq) + H_2O + 2\,e^- \longrightarrow NO_2^-(aq) + 2\,OH^-(aq)$	0.01
$MnO_2(s) + 2\,H_2O + 2\,e^- \longrightarrow Mn(OH)_2(s) + 2\,OH^-(aq)$	−0.05
$CrO_4^{2-}(aq) + 4\,H_2O + 3\,e^- \longrightarrow Cr(OH)_3(s) + 5\,OH^-(aq)$	−0.12
$Cu(OH)_2(s) + 2\,e^- \longrightarrow Cu(s) + 2\,OH^-(aq)$	−0.36
$S_8(s) + 16\,e^- \longrightarrow 8\,S^{2-}(aq)$	−0.48
$Fe(OH)_3(s) + e^- \longrightarrow Fe(OH)_2(s) + OH^-(aq)$	−0.56
$2\,H_2O + 2\,e^- \longrightarrow H_2(g) + 2\,OH^-(aq)$	−0.8277
$2\,NO_3^-(aq) + 2\,H_2O + 2\,e^- \longrightarrow N_2O_4(g) + 4\,OH^-(aq)$	−0.85
$Fe(OH)_2(s) + 2\,e^- \longrightarrow Fe(s) + 2\,OH^-(aq)$	−0.877
$SO_4^{2-}(aq) + H_2O + 2\,e^- \longrightarrow SO_3^{2-}(aq) + 2\,OH^-(aq)$	−0.93
$N_2(g) + 4\,H_2O + 4\,e^- \longrightarrow N_2H_4(aq) + 4\,OH^-(aq)$	−1.15
$Zn(OH)_4^{2-}(aq) + 2\,e^- \longrightarrow Zn(s) + 4\,OH^-(aq)$	−1.22
$Zn(OH)_2(s) + 2\,e^- \longrightarrow Zn(s) + 2\,OH^-(aq)$	−1.245
$Zn(CN)_4^{2-}(aq) + 2\,e^- \longrightarrow Zn(s) + 4\,CN^-(aq)$	−1.26
$Cr(OH)_3(s) + 3\,e^- \longrightarrow Cr(s) + 3\,OH^-(aq)$	−1.30
$SiO_3^{2-}(aq) + 3\,H_2O + 4\,e^- \longrightarrow Si(s) + 6\,OH^-(aq)$	−1.70

Table of Standard Thermodynamic Data (25°C and 1 bar)

Species	$\Delta_f H°$ kJ/mol	$S°$ J/mol·K	$\Delta_f G°$ (note [a]) kJ/mol
Silver			
Ag(s)	0	42.55	0
Ag$_2$CO$_3$(s)	−505.8	167.4	−436.8
Ag$_2$O(s)	−31.05	121.3	−11.2
AgCl(s)	−127.068	96.2	−109.789
AgBr(s)	−100.37	107.1	−96.9
AgI(s)	−61.83	115.5	−66.19
AgNO$_3$(s)	−124.39	140.92	−33.47
Aluminum			
Al(s)	0	28.33	0
AlCl$_3$(s)	−704.2	110.67	−628.8
Al$_2$O$_3$(s)	−1675.7	50.92	−1582.3
Argon			
Ar(g)	0	154.843	0
Boron			
B(s)	0	5.86	0
BF$_3$(g)	−1137	254.01	−1120.35
Barium			
Ba(s)	0	62.8	0
BaCO$_3$(s)	−1216.3	112.1	−1137.6
BaCl$_2$(s)	−858.6	123.68	−810.4
BaO(s)	−553.5	70.42	−525.1
BaSO$_4$(s)	−1473.2	132.2	−1362.2
Beryllium			
Be(s)	0	9.5	0
BeO(s)	−609.6	14.14	−580.3
Bromine			
Br(g)	111.88	174.91	82.429
Br$_2$(l)	0	152.231	0
Br$_2$(g)	30.907	245.463	3.11
HBr(g)	−36.4	198.695	−53.45
Carbon			
C(s, graphite)	0	5.74	0
C(s, diamond)	1.895	2.377	2.9
C(g)	716.682	158.096	671.257
CO(g)	−110.525	197.674	−137.168

Species	$\Delta_f H°$ kJ/mol	$S°$ J/mol·K	$\Delta_f G°$ (note [a]) kJ/mol
Carbon			
$CO_2(g)$	−393.509	213.74	−394.359
$COCl_2(g$, phosgene)	−218.8	283.53	−204.6
$CH_3Cl(g)$	−80.83	234.58	−57.37
$CHCl_3(g)$	−103.14	295.71	−70.34
$CCl_4(g)$	−102.9	309.85	−60.59
$CCl_4(l)$	−135.44	216.4	−65.27
$HCN(g)$	135.1	201.78	124.7
$CH_4(g$, methane)	−74.81	186.264	−50.72
$C_2H_2(g$, acetylene)	226.73	200.94	209.2
$C_2H_4(g$, ethene)	52.25	219.45	68.12
$C_2H_6(g$, ethane)	−84.68	229.6	−32.82
$C_3H_8(g$, propane)	−103.8	269.9	−23.49
$C_6H_6(l$, benzene)	49.03	172.8	124.5
$CH_3OH(g$, methanol)	−200.66	239.7	−162
$CH_3OH(l$, methanol)	−238.66	126.8	−166.36
$C_2H_5OH(g$, ethanol)	−235.1	282.7	−168.49
$C_2H_5OH(l$, ethanol)	−277.69	160.7	−174.78
Calcium			
$Ca(s)$	0	41.42	0
$Ca(g)$	178.2	158.884	144.3
$CaC_2(s)$	−59.8	69.96	−64.9
$CaCO_3(s$, calcite)	−1206.92	92.9	−1128.79
$CaCO_3(s$, aragonite)	−1207.13	88.7	−1127.75
$CaCl_2(s)$	−795.8	104.6	−748.1
$CaF_2(s)$	−1219.6	68.87	−1167.3
$CaH_2(s)$	−186.2	42	−147.2
$CaO(s)$	−635.09	39.75	−604.03
$Ca(OH)_2(s)$	−986.09	83.39	−898.49
$Ca(OH)_2(aq)$	−1002.82	−74.5	−868.07
$CaSO_4(s)$	−1434.11	106.7	−1321.79
Chlorine			
$Cl(g)$	121.679	165.198	105.68
$Cl^-(g)$	−233.13	—	—
$Cl_2(g)$	0	223.066	0
$ClO(g)$	101.219	226.65	97.48
$ClO_2(g)$	102.5	256.84	120.5
Cesium			
$Cs(s)$	0	85.23	0
$Cs^+(g)$	457.964	—	—
$CsCl(s)$	−443.04	101.17	−414.53
Copper			
$Cu(s)$	0	33.15	0
$CuO(s$, tenorite)	−157.3	42.63	−129.7
$Cu_2O(s$, cuprite)	−168.6	93.14	−146
$Cu(OH)_2(s)$	−449.8	—	—
$CuS(s$, covellite)	−53.1	66.5	−53.6
$Cu_2S(s$, chalcocite)	−79.5	120.9	−86.2

Species	$\Delta_f H°$ kJ/mol	$S°$ J/mol·K	$\Delta_f G°$ (note [a]) kJ/mol
Fluorine			
$F(g)$	78.99	158.754	61.91
$F^-(g)$	−255.39	—	—
$F^-(aq)$	−332.63	−13.8	−278.79
$F_2(g)$	0	202.78	0
$HF(g)$	−271.1	173.779	−273.2
Iron			
$Fe(s)$	0	27.28	0
$FeCl_2(s)$	−341.79	117.95	−302.30
$FeCl_3(s)$	−399.49	142.3	−344.00
$Fe(CO)_5(l)$	−774.0	338.1	−705.3
$Fe_2O_3(s, \text{hematite})$	−824.2	87.4	−742.2
$Fe_3O_4(s, \text{magnetite})$	−1118.4	146.4	−1015.4
$Fe(OH)_3(s)$	−823	106.7	−696.5
$FeS_2(s, \text{pyrite})$	−178.2	52.93	−166.9
Hydrogen			
$H(g)$	217.965	114.713	203.247
$H_2(g)$	0	130.684	0
$H^+(g)$	1536.202	—	—
$HF(g)$	−271.1	173.779	−273.2
$HBr(g)$	−36.4	198.695	−53.45
$HCl(g)$	−92.307	186.908	−95.299
$HCN(g)$	135.1	201.78	124.7
$HI(g)$	26.48	206.594	1.7
$H_2O(g)$	−241.818	188.825	−228.572
$H_2O(l)$	−285.83	69.91	−237.129
$H_2O_2(g)$	−136.31	232.7	−105.57
$H_2O_2(l)$	−187.78	109.6	−120.35
$H_2S(g)$	−20.63	205.79	−33.56
Helium			
$He(g)$	0	126.15	0
Mercury			
$Hg(l)$	0	76.02	0
$HgCl_2(s)$	−224.3	146	−178.6
$Hg_2Cl_2(s)$	−265.22	192.5	−210.745
$HgO(s, \text{red})$	−90.83	70.29	−58.539
$HgS(s, \text{red})$	−58.2	82.4	−50.6
$HgS(s, \text{black})$	−53.6	88.3	−47.7
$Hg_2SO_4(s)$	−743.12	200.66	−625.815
Iodine			
$I(g)$	106.838	180.791	70.25
$I^-(g)$	−197	—	—
$I_2(s)$	0	116.135	0
$I_2(g)$	62.438	260.69	19.327
$ICl(g)$	17.78	247.551	−5.46
Potassium			
$K(s)$	0	64.18	0
$K(g)$	89.24	160.336	60.59
$K^+(g)$	514.26	—	—
$KBr(s)$	−393.798	95.9	−380.66
$KCl(s)$	−436.747	82.59	−409.14

Species	$\Delta_f H°$ kJ/mol	$S°$ J/mol·K	$\Delta_f G°$ (note [a]) kJ/mol
Potassium			
$KClO_3(s)$	−397.73	143.1	−296.25
$KF(s)$	−567.27	66.57	−537.75
$KI(s)$	−327.9	106.32	−324.892
$KNO_3(s)$	−494.63	133.05	−394.86
$KOH(s)$	−424.764	78.7	−379.08
Magnesium			
$Mg(s)$	0	32.68	0
$MgCl_2(s)$	−641.32	89.62	−591.79
$MgCO_3(s, \text{magnesite})$	−1095.8	65.7	−1012.1
$MgF_2(s)$	−1123.4	57.24	−1070.2
$MgO(s)$	−601.7	26.94	−569.43
$Mg(OH)_2(s)$	−924.54	63.18	−833.51
$MgS(s)$	−346.0	50.33	−341.8
Manganese			
$Mn(s, \text{alpha})$	0	32.01	0
$MnO_2(s)$	−520.03	53.05	−465.14
$MnS (s, \text{green})$	−214.2	78.2	−218.4
Nitrogen			
$N(g)$	472.704	153.298	455.563
$N_2(g)$	0	191.61	0
$NH_3(g)$	−46.11	192.45	−16.45
$N_2H_4(l)$	50.63	121.21	149.34
$NH_4Cl(s)$	−314.43	94.6	−202.87
$NH_4Cl(aq)$	−299.66	169.9	−210.52
$NH_4NO_3(s)$	−365.56	151.08	−183.87
$NH_4NO_3(aq)$	−339.87	259.8	−190.56
$NO(g)$	90.25	210.761	86.55
$NO_2(g)$	33.18	240.06	51.31
$N_2O(g)$	82.05	219.85	104.2
$N_2O_4(g)$	9.16	304.29	97.89
$N_2O_4(l)$	−19.5	209.2	97.54
$NOCl(g)$	51.71	261.69	66.08
$NOBr(g)$	82.17	273.66	82.42
Sodium			
$Na(s)$	0	51.21	0
$Na(g)$	107.32	153.712	76.761
$Na^+(g)$	609.358	—	—
$NaF(s)$	−573.647	51.46	−543.494
$NaCl(s)$	−411.153	72.13	−384.138
$NaCl(g)$	−176.65	229.81	−196.66
$NaCl(aq)$	−407.27	115.5	−393.133
$NaBr(s)$	−361.062	86.82	−348.983
$NaI(s)$	−287.78	98.53	−286.06
$Na_2CO_3(s)$	−1130.68	134.98	−1044.44
$NaNO_2(s)$	−358.65	103.8	−284.55
$NaNO_3(s)$	−467.85	116.52	−367
$NaOH(s)$	−425.609	64.455	−379.494
$NaOH(aq)$	−470.114	48.1	−419.150
$Na_2O(s)$	−414.22	75.06	−375.46

Species	$\Delta_f H°$ kJ/mol	$S°$ J/mol·K	$\Delta_f G°$ (note [a]) kJ/mol
Neon			
Ne(g)	0	146.328	0
Nickel			
Ni(s)	0	29.87	0
NiCl$_2$(s)	−305.332	97.65	−259.032
NiO(s)	−239.7	37.99	−211.7
NiS(s)	−82	52.97	−79.5
Oxygen			
O(g)	249.17	161.055	231.731
O$_2$(g)	0	205.138	0
O$_3$(ozone)	142.7	238.93	163.2
Phosphorus			
P(g)	314.64	163.193	278.25
P$_4$(s, white)	0	164.36	0
P$_4$(s, red)	−70.4	91.2	−48.4
PH$_3$(g)	5.4	210.23	13.4
PCl$_3$(g)	−287	311.78	−267.8
PCl$_5$(g)	−360.18	364.21	−305
P$_4$O$_{10}$(s)	−2984.0	228.86	−2697.7
H$_3$PO$_4$(s)	−1279.0	110.5	−1119.1
Lead			
Pb(g)	195	175.373	161.9
Pb(s)	0	64.81	0
PbBr$_2$(s)	−278.9	161.5	−261.92
PbCl$_2$(s)	−359.41	136.0	−314.1
PbO(s, red)	−218.99	66.5	−189.93
PbO(s, yellow)	−217.32	68.7	−187.89
PbO$_2$(s)	−277.4	68.6	−217.33
Pb$_3$O$_4$(s)	−718.4	211.3	−601.2
PbS(s, galena)	−100.4	91.2	−98.7
PbSO$_4$(s)	−919.94	148.57	−813.14
Sulfur			
S(s, rhombic)	0	31.8	0
S(g)	278.805	167.821	238.25
SF$_6$(g)	−1209	291.82	−1105.3
SO$_2$(g)	−296.83	248.22	−300.194
SO$_3$(g)	−395.72	256.76	−371.06
SO$_3$(l)	−441.04	113.8	−373.75
SO$_2$Cl$_2$(g)	−364	311.94	−320
H$_2$SO$_4$(l)	−813.989	156.904	−690.003
H$_2$SO$_4$(aq)	−909.27	20.1	−744.53
Silicon			
Si(s)	0	18.83	0
SiBr$_4$(l)	−457.3	277.8	−443.9
SiC(s)	−65.3	16.61	−62.8
SiCl$_4$(g)	−657.01	330.73	−616.98
SiH$_4$(g)	34.3	204.62	56.9
SiF$_4$(g)	−1614.94	282.49	−1572.65
SiO$_2$(s, quartz)	−910.94	41.84	−856.64

Species	$\Delta_f H°$ kJ/mol	$S°$ J/mol·K	$\Delta_f G°$ (note [a]) kJ/mol
Tin			
Sn(s, white)	0	51.55	0
Sn(s, gray)	−2.09	44.14	0.13
$SnCl_4(g)$	−471.5	365.8	−432.2
$SnCl_4(l)$	−511.3	258.6	−440.1
SnO(s)	−285.8	56.5	−256.9
SnO_2(s, cassiterite)	−580.7	52.3	−519.6
SnS(s)	−100	77	−98.3
Titanium			
Ti(s)	0	30.63	0
$TiCl_4(g)$	−763.2	354.9	−726.7
$TiCl_4(l)$	−804.2	252.34	−737.2
Xenon			
Xe(g)	0	169.683	0
Zinc			
Zn(s)	0	41.63	0
$Zn^{2+}(g)$	2782.78	—	—
$ZnCl_2(s)$	−415.05	111.46	−369.398
ZnO(s)	−348.28	43.64	−318.3
ZnS(s, wurtzite)	−192.63	—	—
ZnS(s, sphalerite)	−205.98	57.7	−201.29

Source: The National Bureau of Standards Tables of Chemical Thermodynamic Properties, 1982.
[a] While all of these values are dependent upon temperature, $\Delta_r G°$ for reactions is **particularly sensitive. Use $\Delta_f G°$ only if the temperature is** *exactly* **298 K.** Otherwise, use $\Delta_r H° - T \Delta_r S°$.

Thermodynamic Data for the Evaporation of Liquid Water

T [a] °C	P_{vap} mmHg	$\Delta_{vap}H°$ (kJ/mol)	$\Delta_{vap}S°$ (J/mol·K)	T [a] °C	P_{vap} mmHg	$\Delta_{vap}H°$ (kJ/mol)	$\Delta_{vap}S°$ (J/mol·K)
0	4.58	45.072	122.59	40	55.32	43.473	117.10
1	4.93	45.029	122.44	50	92.51	43.087	115.88
2	5.29	44.986	122.29	60	149.38	42.730	114.79
3	5.69	44.944	122.13	70	233.70	42.341	113.64
4	6.10	44.902	121.98	80	355.10	41.995	112.64
5	6.54	44.860	121.83	90	525.76	41.654	111.68
10	9.21	44.656	121.10	91	546.05	41.621	111.59
15	12.79	44.462	120.41	92	566.99	41.588	111.50
20	17.54	44.274	119.77	93	588.60	41.556	111.41
21	18.65	44.238	119.64	94	610.90	41.524	111.32
22	19.83	44.202	119.52	95	633.90	41.492	111.24
23	21.07	44.166	119.39	96	657.62	41.460	111.15
24	22.38	44.130	119.27	97	682.07	41.428	111.07
25	23.76	44.095	119.15	98	707.27	41.397	110.98
26	25.21	44.060	119.03	99	733.24	41.366	110.90
27	26.74	44.025	118.92	100	760.00	41.335	110.81
28	28.35	43.990	118.80	101	787.57	41.304	110.73
29	30.04	43.956	118.68	102	815.86	41.274	110.65
30	31.82	43.922	118.57	103	845.12	41.243	110.57
31	33.70	43.846	118.32	104	875.06	41.213	110.49
32	35.66	43.804	118.18	105	906.07	41.183	110.41
33	37.73	43.761	118.04	110	1075	41.036	110.02
34	39.90	43.719	117.90	120	1489	40.755	109.29
35	42.18	43.677	117.76	130	2026	40.491	108.62
36	44.56	43.636	117.63	140	2711	40.249	108.03
37	47.07	43.594	117.50	150	3570	39.989	107.40
38	49.69	43.554	117.36	200	11659	38.883	104.90
39	52.44	43.513	117.23	250	29818	38.021	103.20

[a] Values between 0°C and 100°C are based on National Institute of Standards and Technology data; values above 100°C are approximations based on known vapor pressures.

Answers to Selected Exercises

Chapter 1

1.1 (a) 1/6; (b) 1/2; (c) 5/6

1.4 (a) 1/4; (b) 3/16; (c) 3/8

1.7 (a) 1/4; (b) 1/52; (c) 10/13; (d) 3/51

1.10 1.35 mol H_2 + 0.33 mol D_2 + 1.32 mol HD

1.12 0.43 mol $^{35}Cl_2$ + 0.043 mol $^{37}Cl_2$ + 0.27 mol $^{35}Cl^{37}Cl$

1.13 (a) 0.92 mol $^{35}Cl_2$ + 0.026 mol $^{37}Cl_2$ + 0.31 mol $^{35}Cl^{37}Cl$; (b) 0.48 mol $^{35}Cl_2$ + 0.093 mol $^{37}Cl_2$ + 0.42 mol $^{35}Cl^{37}Cl$; (c) 0.32 mol $^{35}Cl_2$ + 0.43 mol $^{37}Cl_2$ + 0.75 mol $^{35}Cl^{37}Cl$

1.17 (a) 0.187 mol $^{191}Ir_2O_3$ + 0.529 mol $^{193}Ir_2O_3$ + 0.629 mol $^{191}Ir^{193}IrO_3$; (b) 4.00

1.19 51.8% ^{107}Ag and 48.2% ^{109}Ag

Chapter 2

2.1 (a) Distribution A, giving one quantum to two different particles, 10 ways. Distribution B, giving two quanta to one particle, 5 ways; (b) twice as probable

2.3 There are five distinct distributions, A, B, C, D, and E. Distribution A: one particle has four quanta of energy, $W = 5$ $W/W_{total} = 0.07$; Distribution B: one particle has three quanta of energy and one particle has one quantum of energy, $W = 20$ $W/W_{total} = 0.28$; Distribution C: one particle has two quanta of energy and two particles have one quantum of energy, $W = 30$ $W/W_{total} = 0.43$; Distribution D: two particles have two quanta of energy, $W = 10$ $W/W_{total} = 0.14$; Distribution E: four particles have 1 quantum of energy each, $W = 5$ $W/W_{total} = 0.07$; $W_{total} = 70$

2.5 (a) Distribution A, giving one quantum to one particle and two quanta to one particle, 20 ways. Distribution B, giving one quantum to three particles, 10 ways; (b) twice as probable

2.9 (a) $W_A = 56$, $W_B = 1$, $W_C = 28$, $W_D = 1120$; (b) D; (c) 20 times more probable; (d) 1120 times more probable

2.12 4

2.14 0

2.17 3.9×10^{-21} J

2.21 (a) 3.9×10^{-21} J; (b) 204 K

2.22 (a) 298 K; (b) 0.24

2.26 (a) 0.28; (b) 0.64

2.29 (a) 0.99998; (b) 0

2.32 (a) 3.6×10^{-10}; (b) 0.013

2.34 (a) 25; (b) 124

2.35   (a) 3.7×10^{-19} J; (b) 7.5×10^{-20} J

Chapter 3

3.1   (a) 7×10^{-10} m; (b) 1.1×10^{-34} m; (c) 1.2×10^{-43} m

3.3   (a) 6.02×10^{-18} J, 2.41×10^{-17} J, 5.40×10^{-17} J; (b) 3.28×10^{-21} J, 1.31×10^{-20} J, 2.95×10^{-20} J

3.6   (a) $-2.1798736 \times 10^{-18}$ J, $-5.449684 \times 10^{-19}$ J; (b) 1.5×10^{15}; (c) 0; (d) 54000 K

3.8   (a) 656.1124 nm; (b) 1874.6 nm; (c) 91.1267 nm

3.12   only statement (d)

3.14   (a) 8.67×10^{13} s^{-1}; (b) 8.65×10^{13} s^{-1}; (c) 6.21×10^{13} s^{-1}; (d) 6.20×10^{13} s^{-1}

3.16   (a) 0.0155; (b) 0.0157; (c) 0.0492; (d) 0.0492

3.18   1460 K

3.20   (a) This statement is true since while $^{19}F_2$ has a larger reduced mass, it has a longer equilibrium bond length than $^{16}O_2$. (b) This statement is true since $^{14}N^{16}O$ has both a larger reduced mass and bond length than $^{12}C^{16}O$. (c) This statement is false since while $^{1}H^{79}Br$ and $^{19}F_2$ have the same bond lengths, $^{1}H^{79}Br$ has a much smaller reduced mass, so it will have more widely spaced energy levels. (d) This statement is true since while $^{1}H^{127}I$ and $^{1}H^{79}Br$ have virtually identical reduced mass, $^{1}H^{79}Br$ has a shorter bond length, giving it the more widely spaced energy levels.

3.22   (a) 113 pm. (b) 7.78×10^{-23} J

3.24   (a) 6.14×10^{-24} J; (b) yes

3.26 3.92×10^{-41} J

3.30   (a) 1.64×10^{-54} J; (b) 9.92×10^{-37} J

3.33   (a) vibrational, rotational; (b) rotational, translational; (c) electronic; (d) translational

3.35   more, translational, the length of the side of the box in which it is contained, d, is lengthened.

3.37   (a) vibrational; (b) vibrational, rotational; (c) translational, rotational; (d) electronic; (e) vibrational

3.42   The spacing between electronic energy levels in the blue light laser is larger than in the HeNe laser.

Chapter 4

4.1   State variables are independent of the path to the current state.

4.4   (a) 1.65 bar; (b) 3.963×10^{-40} J, 7.926×10^{-40} J, 3.963×10^{-40} J; (c) 11.7 K; (d) 218 K

4.6   (a) 99.8 L, 199.6 L, -9.98 kJ; (b) 99.8 L, 24.9 L, $+30.0$ kJ; (c) -3.32 kJ; (d) 0

4.10   (a) 2.2 J/g·K, 1.8 J/g·K, 2.4 J/g·K, 2.5 J/g·K, 4.2 J/g·K; (b) benzene; (c) water

4.11   (a) 200 J; (b) 2400 J; (c) 3140 J; (d) 1003 J

4.15   7.846 kJ/K

4.17   (a) 14.42 kJ/K; (b) -13570 kJ; (c) -27410 kJ

Chapter 5

5.1 -545 ft

5.9 -296.5 kJ

5.11   (a) -107 kJ/mol; (b) -642 kJ/mol; (c) $+856$ kJ/mol

5.14   (a) -456 kJ/mol; (b) $+9$ kJ/mol; (c) -170 kJ/mol

5.18   (a) -579 kJ/mol; (b) -143 kJ/mol; (c) -590 kJ/mol; (d) -229 kJ/mol

5.21   (a) $Xe(g) + 2 F_2(g) \longrightarrow XeF_4(g)$; (b) -212 kJ/mol

Chapter 6

6.1   (a) Case 2; (b) Cases 1 and 3; (c) Case 4

6.3   (a) Case 1: $K = 0$, Case 2: $K \longrightarrow$ infinity, Case 3: $K = 0$, Case 4: $K = 1$; (b) Case 1: $K = 1$, Case 2: $K = 2$, Case 3: $K = 3$, Case 4: $K = 1$

6.5   At low temperatures, the side of the reaction with the lowest ground state will be favored. The equilibrium constant K will increase with increasing temperature if the products have a higher ground state than the reactants. If the reactants have a higher ground state than the products, K will decrease with increasing temperature. The limit of K at high temperature is determined by the relative energy spacings of the products and reactants.

6.7   (a) (b); (b) (a) and (c); (c) (a) and (c); (d) (b)

6.9   (a) K is 0 at $T = 0$. It will increase with increasing T, crossing 1, and approach infinity as T approaches infinity. (b) K is 0 at $T = 0$. It will increase with increasing T, crossing 1, and approach infinity as T approaches infinity. (c) K is infinite at $T = 0$. It will decrease with increasing T, crossing 1, and approach a number between 0 and 1 as T approaches infinity.

Chapter 7

7.2   State A: 0 aJ, 1, 0 J/K; State B: 3 aJ, 56, 5.6×10^{-23} J/K; State C: 8 aJ, 1120, 9.69×10^{-23} J/K

7.6   (a) $SiBr_4(g)$; (b) $GeH_4(g)$; (c) $H_2Se(g)$

7.8   (a) $BaCl_2(s)$; (b) $CCl_4(g)$; (c) $CO_2(g)$

7.11   Entropy generally increases in going from solids to liquids to gases. For a pure substance there is more "disorder" in the gas than in the liquid, and more "disorder" in the liquid than in the solid.

7.13   Expect that the energy levels in P(white) are more closely spaced than in either P(red) or P(black).

7.16   Expect that the energy levels in PbO(s, yellow) are slightly more closely spaced than in PbO(s,red)

7.18   261.41 J/mol K; expect that the vibrational, rotational, and translational energy levels of carbon tetrafluoride would be farther apart thus resulting in a lower entropy.

7.21   (a) A to B: +3 aJ, 5.56×10^{-23} J/K, B to C: +5 aJ, 4.13×10^{-23} J/K; (b) in both cases W increases so ΔS will be positive since the logarithm of a number greater than one is positive; (c) ΔS will be positive since both q and T are positive

7.23   (a) A to B: -28 aJ, -1.0813×10^{-22} J/K, B to C: -8 aJ, $- \times 10^{-23}$ J/K; (b) in both cases W decreases so ΔS will be negative since the logarithm of a number less than one is negative; (c) ΔS will be negative since q is negative and T positive

7.25   (a) very positive; (b) very negative; (c) very positive

7.27   (a) +10 J/K, small; (b) +89 J/K, very positive; (c) -20 J/K, small

Chapter 8

8.1   (a) 0.250; (b) 2.88 J/K; (c) -0.60 J/K

8.3   (a) 0.286; (b) 2.60 J/K; (c) -0.70 J/K

8.4   (a) 9.41 J/K; (b) 9.41 J/K; (c) 9.41 J/K; (d) this will not work as in Equation 8.6 since the pressures in the two flasks are *not* the same

8.9   (a) 35 J/K; (b) 0.0790 J/K

8.13   (a) 189 J/K; (b) 18.9 J/K; (c) 378 J/K

8.15   283 J/K

8.17   160 J/K

8.20   (a) -153.53 J/mol·K; (b) -187.14 J/mol·K

8.23   (a) $Q = \dfrac{P_{NH3}^2}{P_{H2}^3 P_{N2}}$; (b) $Q = \dfrac{P_{PCl5}}{P_{PCl3} P_{Cl2}}$

Chapter 9

9.1   (a) both positive; (b) both negative

9.3   (a) both positive; (b) both negative

9.5   (a) 1.47 kJ; (b) 8.35 kJ

9.7   4.54 kJ

9.9   (a) -452.176 kJ/mol; (b) $+176.20$ kJ/mol; (c) -397.32 kJ/mol

9.11   (a) -146.4 kJ/mol; (b) $+287.78$ kJ/mol; (c) $+667.87$ kJ/mol

9.15   (a) $-1/2$ mol, -146.0 kJ, -144.8 kJ; (b) 0 mol, -104 kJ, -104 kJ; (c) $+1$ mol, -72.8 kJ, -75.3 kJ

9.17   (a) $+3.895$ kJ, -0.248 kJ; (b) $+38.95$ kJ/mol, $+36.47$ kJ/mol

9.20   (a) -291 kJ/mol; (b) -138 kJ/mol

9.22   (a) -771.04 kJ/mol; (b) -132.44 kJ/mol

9.23 $+506.16$ kJ/mol

Chapter 10

10.2   The slope of a G-T graph is determined by S since $G = H - TS$. S generally increases, making the curve steeper, as T increases because more particles move to higher energy levels. Having more levels populated in a system allows more ways to distribute the energy, which increases S.

10.3   (a) Curve for A starts below the curve for B, the curve for B decreases faster so the two curves cross. (b) Curve for A starts below the curve for B, the curve for A decreases faster so the two curves do not cross. (c) Curve for B starts below the curve for A, the curve for A decreases faster so the two curves cross.

10.5   (a) Curve for B starts below the curve for A, the curve for B decreases faster so the two curves do not cross. (b) Curve for B starts below the curve for A, the curve for A decreases faster so the two curves cross. (c) Curve for B starts below the curve for A, the curve for A decreases faster so the two curves cross.

10.8   (a) Curve for $I_2(s)$ starts below the curve for $I_2(g)$, the curve for $I_2(g)$ decreases faster so the two curves cross at 432 K. (b) Curve for $Br_2(l)$ starts below the curve for $Br_2(g)$, the curve for $Br_2(g)$ decreases faster so the two curves cross at 331 K. (c) Curve for C(*graphite*) starts below the curve for C(*diamond*), the curve for C(*graphite*) decreases faster so the two curves do not cross. (d) Curve for P(*red*) starts below the curve for P(*white*), the curve for P(*white*) decreases slightly faster so the two curves cross at 962 K.

10.12   (a) 25°C, 298 K, 62438 J, -62438 J, 144.5 J/mol·K, -209 J/K, -65.0 J/K; (b) 125°C, 398 K, 62438 J, -62438 J, 144.5 J/mol·K, -156.9 J/K, -12.3 J/K; (c) 159°C, 432 K, 62438 J, -62438 J, 144.5 J/mol·K, -144.5 J/K, 0.0 J/K; (d) 175°C, 448 K, 62438 J, -62438 J, 144.5 J/mol·K, -139.4 J/K, 5.2 J/K;

10.15   79.86 J/mol·K, 18.77 kJ/mol, 272.04 K, 83.31 J/mol·K, 84.38 J/mol·K

10.18   (a) 72.9 J/K; (b) $\Delta_{vap}H°$ is much larger for water while $\Delta_{vap}S°$ for methane and water are similar.

Chapter 11

11.1   (a) -57.2 kJ, -175.83 J/K, -4803 J, 6.95; (b) -108.3 kJ, -137.21 J/K, -67411 J, 6.544×10^{11}

11.4 1.25×10^{-15}, 1.00×10^{-14}, 1.06×10^{-13}, 9.23×10^{-13}

11.7   179 kJ/mol, 89.46 J/mol·K, 2007 K

11.9   (a) K_{eq} decreases with increasing T so $\Delta_r H$ is negative, K_{eq} crosses 1 so $\Delta_r S$ is negative; (b) −50.2 kJ/mol, −63 J/mol·K

11.11   (a) 4.05×10^{18}, (b) 0.289

11.14   (a) Curve for $I_2(s)$ starts below the curve for $I_2(g)$, the curve for $I_2(g)$ decreases faster so the two curves cross at 432 K. (b) Curve for $Br_2(l)$ starts below the curve for $Br_2(g)$, the curve for $Br_2(g)$ decreases faster so the two curves cross at 331 K. (c) Curve for Sn(*gray*) starts below the curve for Sn(*white*), the curve for Sn(*white*) decreases faster so the two curves cross at 282 K.

11.17   Curve for $CO_2(aq)$ starts below the curve for $CO_2(g)$, the curve for $CO_2(g)$ decreases faster so the two curves cross. When the temperature is raised the reaction $CO_2(g) \longrightarrow CO_2(aq)$ is no longer the favored reaction, so to keep the champagne bubbly, we should keep the bottle cold.

Chapter 12

12.2   Curve for C(*graphite*) starts below the curve for C(*diamond*), the curve for C(*graphite*) decreases faster so the two curves do not cross. So at least at 1 bar, graphite and diamond are never in equilibrium since their curves never cross on the G−T graph.

12.3   195 K

12.5   176 K

12.7   10.76 kJ/mol, 40.5 J/mol·K

12.10   (a) 348 K; (b) 353 K

12.13   (a) 370 K; (b) 352 K

12.15   (a) 371 K; (b) 400 K

12.17   In Denver, the atmospheric pressure is lower than in Northfield. The curve for $H_2O(g)$ in Denver decreases faster than the curve for $H_2O(g)$ in Northfield so the curve for $H_2O(g)$ in Denver crosses the curve for $H_2O(l)$ at a lower temperature. As a consequence, water will boil at a lower temperature in Denver. This means that the water will not get as hot before it boils and it will take longer to cook an egg.

12.20   Since the actual vapor pressure is usually much less than the equilibrium vapor pressure, it is impossible for enough water to evaporate from the puddle to significantly alter the vapor pressure of water in the entire atmosphere. So the puddle simply completely evaporates without ever establishing the equilibrium vapor pressure.

12.23   34°F, yes

12.27   At pressures less than 5 bar, the gaseous form of CO_2 is favored. Only above 5 bar is $CO_2(l)$ more stable than $CO_2(g)$.

12.29   The critical point occurs when the pressure of a gas is raised to the point where the entropy of the gas is identical to that of the liquid. At this point, the curvature of the G–T curves for the gas and liquid curves are the same, and the two phases are indistinguishable.

12.33   (a) Curve for $C_2H_5OH(s)$ starts below the curve for $C_2H_5OH(l)$, the curve for $C_2H_5OH(l)$ decreases faster so these two curves cross at 159 K. Curve for $C_2H_5OH(l)$ starts below the curve for $C_2H_5OH(g)$, the curve for $C_2H_5OH(g)$ decreases faster so these two curves cross at 352 K. The curve for $C_2H_5OH(g)$ crosses the curve for $C_2H_5OH(g)$ at 312 K, but this crossing point occurs above the $C_2H_5OH(l)$ curve;. (b) Above 1 bar; (c) less than 1 bar; (d) 159 K, 2.3×10^{-8} bar

12.35   (a) Curve for $C_3H_7OH(s)$ starts below the curve for $C_3H_7OH(l)$, the curve for $C_3H_7OH(l)$ decreases faster so these two curves cross at 148 K. Curve for $C_3H_7OH(l)$ starts below the curve for $C_3H_7OH(g)$, the curve for $C_3H_7OH(g)$ decreases faster so these two curves cross at 475 K. The curve for $C_3H_7OH(g)$ crosses the curve for $C_3H_7OH(g)$ at

370 K, but this crossing point occurs above the C3H7OH(l) curve. (b) Above 1 bar; (c) less than 1 bar; (d) 148 K, 1.7×10^{-9} bar

12.38 (b) 5.06 kJ/mol, 30.8 J/mol·K

12.40 In a solution that is saturated with gas, raising the temperature will not allow more to dissolve. On the G-T graph moving from the saturation point to higher temperature makes the reaction $gas \longrightarrow aq$ an unfavorable one. This is mainly a function of $\Delta_r H$ not $\Delta_r S$. In the usual case where ΔH is negative for the dissolving of a gas, raising the temperature will not allow more gas to dissolve. This can also be seen by examining the Clausius-Clapeyron equation.

12.42 (a) 369.8 K; (b) 369.8 K; (c) 369.4 K

Chapter 13

13.1 (a) 12960 C; (b) 0.1343 mol e$^-$

13.3 (a) 8640 C; (b) 0.0895 mol e$^-$

13.5 (a) 10.5 hr; (b) 3.35 hr; (c) 1862 hr

13.8 19440 J

13.10 (a) $NiO_2(s) + Cd(s) + 2\ H_2O\ (l) \longrightarrow Ni(OH)_2(s) + Cd(OH)_2(s)$; (b) 6.09 g; (c) 9.831 g

13.11 (a) spontaneous, −1.949 kJ/mol; (b) spontaneous, −139 kJ/mol; (c) spontaneous, −233.5 kJ/mol; (d) spontaneous, −150 kJ/mol

13.15 (a) spontaneous, oxidizing agent: Ce^{+4}, reducing agent: Tl; (b) not spontaneous, oxidizing agent: Fe^{+2}, reducing agent: Cu; (c) spontaneous, oxidizing agent: H^+, reducing agent: Zn; (d) not spontaneous, oxidizing agent: Sn^{+2}, reducing agent: Ag

13.17 (a) $3\ Fe(s) + 8\ H^+(aq) + 2\ NO_3^-(aq) \longrightarrow 2\ NO(g) + 4\ H_2O(g) + 3\ Fe^{2+}(aq)$; (b) +1.40 V, spontaneous; (c) 1.14×10^{142}

13.18 (a) +1.10 V, −212 kJ, 1.61×10^{37}; (b) +1.04 V, −801 kJ, 5.1×10^{70}; (c) +0.52 V, −400 kJ, 1.54×10^{70}; (d) −1.10 V, +212 kJ, 6.22×10^{-38}

13.20 (a) +1.10 V; (b) +1.22 V; (c) +0.923 V; (d) +1.04 V

13.21 6.2×10^{-38} M

13.22 2.1×10^{-24} M

Page numbers often refer to the first page of the section in which the topic is discussed.